THE MECHANICS OF ROLLERS

David R. Roisum, Ph.D.

TAPPI
Atlanta,GA

TAPPI PRESS

TAPPI PRESS
Technology Park/Atlanta
P.O. Box 105113
Atlanta, GA 30348-5113 U.S.A.

ISBN 0-89852-313-3
TP 0101R255
Printed in the United States of America

Library of Congress Cataloging-in-Publication Data

Roisum, David R.
The mechanics of rollers / David R. Roisum.
p. cm.
Includes bibliographical references and index.
ISBN 0-89852-313-3
1. Papermaking machinery. 2. Rolling contact. 3. Cylinders.
I. Title.
TS1117.R65 1996
681'.7676--dc20 95-52109
CIP

Preface

Rollers seldom have the attention they deserve. They are often regarded as simple, unimportant and uninteresting. How often when viewing a process do we see the web, but not the roller underneath? How often when considering a process do we think in terms of calendering, coating, printing and so on, rather than the rollers which provide those process functions?

It is all too easy to view rollers as trivial components because their shape is but a cylinder and their action merely rotation. However, as demonstrated in these pages, there is more than enough complexity to fill a book. Indeed, many of these chapters could have been expanded into books unto themselves.

It is all too easy to regard rollers as unimportant when there is no obvious defect or delay associated with them. How could a roller be at fault if it continues to perform its primary mission of spinning on its axis? We all know when our web is bad because it looks bad or tests bad. But when is a roller bad?

We can better understand the importance of rollers if we change our perspective to see the world from the web's viewpoint. Almost everything in a web's universe is rollers and the spans between rollers. From this vantage it now becomes clear that not only are rollers important, they are the most important element in a web process.

Rollers are the foundation of all web manufacturing and converting processes. They are the building blocks of all web machines. Rollers allow us to make products by an economic continuous fashion instead of by an inefficient batch process.

Thus, the position of the roller can now be properly lifted to one of vital and central importance. Even the lowly idler must be understood and attended to, or else we may suffer as one converter who lost $1M of product due to an undersized $1K idler.

While rollers are important, they are sometimes attributed to have powers they do not possess. For example, there is a common fallacy in the web industries that spiral grooving will spread a web. The reality, however, is quite different. Grooving of any type has the tendency to contract rather than spread a web. In this book we show that the purpose of grooving is to retain traction despite air/fluid entrainment. However, common grooving patterns are like using tractor tires on an automobile, very much inappropriate for the duty.

Roller peripherals, such as drive controls, are not always understood as well as they should be. A common misconception is the "isolation nip" driven under draw or speed control. Here, the thought is that the nip will prevent tension disturbances from traveling down the line. The reality, however, is quite different. Draw or speed control will pass all disturbances, and the presence of a nip merely ensures it. Indeed, draw/speed is the only control scheme that does not attenuate upstream tension fluctuations. Had the designer known this, the isolation nip would not have been included, the machine cost would be less, and the process performance would be improved.

I do not expect that everyone will find rollers as interesting as I, because it is a matter of personal taste. I do not expect stories to be written of their deeds, or songs to be sung of their virtues. My expectations are much more modest.

Perhaps some of the readers will better appreciate the role of rollers in the health of their processes. Perhaps more machines will be built with fewer rollers, but the rollers that are supplied will be of better integrity. Perhaps a few charged with improving process performance will include rollers in their program priorities. Perhaps a very few may be inspired to research rollers and expand our understanding of their mechanics. Only when we understand can we appreciate, and only when we appreciate can we apply.

Acknowledgments

My first book in this series, *The Mechanics of Winding*, was published by TAPPI PRESS in 1994. You might think that this second book would be easier to write than the first because of the experience gained. Unfortunately, it proved harder and longer, mainly because there is so much to say about rollers. Indeed, there is much truth to the saying that "you never really finish a book, you abandon it."

Hopefully this should not dampen my enthusiasm for the next project, *The Mechanics of Web Handling*, though it may be completed on a slightly longer interval. Briefly, web handling is the science of moving a web through a machine with a minimum of trouble.

Now I know very well why authors include an acknowledgment page. They sincerely appreciate any assistance or help with this almost herculean task. First, I would like to reiterate my appreciation for TAPPI. If it were not for TAPPI, our knowledge of rollers would be more limited and less accessible. All one has to do is to look at the bibliography listings at the ends of the chapters to see the major role that TAPPI has played. Specifically, I would like to thank Tomya Moore, Steve Yaeger and others of TAPPI PRESS for their assistance, guidance, patience and tolerance.

Second, I again owe a debt to the Web Handling Research Center at Oklahoma State University. Their theoretical and experimental studies of the science of web handling, of which web-roller interaction plays a central role, has provided the foundation for many of the topics covered in this book.

Finally, I would like to express my gratitude to the following individuals and their companies who have provided technical information and technical review of one or more of the chapters in the book. It is my belief that it is always wise to have a "sanity check" by others, especially when presenting new, controversial or subjective material.

A sincere thank you to:

Peter Werner
Allen Bradley Co.

Ron Buono
American Roller Co.

Bruce Lindstrand
Beloit Corp.

Duane Smith
Black Clawson

Eric Westerheim
Epoch Industries

Lou Bargatti
Magnat Rolls

Doug Liescheidt
Magnetic Power Systems

Gary Teeter
Mark Andy Inc.

Matt Oliver
Montalvo Corp.

Jim Dunn
Oasis

Gerry Buxton
PCMC

George Maniatty
Precision Roll Grinders

James Stoddard
Republic Roller Corp.

Joe Vosters
Spencer Johnston Co.

Allen Killeen
Stowe Woodward Co.

James Franklin
Tidland Corp.

John Slotten
Valley Roller Co.

Mel Kruse
Valmet Paper Machine

John Shelton
Web Handling Research

Jack Roberts
Webex Inc.

Table of Contents

20 Maintenance

Appendices

List of Illustrations

1 Introduction

2 Tension and Tension Control

3 Deflection

4 Traction, Torque and Power

5 Nipped Rollers

6 Shells

7 Journals and Shafts

8 Bearings

9 Coatings and Covers

10 Surface Topology

11 Vibration and Balance

12 Alignment

13 Mechanical Drives

14 Electric Drives

15 Temperature Control

16 Cores, Chucks and Shafts

17 Spreaders

18 Specialty Rollers and Elements

19 Wound Rolls

20 Maintenance

List of Tables

1 Introduction

2 Tension and Tension Control

3 Deflection

4 Traction, Torque and Power

5 Nipped Rollers

6 Shells

8 Bearings

9 Coatings and Covers

10 Surface Topology

11 Vibration and Balance

12 Alignment

17 Spreaders

18 Specialty Rollers and Elements

19 Wound Rolls

Chapter 1

Introduction

This chapter describes the usage and importance of rollers in web manufacturing, processing and converting. Also included is an introduction to their construction and selection.

The Foundation of Web Processes

The most common web materials include paper, film, foils, nonwovens and textiles. However, within each of these classes are a wide variety of consumer products as given in Table 1.1. Web products are also commonly used in the construction of buildings, such as insulation and tiles, as well as in the construction of automobiles and aircraft. Our modern way of life is made possible, in large part, by the products of the web industries.

As might be expected, the manufacturing processes for making paper are quite different than those for making film, and nonwoven machinery is quite different from textile machinery. Also, the widths of webs varies considerably from as little as 1 millimeter for pinstriping tape to more than 10 meters for wide paper machinery. Finally, the strengths and weights of web materials varies from lightweight tissues and cigarette papers to heavy coils of metals.

Despite the tremendous variety of webs and their manufacturing processes, however, they all have one thing in common. Rollers! Rollers are the foundation of all web manufacturing and converting machinery. Most of the elements that touch the web during manufacturing are rollers. Seldom does a web touch less than 10 rollers before it is complete, and it may touch more than 100 rollers for complex laminates.

Without rollers, webs could not be made continuously but would instead need to be cast in batches. It is this continuous processing made possible by rollers that makes web materials economical.

Table 1.1
Web Material and Product Examples

Film

- bags
- adhesivie tape
- magnetic tape
- photographic
- wrap, food
- wrap, shrink
- wrap, stretch

Foil

- aluminum food wrap

Nonwovens

- filters
- insulation
- medical garments and covers
- personal hygiene products

Paper

- bags
- boxes
- currency
- labels
- magazines
- newspapers
- tape
- tissue
- toweling
- waxed
- wrapping
- writing

Textiles

- carpet
- clothing
- upholstery

Roller Applications

Roller applications can be broken into two distinct categories: transport or processing. Rollers used for transport are not intended to modify the properties of the web in any permanent way. Rather, they are used merely to get the web from one processing element to another. Unfortunately, they can modify the web if it is scratched due to sliding or if troughs in the web fold over into creases going over a roller.

Because transport rollers can only damage rather than improve the web, they should be reduced to an absolute minimum. It is unfortunate that most machines have unnecessarily high roller counts. Not only does every roller present the risk of web damage, but they also increase equipment and maintenance costs, make threading and startup more difficult, and cause the machine to take up more precious floor space.

Transport rollers, as seen in Table 1.2, can be web driven (idler rolls) or externally driven (pull or draw rolls). Indeed, a major roller design and selection criteria is whether it needs to be driven externally such as by an electric motor. It must be driven if the tension upset caused by inertial speed changes and/or the bearing torque and windage is excessive for the web. It must also be driven if there is insufficient web-roller traction.

Processing rollers, on the other hand, are intended to modify the web in some way as again seen in Table 2. Obviously, processing rollers must also transport so that all of those considerations hold as well. However, processing is usually much more demanding than mere transport so that those rollers must be selected, manufactured and maintained with somewhat more care.

Finally, there are many other web machinery elements that share essentially the same design considerations as rollers. These include cores, mandrels, shafts and so on. Indeed, the web does not know or care by what name we call the element. It cares about the geometrical details of the cylinder it is wrapping and the forces that are transmitted.

Table 1.2
Roller Applications

Transport Rollers

- bars and pipes
- externally driven
- web driven (idler)

Transport Rollers (special purpose)

- accumulators
- dancer rollers
- guide rollers
- load cell tension rollers
- spreader rollers
  - bowed roller
  - concave roller
  - dual roller
  - edge pull
- squaring rollers
- skewing rollers
- winder drums
- winder lay-on rollers

Processing Rollers

- calender rollers
- chill rolls (film mfg)
- coater rollers
- couch rollers (paper mfg)
- dryer cans
- embossing rolls
- press rolls (paper mfg)
- printer rollers
- laminater rolls
- suction rolls (paper mfg)
- treater rolls (film)

Related Elements

- cores
- corechucks
- coreshafts
- lineshafts
- mandrels
- pulleys
- spools
- wound rolls

Dead and Live Shaft Arrangements

Rollers are classified as either dead shaft or live shaft arrangements as determined by whether the central shaft is fixed or free to rotate. Many rollers less than 1 meter in width are of a dead shaft design, while larger rollers will be (with a few exceptions) of the live shaft design.

The dead shaft arrangement is shown in Figure 1.1. The shaft is usually machined from bar stock of high strength steel such as 4140HT. The machining consists of cutting to length and (usually centerless) grinding of the ends to match the required precision fit on the ID (inner diameter) of the bearings. The shell could be made of a variety of materials but will be typically bored on the ID to fit the bearings and may be ground on the OD (outer diameter) to get good concentricity. The design of the shaft mount varies considerably and need only hold the shaft from shifting or rotating.

The live shaft arrangement is shown in Figure 1.2. While the construction techniques are similar, the live shaft roller has a much greater complexity. For example, the heads and journals require considerably more machining than does the corresponding shaft of a dead shaft roller. However, the greatest complexity will be in the bearing housings which are typically cast and then machined. This is much more involved than the corresponding shaft mount for the dead shaft roller.

As might be expected, the small dead shaft roller will be less expensive than its live shaft counterpart. However, cost cutting has often gone too far because some of these rollers have very poor tolerances and quality. This is because a dead shaft roller can be made from raw stock with little or no machining cleanup.

The dead shaft roller also has no provision for an external drive coupling. Thus, the torque to drive the roller must come from the web itself. This is done as a web tension buildup or difference across the roller sufficient to overcome bearing drag, windage (pumping of air), and inertia during speed changes.

Figure 1.1
Dead Shaft Roller

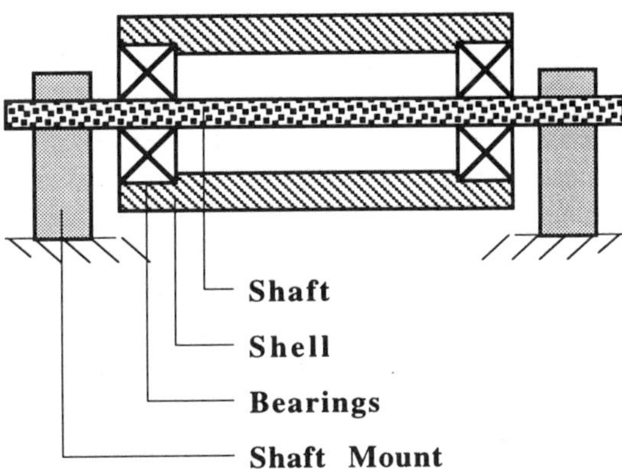

Figure 1.2
Live Shaft Roller

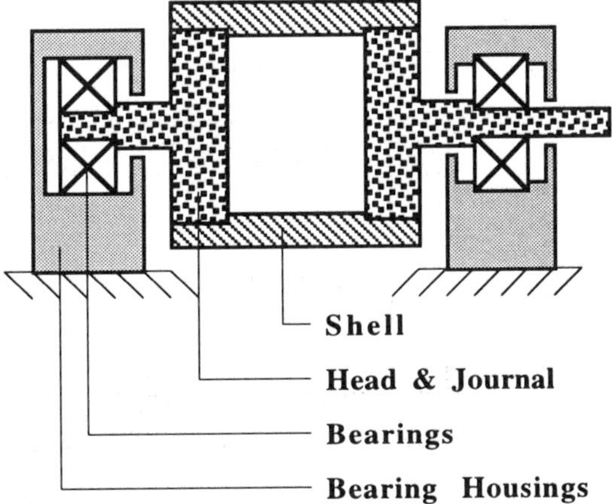

Similarly, the dead shaft roller has no provision for directly attaching a footage counter or a speed measuring sensor such as an encoder or tachometer. Rather, these sensors will often be added as a wheel which bears on the outer shell. The problem, however, is that shell driven sensors are high maintenance items and often have poor accuracy due to slippage and wear.

A better approach for dead shaft footage or speed measurement is to use a noncontacting photoeye or proximity switch to detect and count the passage of a target on the shell. This has the additional advantage of not increasing the drag on the roller.

Heads and Journals

There are a number of ways of attaching the head/journal to a shell as seen in Figure 1.3. Welding may be the most economical because very little machining or preparation is required. Press fits are also economical, but require very precise control of the OD of the head and ID of the shell so that the fit is neither too loose so that the joint comes apart or too tight so that the head can't be pressed in. Bolting is somewhat more involved because of the large amount of drilling and tapping required in addition to machining the heads and shell. Bolting is most often used for large pressure vessels such as dryer cans.

Other means of mounting heads include brazing (metals) and adhesives (composites and polymers).

Figure 1.3
Fixing Heads/Journals to Shells

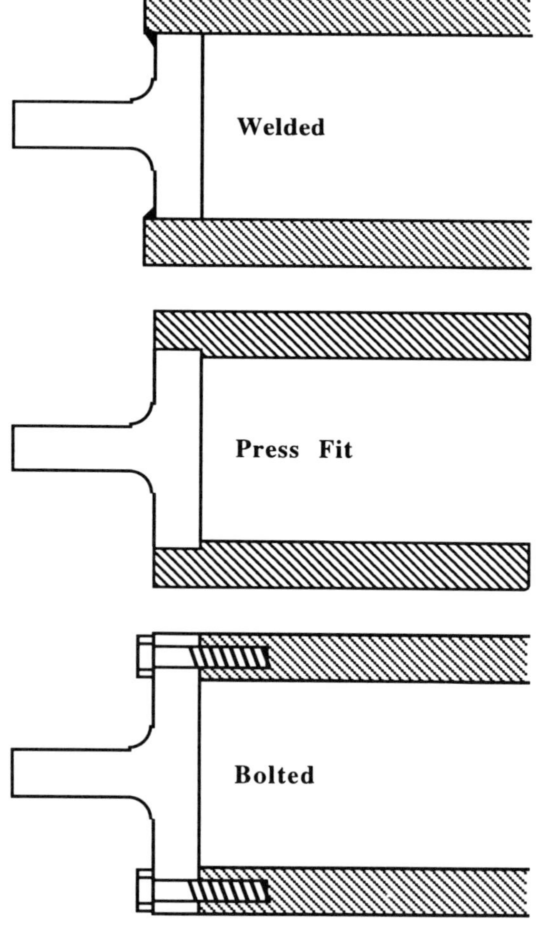

Bearings

There are a variety of bearing styles used in rollers as seen in Table 1.3. However, the spherical roller bearing is by far the most common.

Bearings must withstand a radial load due to the combined effects of roller weight, web tension, nips and other loads. Additionally, many bearings, such as on nipped rollers, must have a substantial thrust load capability. The bearing must survive these loads for tens or even hundreds of millions of revolutions. Bearings also have RPM (revolutions per minute) and misalignment ratings that must not be exceeded. Selecting the most appropriate bearing for any application can be bewildering because of the large variety of options. However, sizing them is truly a design calculation that will be covered in more detail in Chapter 7.

There are also a number of bearing lubrication options with the most common given in Table 1.4. Grease is common, but many bearings get overlooked by the oiler. Also, grease adds considerably to bearing friction. Oil is also common but requires more expensive bearing housings. An evolution of oil is continuous lubrication where oil is metered into a bearing and the excess drained and returned to tank. This helps keep the bearing cooler and cleaner, and thus increases its life.

Table 1.4
Lubrication

Grease (common, but high friction)
Oil - static (requires housings and seals)
Oil - continuous lube (very long life)

Table 1.3
Bearing Options

Babbitt Bearing (very old equipment)
Ball Bearing (small rollers)
Bushing (oil impregnated bronze, short life)
Gas Bearing (nearly frictionless)
Journal Bearing (high rpm, forced oil lube)
Spherical Roller Bearing (most common)

Roller Geometry - Axial

There are several widths that determine the axial geometry of a roller as seen in Figure 1.4.

Minimum Deckle - the minimum width of product to be run.

Maximum Deckle - the maximum width of the product to be run, which is important for roller sizing.

Face - the effective and useful width of a roller needs to be somewhat greater (typically 2-4") than the widest product to allow for edge position wander.

Width - the width of the shell structure which may be equal or greater than the face.

Bearing-to-Bearing Centerline - the distance between bearing centers which should be held common to most rollers in a machine for ease of framework design.

Overall Width - the width used for designing clearances and machine layout.

The axial quality tolerance of greatest importance in a roller is its straightness, which we may expect to be around 0.002" for typical non-precision rollers.

Figure 1.4
Axial Roller Geometry

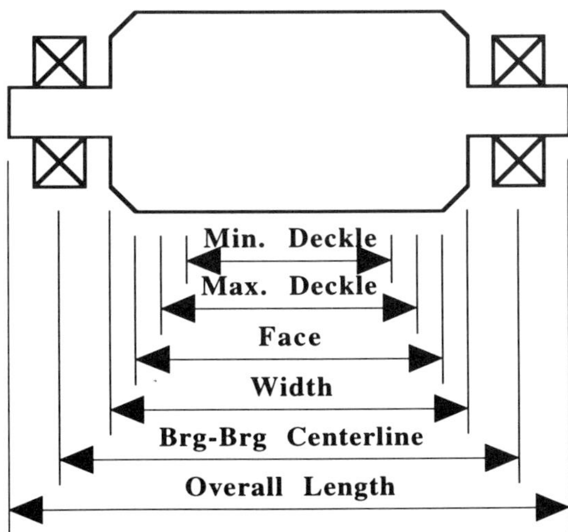

Roller Geometry - Diametral

There are several diameters that need to be specified for a roller as seen in Figure 1.5.

The OD (outer diameter) is determined by a deflection standard (Chapter 3). However, a safe rule of thumb for nipped rollers would be a slenderness ratio (length/diameter) of no more than 10 and unnipped rollers of not more than 15. The OD would be selected as the next stock size greater than the calculated requirement. OD's are nominally available in 1/2" increments for small rollers, 1" increments for rollers 4-10" in diameter, 2" increments from 12-20" diameter and so on.

Wall thickness and shell material have a much smaller contribution to inertia and deflection than diameter. Walls must be sufficient to avoid ovaling or oil-canning.

The journal diameter must be carefully sized to withstand bending and torque stresses without breakage (Chapter 6). Also, generous radii must be provided at journal diametral changes (e.g., inside the bearing and going into the head) to reduce the stress concentration factor. Very conservative safety factors are required here.

The bearing fit on the journal is usually a snug press so that the bearing may have to be induction heated to slip it on. The fit on the housing is a little looser but must still be tight enough to avoid any slippage of the outer race.

The important diametral quality tolerance is TIR (total indicated runout) which may be 0.0001" for the best precision rollers, 0.002" for other metal rollers and about 0.005" for covers.

Figure 1.5
Diametral Roller Geometry

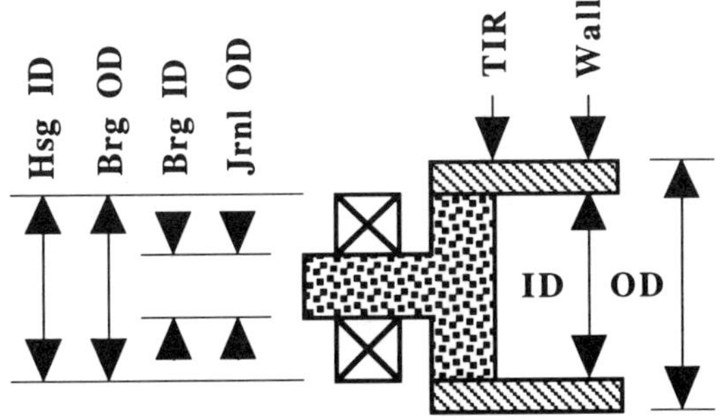

Shell Materials

There are a variety of materials used for roller shells.

Aluminum Tubing - Aluminum tubing (such as 2024 or 6061T6) is ubiquitous for small rollers because it is easy to machine, relatively inexpensive and may have a lower inertia (for a given deflection). However, aluminum tends to wear, scratch and mark easily. To avoid this, some shells are anodized, or plated with a harder material. Also, aluminum has no fatigue endurance limit, as do most steels, so that even a conservative design could result in fatigue failure.

Cast Steel - is centrifugally cast into large diameters, wall thicknesses and lengths.

Composite - graphite epoxy or other composite tubes are sometimes used on rollers where low weight or inertia is required.

DOM Steel Tubing - drawn-over-mandrel (such as UNS G10260) steel tubing is used for many roller shells smaller than 10" in diameter because it has a uniform wall, circularity and straightness, and thus requires less machining and is easier to balance.

Stainless Steel - is used in environments that are acidic, caustic or where good release is required.

Other materials that have been used for shells of roller-like components include fiber cores (spiral wound paper and adhesives), wood mandrels, PVC (poly-vinyl chloride) sewer pipe, and even heavy rubber bands (on a specialized spreader roller).

The economic factors involved in shell material selection include raw material and machining costs and availability in the required length, diameter and wall thickness. The performance factors involved in shell material selection include strength-to-weight, stiffness-to-weight, abrasion and dent resistance, fatigue strength and fracture toughness, corrosion resistance and release.

Coverings and Coatings

There are a variety of materials used in covering rollers as given in Table 1.5. Coverings are used to give nipped rollers more radial compliance and thus soften and even out the stresses, and for unnipped rollers for increasing their traction with the web. Cover materials are selected based on hardness as well as solvent, abrasion and temperature resistance.

Metal rollers are sometimes coated with more durable metals to increase abrasion resistance as given in Table 1.6. Anodized aluminum oxide is used to increase the hardness of aluminum rollers and is common in printing. Tungsten carbide is also commonly flame sprayed on steel rollers to provide a controllable traction that can vary from near mirror smoothness to an extremely rough grit. Both chrome and tungsten carbide may also provide a better release from some webs.

In some cases, rollers are covered with sleeves that slip over axially instead of being formed onto the shell. Some sleeves are made for easy removal and replacement.

Table 1.5
Roller Covers

Acrylic
Buna N (Nitrile)
Butyl
Cotton
EPDM
Felt
Fluorocarbon Elastomers
Hypalon
Natural Rubber
Neoprene
Paper
Polyurethane
SBR
Silicone
Tapes such as cork or polymers
Thiokol
Urethane

Table 1.6
Roller Coatings

Anodized Aluminum Oxide
Ceramic
Chromium
Nickel
Tungsten Carbide

To Drive or Not to Drive

The question of whether and how to drive a roller is asked after roller width is sized based on the maximum web width and the diameter sized based on a maximum deflection (Chapter 3). Typically a large diameter roller, or a light weight web, or a fast accelerating/decelerating machine will require an external drive on some or all rollers.

There is a web tension increase going over an undriven roller in the direction of web travel required to overcome bearing torque, windage and rolling resistance for nipped rollers. Also, speed changes will cause an increase in tension during acceleration and a decrease during deceleration. If this tension buildup across a roll or a series of rolls is too high, the roller must be driven externally (Chapter 4).

Drives may also be required to intentionally step a tension up or down going through a process line. For example, it may be desirable to taper the tension during the buildup of a roll on the winder at the end of a line, but maintain a constant tension upstream through some other critical process. In this case, the winder would be isolated from the rest of the machine with an intermediate driven roller. Similarly, an additional drive point may be required if the web expands or contracts at some process such as during calendering, coating, embossing, heating, and so on.

A driven roller may also be required in an unusual case of an exceptionally slippery web with a light wrap angle. Here, the web may not have enough traction to turn/accelerate the roll even if the tension differential were acceptable. However, a better approach would be to increase the wrap angle and give the roller a more tractive surface (texture and/or cover) rather than hang a drive onto it.

Whether any roller needs a drive or can be driven by the web alone is a fallout from engineering design calculations. Drive points must be justified because they greatly increase the cost, complexity and maintenance of a web handling line.

Roller Drives

A roller can be driven by a variety of mechanical and electrical means as given in Table 1.7. In almost all cases the line is somehow powered by an electric drive motor(s). However, power may be transmitted indirectly from the motor to the roller such as by belting off of a lineshaft or another roller.

The most common electrical drive is a DC motor because it provides a great deal of control flexibility. However, the recent appearance of AC vector drives promises even better control, reduced maintenance and other features, but at an increased cost.

The most common mechanical drive component would probably be the V-belt. The caution here is that the effective pitch, and thus roller speed, varies on V-belts due to temperature, wear, and tension. Chains and sprockets should never be used because they give an oscillatory input to rollers.

If a roller need only be braked instead of requiring power input (motoring), such as with many unwinds, mechanical brakes are a very acceptable and economical alternative. Of these, the pneumatically control disk brake is the most common.

Table 1.7
Roller Drives

Electrical
- AC motor
- DC motor
- Variable Frequency Vector Drives
- Stepper Motors

Mechanical
- Chains and Sprocket (avoid entirely)
- Lineshaft and Gearbox
- Timing Belt
- V-Belt (don't use for draw control)

Brake
- Band Brake
- Eddy Current Brake/Clutch
- Magnetic Particle Brake/Clutch
- Pneumatic Disk
- Pneumatic Shoe

Roller Issues

There are a number of issues with rollers, indeed a whole book full. Here I merely highlight some consistently recurring problems based on my field experiences.

Too Many Rollers!

Most machines have an excessively high roller count. It is true that process rollers add value to a web, such as those used for coating, printing, laminating and so on. Transport rollers, on the other hand, do not add value to a web and in some cases can harm it. In my machine audits, I will typically find that about 10-50% of the rollers could or should be removed

Every extra roller adds to the installation costs of the machine. The additional expenditure is not only for the roller itself, but also for the framework to hold the roller, peripherals attached to it such as drives, as well as for additional floor space. Recurring costs are increased as rollers will need to be realigned periodically. Also, rollers may need to be reground or resurfaced or may need a bearing replacement during their lifetime. Finally, every extra roller increases operational costs for threading and cleanup. Large roller counts also make maintenance and troubleshooting more difficult.

Every additional roller increases the chances of damaging a web. If a roller is misaligned, the skewed web stresses may stretch or break the web on one side, or cause a foldover or crease to form. If a web slips on a roller, the web may be scratched and web control for tension, width and edge position may be compromised. Tension, width and edge position may also be compromised during run if the roller has excessive bearing drag or during speed changes if the roller has excessive inertia. The worst offender may be a nipped roller, which is very unforgiving of roller tolerances and web uniformity. This is why the nipped pull roller (a.k.a., the isolation nip) is to be avoided at every opportunity.

The best time to reduce roller count is at the design layout stage of a machine project. Since even a modest idler roller can cost thousands of dollars, we would expect that the designer would spend an appropriate effort to see if the machine can be arranged to avoid it without functional compromise. Adding a roller to increase a wrap angle on another, to reorient the web, or to reduce span lengths are weak justifications. Indeed, you may be able to judge the skill of the builder merely by counting the number of rollers used to provide a certain set of web functions.

It may be idealistic to assume that all machines are designed and arranged in an optimal fashion. Thus, after startup the plant process engineer with solid web handling skills should reevaluate every roller to see if it was truly required. At this time the engineer should also evaluate whether the supplied tolerances were appropriate for the application. Minimal roller count machines will tend to run smoother and more reliably, and at a reduced risk of damage to the web caused by rollers. In fact, we will show in Chapter 4 that there is a maximum number of idlers for any line.

Complexity

On some machines you will find a plethora of different roller diameters, lengths, mountings and surfaces within a single machine. For example, some of the idlers might have a small diameter, plain surface and be mounted from underneath; while their neighbors might be larger and grooved or cork taped and be mounted from the ends. However, in most cases all of the idler rollers in a line should be identical and sized for the worst condition (180 degree wrap pulling down).

Process rollers are a little more difficult to make interchangeable, but may be worth the effort. For example, I once designed a pilot line where embossing rollers, calender rollers and winding drums were interchangeable. The embossing pattern, calender cover and winder grooves were merely added to a common base cylinder. A machine like this would require fewer spare parts and would be simpler to maintain.

Slender Roll Deflection

Rollers can't be allowed to deflect excessively due to the combined effects of gravity, tension or nip. Deflection tends to cause web contraction, gathering and wrinkling on unnipped rollers. Deflection is even more serious on nipped rollers because the web may wrinkle or the process may be applied nonuniformly at the center with respect to the ends.

In Chapter 3, we show how to calculate deflection and also give deflection tolerances for various grades of precision. However, a crude rule of thumb to avoid deflection problems is to have the length no more than 10X the diameter for nipped rollers and 15X for unnipped rollers.

Alignment

All rollers must be aligned to within the tolerances of about the thickness of a human hair to avoid wrinkling, skewed tension distributions and other web handling problems. The builders of large paper equipment have long taught and practiced precision alignment using optical transits. Their customers, the paper mills, also recognize the need and regularly practice realignment. This awareness has been spreading in recent years to some film, printing and other web applications. However, a lot of web equipment, particularly the smaller machines, are aligned using tape measures and levels, if at all. Unfortunately, hand tools are about an order of magnitude too crude for most applications.

We should expect that the machine builder provide alignment and realignment tolerances in their maintenance manual. All too often, however, the builder may not even be aware of the consequences of roller misalignment or how to calculate tolerances such as given in Chapter 12. Perhaps the largest problem is those builders who know that rollers need to be in alignment, but assume that initial machining can satisfy the requirements. Unfortunately, initial machining alignment is lost due to framework creep and mounting movement, even before the machine is first bolted to the floor. Later, additional framework and foundation movement will further degrade alignment over a period of time.

Thus, we must be ever vigilant to builders, who through arrogance or ignorance, do not make provision for periodic re alignment. The common example here is the machined sideplates where roller mountings are bolt bound to a realignment movement. The irony here is that the builder who takes the care to machine roller mountings to close tolerance may also be the one that precludes an easy realignment when necessary.

Finally, the maintenance department must be careful to preserve an expensive and hard fought alignment during roller changes. For example, the shims, pins and other alignment aids must be tagged and kept with the roller, and reinstalled in exactly the same order and orientation.

Precision

Rollers need to be cylinders which are straight and true to the tolerances of the thickness of a human hair or, in many cases, better. Transport rollers that are bent or have runout transmit tension surges into the web and may encourage loss of traction. Precision is even more important on process rollers because the uniformity of the process depends largely on the geometrical tolerances of rollers. A side benefit to avoiding excessive roller deflection is that the stout roller is much more robust to the occasional accident or overload. Slender rollers, on the other hand, are difficult to machine precisely and will lose their precision when they get bent.

Web/Roller Slippage

Traction can sometimes be lost on rollers so that the web slides across part or all of its width. Factors that aggravate slippage are lightly wrapped rollers, slippery web and roller surfaces, air/fluid entrainment and high inertial or drive torques. If an idler roller can't be configured to avoid slippage, it must be driven.

The obvious result of slipping is roller surface wear and web marking (of some grades). However, a more serious universal result is the loss of web control. When a web changes between floating, sliding or traction, there will be a tension and edge position upset. This becomes particularly debilitating with processes that require registration, or have those with tight tolerances for length, width or thickness.

Excessive Clearance and Compliance

Rollers and the mountings can have excessive clearance and compliance. The resulting problems include machine vibration and alignment difficulties. Bearings can have more clearance than the few ten thousandths of an inch needed for proper operation. An extreme example is some self-aligning bearing styles which may have many mils (0.001") of clearance, even when new. More often, it is a pivoting or sliding roller mounting that has excessive clearance. However, even a tight bearing and mounting can be compromised if it is attached to a flimsy framework, such as a cantilevered beam. In general, the total system clearance and framework compliance under applied loads should be much smaller than values for allowable deflection or misalignment. Framework design is covered in more detail at the end of Chapter 11.

Balance

All rollers, even those running at very low speeds, need to be balanced after machining. Imbalanced rollers cause web surging at any speed, and framework vibration and reduced bearing life beginning at speeds as low as a few hundred FPM (feet per minute). However, many small converting machine rollers are not balanced. A quick check is to spin an idler roller to see if it pendulums to the heavy spot when it stops. Balance grades and balancing are covered in Chapter 11.

Cover Life

The life of a cover may be an order of magnitude shorter than a comparable roller with a metal exterior. Indeed, some roller covers may last only days before they wear out, get cut, swell due to solvents, or the cover comes loose.

Roller Fouling

Rollers seem to collect tape, adhesive, coatings, dust and other contaminants. This can be reduced by using surfaces with better release, such as coatings containing Teflon™ or related polymer.

Web Marking

Rollers with scratches or sharp grooving tend to mark delicate webs, especially when traction is partially or entirely lost. Also, typical grooving patterns used in the industry are too wide for most thin webs. Thus, many films will retain the marks caused by webs deforming into the grooving pattern. In Chapters 4 and 10 we show how to size grooving.

Static Electricity

Static is both a safety and a web handling consideration. Static is often caused by micro or macro sliding of the web across a roller. The generation of static electricity can be minimized by selecting a roller surface that is close to the web material on the triboelectric series, and static can be reduced by methods given at the end of Chapter 18.

Articles - introduction to rollers

1. Roisum, David R. *A Guide to Roller Sizing and Selection.* Converting Magazine, pp 48-50, March 1995.
2. Roisum, David R. *The 10 Commandments of Web Machine Design.* Proc. 3rd Int'l Conf. on Web Handling, Oklahoma State Univ., June 1995.
3. Roisum, David R. *The 10 Commandments of Web Machine Design.* TAPPI Finishing and Converting Conf. Proc., October 1995.

Books - with significant roller coverage

1. Donatas. *Web Processing and Converting Technology and Equipment.* Van Nostrand Reinhold Co., New York, 1984.
2. Weiss, Herbert L. *Coating and Laminating Machines.* Converting Technology Co., Milwaukee WI, 1977.
3. Anon. *Gravure Process and Technology.* Gravure Association of America, Rochester, NY, 1995.

Chapter 2

Tension and Tension Control

In this chapter we show why web tension is needed to size rollers, how to determine a web tension setting for a process, how web tension is controlled and how precisely the tension must be controlled.

Why Web Tension Is Important

Tension is a very important web manufacturing and process setting. If the tension is too low: the web will tend to flop and flutter, will tend to shift and wander sideways going through the machine, and may lose traction on rollers. Conversely, if the tension is too high: portions of the web may be damaged due to local yielding, or web breaks[1] may increase if the material is brittle. However, web tension is also important for sizing rollers.

First, web tension deflects rollers (Chapter 3). True, the magnitude of nip forces (if present) and most roller weights have a higher value than tension. However, the tension contribution is seldom negligible except for very large rollers.

Second, web tension (as well as wrap angle and coefficient of web/roller friction) determines how much torque can be transmitted into the roller to overcome drag, windage and inertia for web driven idler rollers or how much torque can be applied by an externally driven roller before the web breaks loose from its roller (Chapter 4).

Third, web tension determines how much air is pumped in between the web and roller. At higher speeds, a low tension web will tend to lift off a large diameter roller. This results in ever decreasing web/roller traction as speeds increase beyond as little as a few hundred FPM (ft/min). At speeds beyond a few thousand FPM, rollers will often need to be grooved or heavily textured to provide a passage for the air to improve traction. This is similar to the tread on car tires which is needed to reduce hydroplaning when braking on a wet surface.

Fourth, web tension determines changes in length,[2] width[3] and registration.[4] Because web tension is an important process setting, it must be held to within some tolerance. For most web manufacturing and converting lines, we should expect temporal tension (versus time) to be held within ± 10% of its setpoint at and as read by responsive load cells. For example, if the desired nominal tension for a process was 1.0 PLI (pounds per lineal inch), then we might wish to hold tension to within ± 0.1 PLI. Precision tolerances such as this obviously influence how drive controls and sensor feedback must be designed and operated.

However, tension precision directly determines how many nondriven rollers can be used in a web line because tension must be controlled at all locations, not just at a load cell or other point. Thus, we may expect that the downweb tension (versus MD position) should not vary more than say ± 10% of its setpoint within a tension zone. Thus, using the above example, we could only have 10 idlers if the total bearing drag and torque to accelerate an idler roller were only 1/100 PLI!

In summary, the maximum web tension of the heaviest product to be run on any line determines, in part, roller diameter to avoid excessive deflection, as well as drive torque and power. Minimum web tension on the lightest product, on the other hand, helps determine how many rollers can be undriven (or the drive/control precision required for driven rollers), and the minimum wrap angle. More succinctly, maximum tension determines the strength and minimum tension determines the finesse required of a machine design.

In the remainder of this chapter, we describe how to set, measure and control web tension at a sensor (dancer or load cell) or at a control point (drive). We also introduce the concepts of temporal, downweb and crossweb tension variations which will be covered in more detail in later chapters.

Choosing a Process Tension

There are several means of determining a process tension setting for a particular web material. These include previous experiences with similar materials, generic charts and the 10-25% rule. What is most important here is the material rather than the process. In other words, for a given material the optimum tensions for calendering, printing and winding will be quite similar.

The best guide for tension settings are based on experiences with similar materials. If these operations were successful, they will have struck a good balance between running too loose with the attendant difficulties of keeping the web flat and stable throughout the machine, and too tight with the attendant difficulties of web yielding or breaks.

However, the experience need not necessarily be with the same caliper, gauge, thickness or weight. A good estimate of a running tension based on a similar material but of a different **thickness** would be

$$T_2 = T_1 \frac{t_2}{t_1} \tag{2.1}$$

where
T_1 = old tension (lb/in)
T_2 = new tension (lb/in)
t_1 = old thickness (in)
t_2 = new thickness (in)

Here, the units were for lineal tension and thickness in English units. However, any consistent set of units will work equally well.

Lacking any previous experience with a similar material, one can turn to guidelines suggested by machine builders, technical organizations (such as TAPPI for paper), or published in the literature.[5,6] These guidelines will typically be a chart or family of curves of tensions versus thickness/weight for several nominal classes of materials.

Previous experience and generic guidelines work quite well provided that your material has similar MD strength (load to failure), MD stretch (strain at failure), or better yet, yield point.

However, even if you've found no similar tabulation for the material you've developed, the 10-25% tension rule will serve well as a starting point for most web materials.

Tension
Run a tension of 10-25% of the MD strength (or better yet, yield strength) of the material.

Speed
Run a speed differential (often called draw) of no more than 10-25% of the MD stretch (or better yet, yield strain) of the material.

Drawing Operation Exception
True drawing (permanent stretching or elongation) will require tensions equal to strength and speed differentials as determined by the process/product requirements.

Figure 2.1 shows a generic stress-strain diagram for a ductile material. By web convention, we've plotted strength in force/length instead of the mechanics units of force/area. The initial part of the stress-strain curve is linear and elastic (web returns to previous geometry when unloaded) for many materials. At higher stresses, the ductile webs will yield and brittle webs will break. **Yielding** means that when the stresses are removed, the web will retain a permanent increase in length. The near horizontal portion of the curve is typical of many polymers which extend without further increases in load. Notice that the yield strength is only slightly lower than the failure strength as measured ubiquitously in test labs. However, the yield strain is much less than the failure strain for ductile materials.

Figure 2.1
10-25% Rule for Tension Settings

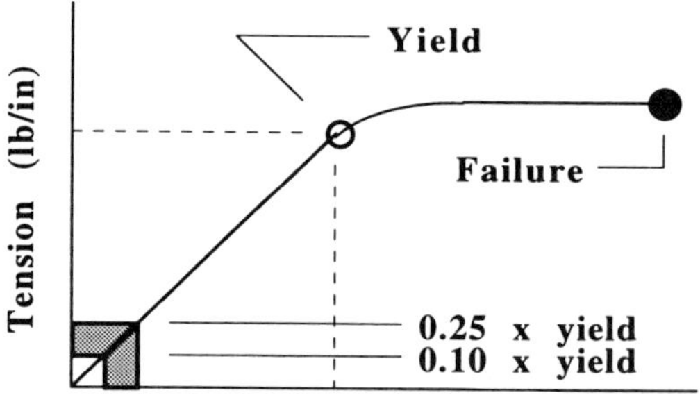

Load Cell Tension

The most direct measurement of web tension is by the use of rollers mounted on load cells.[7,8,9] A load cell is scale which measures a force component produced when a web wraps a roller under tension as seen in Figure 2.2. Most commonly, the load cell is an electronic strain gauge. However, LVDT's (linear variable displacement transformer), spring loaded pneumatic regulators and other force sensors have been used.

In addition to the web tension force, the load cell also must bear the weight of the roller, bearings and mountings as seen in Figure 2.3. Thus, the first rule of load cell sizing is that the **gross** (vector sum of tension and weight) must not exceed the capacity of the cell. Note that some cells have different ratings in the sensing and cross sensing directions.

The second rule of load cell sizing is that the **net** force (tension component in the direction of the sensing) should be at least 10 times the accuracy of the cell for the lowest tension product to be run. For example, if we have a 100# cell with a 1% accuracy, then we would want the web tension force component in the sensing direction (÷2 for a two cell roller) to be at least 10#.

To avoid the proverbial problem of trying to "weigh a pea in a dump truck," the following steps will give an improved sensitivity by increasing the ratio of net to load cell capacity.

1. Select the smallest load cell capacity that meets the first rule of gross weight sizing.
2. Wrap the load cell roller as much as possible.
3. Orient the wrap (or cell) to generate a tension force acting as closely as possible to the sensing direction of the cell.
4. For cells with a vertical sensing direction, orient the wrap so that it pulls upward to relieve roll weight and thus not increase the gross load.
5. Use light rollers and mountings. Composites can be very helpful here.
6. In general, the load cell roller must be an undriven idler so that drive forces and torques are not superposed onto the tension signal.

Figure 2.2
Roller Force from Web Tension

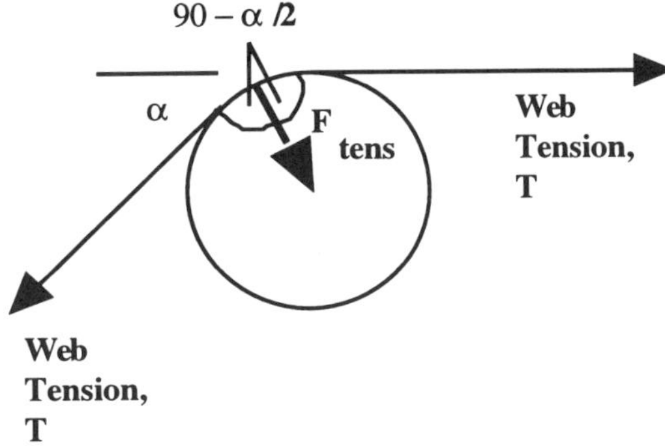

Web tension induces a **force on a wrapped roller**, bar or other element. The direction of the force is oriented with the bisector of the wrap angle. The magnitude of the force is

$$(2.2) \qquad F_{tension} = 2\,T\,W\,\sin(\alpha/2)$$

where
$F_{tension}$ = web force (kN) or (lb)
T = web tension (kN/m) or (lb/in or PLI)
W = web width (m) or (in)
α = wrap angle (degrees)

For a load cell on each end of a roller, the cell will see a gross load that is the vector sum of

(2.3)
$$\mathbf{F}_{\text{load cell}} = \frac{1}{2}\left(\mathbf{F}_{\text{tension}} + \mathbf{F}_{\text{wgt (roller, brgs, mtgs, etc.)}}\right)$$

Figure 2.3
Forces Acting on a Load Cell

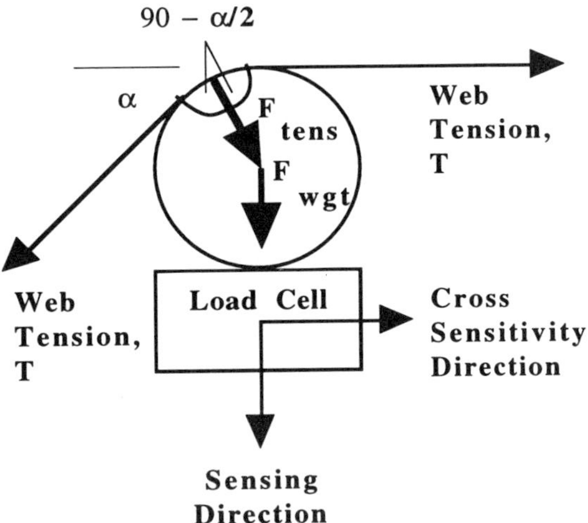

Load Cell Sizing Example

Metric Version

Given a 2 load cell system with the orientation of Figure 2.3 with a 45 degree wrap angle and

F_{wgt} = 50 kg = weight of roller etc.
T = 0.2-0.5 kN/m = web tension range
W = 1.5-1.8 m = web width range

Then,

$$F_{\text{tension max}} = 2\,T\,W\sin(\alpha/2)$$

$$= 2 \times 0.5\frac{\text{kN}}{\text{m}} \times 1.8\text{m} \times \sin(45/2)$$

$$= 0.689\ \text{kN}\ \frac{102.0\ \text{gf}}{\text{N}} = 70.3\ \text{kgf}$$

$$F_{\text{lc vert max}} = \frac{1}{2}\left(F_{\text{tens max vert}} + F_{\text{wgt vert}}\right)$$

$$= \frac{1}{2}\left(70.3\cos(45/2) + 50\right)\text{kgf}$$

$$= 57.5\ \text{kgf}$$

Given mechanically overload protected cells available in 20, 50 and 100 kg capacities, the 100 kg cell would be required. Now let's check sensitivity assuming the cells are accurate to 1% of full load.

Then, the accuracy of the 100 kg cell would be 0.01 x 100 = 1 kg. Then, the tension component in the sensing direction for even a crude application must be at least 10 x 1 = 10 kg.

$$F_{\text{tension min}} = 2\,T\,W\sin(\alpha/2)$$

$$= 2 \times 0.2\frac{\text{kN}}{\text{m}} \times 1.5\text{m} \times \sin(45/2)$$

$$= 0.230\ \text{kN}\ \frac{102.0\ \text{gf}}{\text{N}} = 23.4\ \text{kgf}$$

$$F_{\text{tens sense min}} = \frac{1}{2}\left(23.4\cos(45/2)\right)\text{kgf}$$

$$= 10.81\ \text{kgf}$$

This load cell system just meets the crude sensitivity requirements, but would probably compromise the tension control accuracy of a nice drive.

English Version

Given a 2 load cell system with the orientation of Figure 2.3 with a 45 degree wrap angle and

F_{wgt} = 100 lb = weight of roller etc
T = 1-3 lb/in = web tension range
W = 60-70 in = web width range

Then,

$$F_{\text{tension max}} = 2\,T\,W\sin(\alpha/2)$$

$$= 2 \times 3\frac{\text{lb}}{\text{in}} \times 70\text{in} \times \sin(45/2)$$

$$= 160.7\ \text{lb}$$

$$F_{\text{lc vert max}} = \frac{1}{2}\left(F_{\text{tens max vert}} + F_{\text{wgt vert}}\right)$$

$$= \frac{1}{2}\left(160.7\cos(45/2) + 100\right)\text{lb}$$

$$= 124.2\ \text{lb}$$

Given mechanically overload protected cells available in 50, 100 and 200# capacities, the 200 lb cell would be required. Now let's check sensitivity assuming the cells are accurate to 2% of full load.

Then, the accuracy of the 200 lb cell would be 0.02 x 200 = 4 lb. Then the tension component in the sensing direction for even a crude application must be at least 10 x 4 = 40 lb.

$$F_{\text{tension min}} = 2\,T\,W\sin(\alpha/2)$$

$$= 2 \times 1\frac{\text{lb}}{\text{in}} \times 60\text{in} \times \sin(45/2)$$

$$= 45.9\ \text{lb}$$

$$F_{\text{tens sense min}} = \frac{1}{2}\left(45.9\cos(45/2)\right)\text{lb}$$

$$= 21.2\ \text{lb}$$

Since the tension signal is too low, we would need a 1% accurate cell or try wrapping the cell nearly 90 degrees, or wrap pulling up or lighten the roller so that we could drop down to the next lower capacity. Then we would reiterate on both the capacity and sensitivity calculations. This typical example illustrates why load cell rollers usually need a heavy wrap angle.

Other Load Cell Considerations

Selecting a Load Cell

While there are many considerations for load cell selection beyond those discussed earlier, we will focus on four that are most commonly overlooked. First and most important, the cell should have a 10X or better mechanical (not electrical) overload protection. Mechanical protection is an internal stop that keeps accidental overloads from excessively deflecting and damaging cells. Unfortunately, some vendors do not provide this feature. Second, have spare cells because they will fail eventually.

Third, high speed applications (>1000 mpm) may require high mechanically damped load cells. The is because roller/cell/framework system resonance is almost inevitable, causing high amplitude imbalance noise to compromise the signal. Unfortunately, I am only aware of one load cell vendor that provides this feature. Fourth, high speed load cell rollers must be well balanced as covered later in the book.

Amplifier & Display

Raw load cell signals are amplified and then sent to a display and/or controller. Among many other important electrical considerations, the black box should have the following features.

1. Be able to sum 2 or more cells for tension control use.
2. Have a selector for front, back and difference. Being able to select either a front or a back makes troubleshooting a bad cell much quicker. Being able to select a difference can be used to indicate how level the tension is across the web. This could also be used to automatically position a guide or squaring roller.
3. Have an adjustable filter to clean up noisy signals coming from web bounce or roller imbalance.
4. Be able to scale the display to engineering units of kN/m or lb/in (PLI). An analog meter is much easier for the operator to use than a digital display.
5. Have an additional 0-5 VDC signal terminal to allow hookup of a strip chart recorder or other device for troubleshooting.
6. Lastly, it is sometimes easier to work with a vendor who supplies both load cells and drives (electric motors or mechanical brakes).

Calibrating Load Cells

Calibrating load cells involves three steps: set zero, set gain, recheck zero. The zero is adjusted without any load applied to the roller. The gain is calibrated by stringing a rope through the machine upon which weights are hung. The position of the rope is centered on the upstream, load cell and downstream rollers following the same path as the web as shown in Figure 2.4. Weights should be hung to give a load near the maximum of the cells to set gain, and half that to crudely check linearity.

Troubleshooting Load Cells

1. See if the load cell display moves when you push on the roller. If not, consult the load cell manufacturer's troubleshooting procedure.
2. Check to see if the zero has moved since the last calibration. An unexplained movement of zero on a single cell is a symptom of the most common load cell problem: it has been mechanically damaged because it was overloaded or struck by something.
3. Change out a cell with a spare: did the new cell solve the problem? If not, it might be the amplifier box. Try changing it out.
4. Noisy load cell readings come from two major sources. The first is mechanical vibration. The second is electrical noise. Systems where the amplifier is remote from the cell are most prone to electrical noise.
5. A 4-20mA signal is much more robust than a 0-5V signal.

Figure 2.4
Calibrating a Load Cell with a Weighted Rope

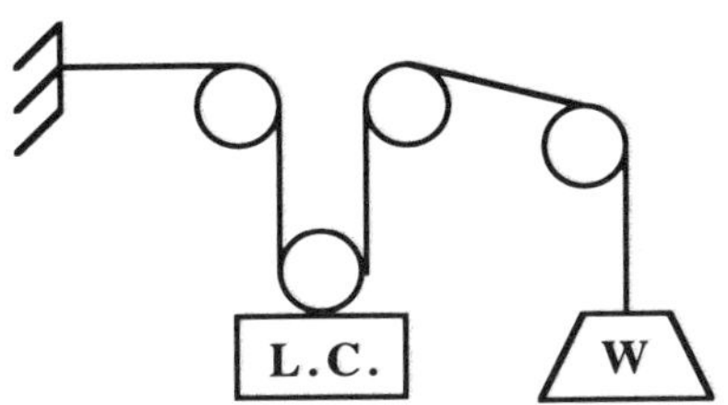

Dancer Controlled Tension

The most common design arrangement of a dancer is shown in Figure 2.5. The dancer roller is connected to a pivoting arm whose position is measured by a rotary or linear pot (potentiometer).[10,11] If the tension is high, the dancer is pulled upward and the pot sends a signal to the downstream drive to slow down (or the upstream drive to speed up). If the tension drops, so does the dancer roller.

A minimum tension is provided by the weight of the roller and a portion of the arm. To change tension, pneumatic cylinder pressure might be varied or weights might be added or removed.

Dancers are one of the oldest web tension sensors, and are still commonly used today. Their primary advantage is that they can accommodate drives that are sloppy (as were most before digital drives) or for applications which have extremely fast tension transients (such as on flying splices). The dancer is tolerant of fast tension transients and slow drives because of the length of stored material in the loop that can be drawn from or fed into. A less common arrangement of dancer is a festoon as seen in Figure 2.6. This provides even more stored material and is used for zero speed splicing of unwinding or winding rolls.

However, there are several disadvantages and misconceptions with dancers. The first disadvantage is that, while it does respond to changes in tension, it does not display tension as does a load cell. Thus, there is no output of average or instantaneous tension to be used for process design, quality control or troubleshooting. The best that can be done is to calculate the (average) tension as a function of cylinder pressure or arm counterweight. Even then, however, one would not be able to determine tension excursions to gauge the health of the drive control system.

Another disadvantage of the dancer is the ubiquitous use of pots (variable potentiometers) for position sensing. An example of a pot is the volume knob on a radio. It is not durable to continuous turning. Also, the pot fails in the worst manner possible as occasional and erratic control glitches as the slider gets dirty.

Figure 2.5
A Pivoting Dancer

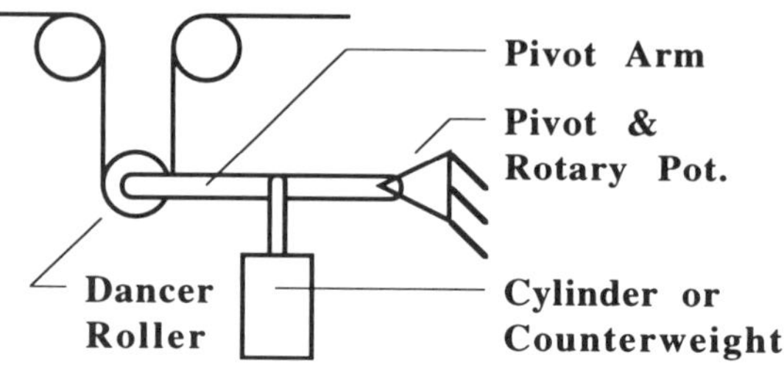

Figure 2.6
A Festoon Dancer

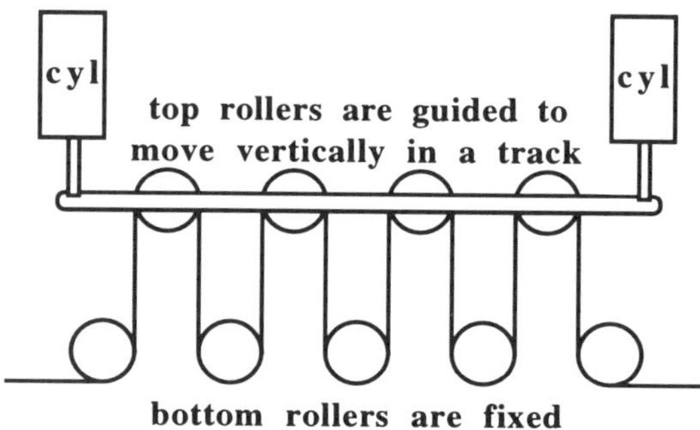

However, the best evaluation of dancer/drive performance is to measure tension excursions after the dancer with a load cell. As seen in Figure 2.7, tension is poorly held, especially during speed changes. Much of the poor performance of dancers is excessive **mechanical friction** of the pivot and cylinder assembly which should be much less than 1/10th of the force of the lightest tension to be run. It is also important to keep dancer **mass** to a minimum to improve response.

Figure 2.7
Tension (Im)Precision of a Dancer

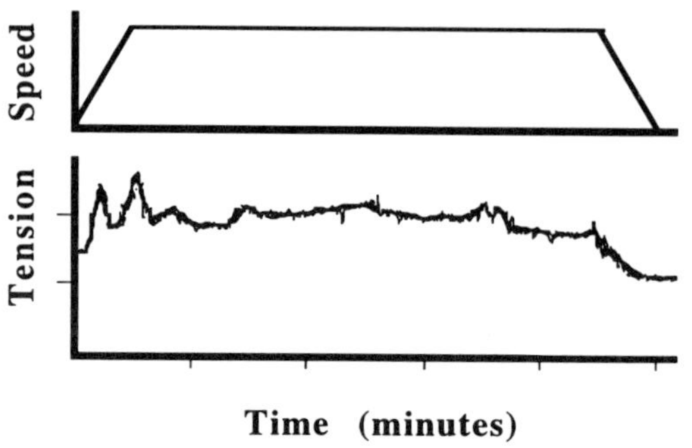

Dancer Design

The first step in dancer design, as it is for load cells, is to size the **roller**. As discussed in more detail in Chapter 3, the nominal diameter is primarily determined by deflection standards which tend to yield slenderness ratios (length/width) of around 15. Then, the minimum roller weight is largely determined because wall thickness and shell material are smaller factors than diameter. Of course, the roller can be made heavier by increasing wall thickness or adding weight.

Second, we must choose a dancer **configuration** that minimizes the required cylinder size so that cylinder friction which compromises sensitivity is also reduced. Indeed, the cylinder friction alone will often put a dancer out of spec without even considering pivot/slide friction or regulator hysteresis. The two primary configurations are the horizontal and vertical arrangements as seen in Figure 2.8. The vertical arrangement is suggested for materials whose nominal running tension is similar to the already calculated roller weight. The horizontal dancer is more appropriate for light webs whose tension is much lower than roller weight. Note that translating dancers on four bar linkages and particularly slides (festoons) should never be used because of the friction penalty.

Third, we need to size and position an air cylinder to provide coverage of the required **tension range** such that the minimum tension is achieved with no less than 10 psi, and the maximum tension is achieved with no more than 60 psi. Often, we may need to mechanically weight or counterweight the dancer to get the assembly to balance at about the midpoint of the design tension range. If the tension range is quite large, then we may need to provide adjustable counterweights.

Fourth, we must choose a position **sensor**. Since the dancer is in near constant motion, contacting sensors such as rotary pots don't tend to hold up well and can add additional friction of their own. Thus, non-contacting sensors such as servo-pots, LVDT's or ultrasonics may be preferred. Optical encoders may not work well if the dancer doesn't stay put long enough to make the read.

Figure 2.8
Dancer Configurations

vertical dancer for heavier webs

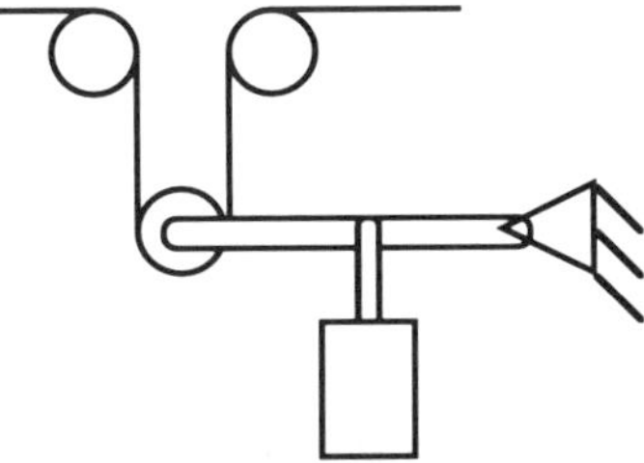

horizontal dancer for lighter webs

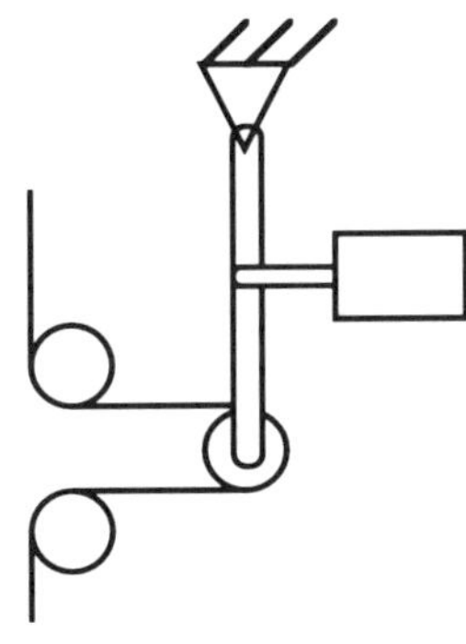

Finally, we need to **calibrate** the dancer which is a graph of tension versus cylinder pressure. This is done by a weighted rope following the web path much like already described for load cells. With a given weight, the cylinder pressure is raised until the dancer moves in one direction, and lowered until the dancer moves in the other. The average pressure is used to make the chart, while the difference is a good measure of sensor **sensitivity**. The hysteresis of the dancer should be <<10% of the minimum design tension. The process is repeated for several weights corresponding to the design tension range.

The **vacuum dancer** is a design variation for tensions that are light, but heavier than can be provided by a catenary. The vacuum dancer uses a light weight puck instead of a pivoting roller to hold loop geometry. The loop and puck are contained in a vacuum chamber to provide additional tension. Loop length feedback is often provided by photoeyes. The most common application is for infeed/outfeeds of magnetic tape drives on some mainframe computers.

Dancer Dynamics

A common misconception is that the dancer provides a damping of tension disturbances as might be generated by an out-of-round supply roll, unstable drive, or vibrating roller. While this can happen, the realities show much more mixed results.

Figure 2.9 shows the **frequency response** of a typical dancer system. This is a plot of the ratio of tension disturbances before and after the dancer as a function of the frequency of the disturbance. A value of one means that the tension disturbance passes through unaltered. A value less than one means the dancer is attenuating or reducing tension fluctuations, while a value greater than one means the dancer is exaggerating or increasing tension fluctuations.

As seen in the figure, the dancer does little to change web tension excursions at low frequencies. However, at intermediate frequencies the dancer system goes into resonance and actually magnifies tension disturbances. The position and severity of the peak depends on the dancer's inertia and other dynamic factors. It is only at very high frequencies that the dancer provides a damping or reduction of tension disturbances. Thus, it might be helpful for shock loads coming from a flying splice or rotary die-cutter where many of the frequency components are high.

Those familiar with vibration will note the similarity of the graph to the response of a single spring mass system. Those who think in electrical terms will classify this system as a low pass filter. Though perhaps surprising to some, the dancer has the same frequency response character as an undriven idler roller. Thus, if damping is the only criteria, a high inertia roller may give similar results.

There is an exception to the mixed benefits of the typical dancer. This is the **inertia compensated dancer roller** that was developed and patented[12] (expired) by Martin Automatic Inc. The frequency response shape is still similar but lies mostly on the attenuation side.

Figure 2.9
Dancer Damping of Tension Fluctuations

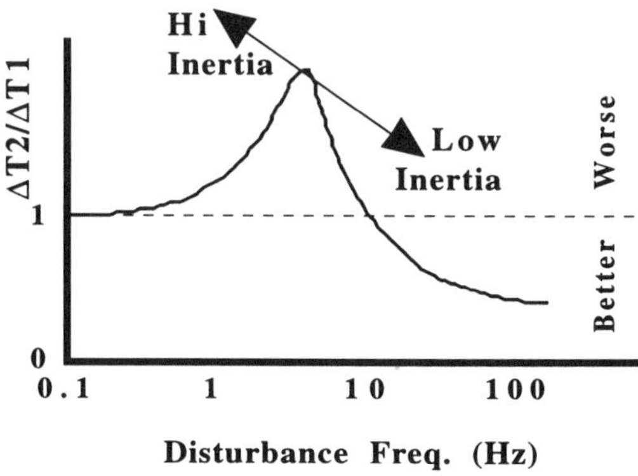

According to the patent, a complete attenuation of tension disturbances is achieved if

$$J = M r^2 \tag{2.4}$$

where

J = polar inertia of the dancer roll(ers)
M = dancer mass
r = dancer roller outer radius

For a typical single translating or pivoting dancer roller, this will require the addition of a flywheel (rotary inertia) to the ends of the roller. Conversely, multiple roller festoon systems may require the addition of mass (translational inertia) to the moveable carriage.

While a desktop demo and the patent are convincing, there were a few higher order terms left out of the derivation, as was the case in other early models.[13,14] The realities of the inertial compensated dancer roller are generally beneficial, but short of the complete attenuation sometimes claimed. This has been verified both experimentally and analytically[15,16] by several independent researchers. While the inertia compensated design might lay a serious claim to smoothing out tension disturbances, the question must be asked if it is necessary and sufficient. Indeed, even idler rollers can significantly reduce the transmission of tension disturbances downstream.

Draw Control

Draw control is the speed regulation of several driven rollers in a process line. While the equipment and controls are quite simple, draw or speed control is **limited to very stretchy materials** such a tissue paper, stretch film, nonwovens and some textiles. (True drawing operations will permanently stretch the web.)

Figure 2.10 is an example of draw control on a roll-to-roll converting line. Here, the unwind, a converting process (calender, emboss, print, slit, etc.), and a windup are all driven at a controlled surface speed. However, the tach or encoder connected on the back end of a motor measures RPM (revolutions per minute), and must calculate surface speed from the equation for rigid body motion,

(2.5) $V = \omega r$

which can be given in conventional units as:

(2.6a) Metric Units

$$V = 0.031416 \times D \times RPM$$

(2.6b) English Units

$$V = 0.26180 \times D \times RPM$$

where the speed V is in MPM (m/min) in the metric system and in FPM (ft/min) in the English system, and while D is the outer diameter of the roll(er) in (cm) for the metric system and (inches) for the English system. The **diameter** of rollers will be measured by a caliper, while the changing diameter of unwinding and rewinding rolls may be measured by a variety means such as ultrasonically.

The draw between any two rollers is expressed as a percent speed change with respect to some reference which should be stated. For example if the converting unit were the speed reference for the system, the windup draw would be

$$\%Draw_{3/2} = 100 \times (V_3 - V_2) / V_2$$

Web strains (elongations) can be calculated from speeds or draws provided that a strain is known somewhere. For example,

Figure 2.10
Draw Control in Roll-to-Roll Converting

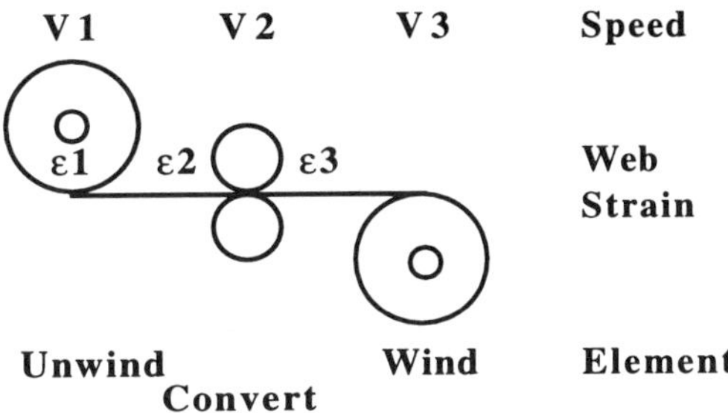

Figure 2.11
Draw Control in a Manufacturing Line

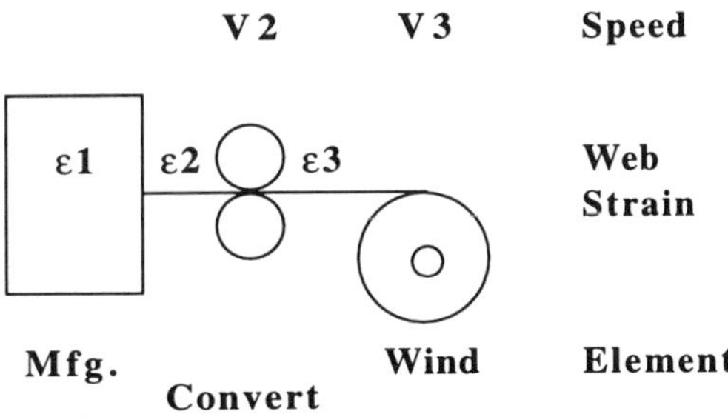

(2.7a) $$\varepsilon_2 = \varepsilon_1 + \frac{\%Draw_{2/1}}{100} \text{ and}$$

(2.7b) $$\varepsilon_3 = \varepsilon_2 + \frac{\%Draw_{3/2}}{100} \text{ etc.}$$

Notice, however, that in the roll-to-roll converting line of Figure 2.10, we would generally not know the strain present in the unwind (ε_1). Similarly in the manufacturing line of Figure 2.11, we would generally not know the strain coming out of the maker (ε_1). Thus, **draw control is not closed loop** on the web because there is no web sensor. This means it is, by itself, unable to measure or compensate for upstream web fluctuations.

Finally, **draw control precision** needs to be appropriately sized. If for example, a typical quality speed control can achieve a 0.1% accuracy, then anything less than a 1% draw setting would be imprecise (10% variation), which means the MD stretch (or better yet, yield) should be greater than 10% (minimum tension 10% of breaking value).

Tension Variations in a Draw Zone

Draw control is required in areas where the web is being yielded because the stress-strain curve is relatively flat there. In other words, small changes in tension can cause large changes and often unpredictable changes in anelstic strain (i.e., permanent length changes). Thus draw is more suitable than tension control in stretching and true drawing operations. Examples include the wet end of a paper machine, tenting of plastic film, and in some oven applications.

However, draw control is also widely used as a cheap but crude form of tension control in the elastic region of a material's response. Here, the objective is to control tension indirectly through draw. Unfortunately, the consistency with which the tension (or **registration**) is held in any particular draw zone is affected by a number of parameters. These factors vividly demonstrate the common fallacy of draw "**tension isolation**."

One strong factor in the variable effects of draw is the **pre-draw zone tension**. In other words as we discussed in the previous section, any changes in the upstream tension will be passed down through the next zone.

Another factor is the **percent draw**, which is the nominal setpoint. Note that it is completely acceptable to run zero or slightly negative draws as that only means that web tension is maintained or drops slightly in the zone. However, we also need to be concerned with how closely the drive system holds the target speed. Clearly, the speed increments and tolerances need to be <<10% of the web's strain.

Web modulus also affects tension such that an increase in modulus will increase the tension at that point. However, if the modulus is increasing because the material is thicker, the average stress will be similar even though tension has changed.

Any **hygrothermal** process that changes the length of the material also changes the tension. For example, if the moisture (paper and some polymers), solvent loading (if it causes the material to swell), or temperature (polymers) increases, the tension will drop and vice versa. This, of course, is in addition to any modulus changes that will also frequently result.

Traction must be maintained on draw controlled rollers. If the web slips on a positive draw zone, the tension will not increase as much as calculated and vice versa. However, as will be discussed in detail in Chapter 4, we can avoid slippage by proper design.

Covered nipping rollers will have an effective radius that increases with increasing nip as discussed in Chapter 5. Conversely, lowering the **nip loading** on a soft pull roller drive section will effectively decrease the draw and hence tension on that zone.

Finally, there is a **time constant** involved in draw control that makes tension react sluggishly to draw changes. For example, if the downstream roller is immediately sped up, the tension does not immediately increase. Rather, it increases as an exponential as shown in Figure 2.12 with a time constant τ = length/speed. Note that similar issues are involved with **registration** (strain) using tension control.

(2.8a) $$T_2 = T_1 + (E\,\Delta\varepsilon\, c)\left(1 - e^{-t/\tau}\right)$$

where

(2.8b) $$\Delta\varepsilon = \frac{V_2 - V_1}{V_1}$$

Figure 2.12
Time Constants and Draw Control

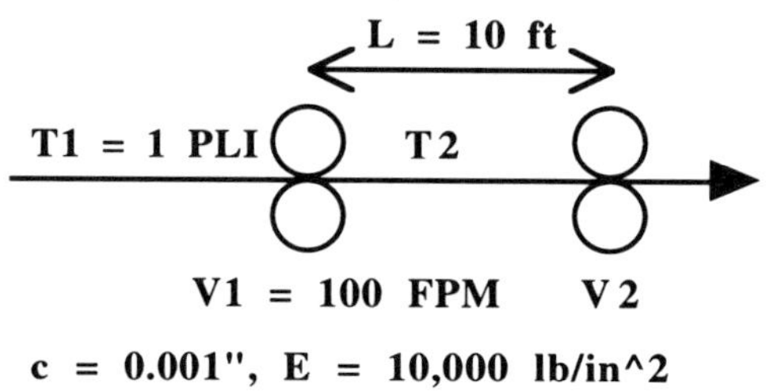

c = 0.001", E = 10,000 lb/in^2

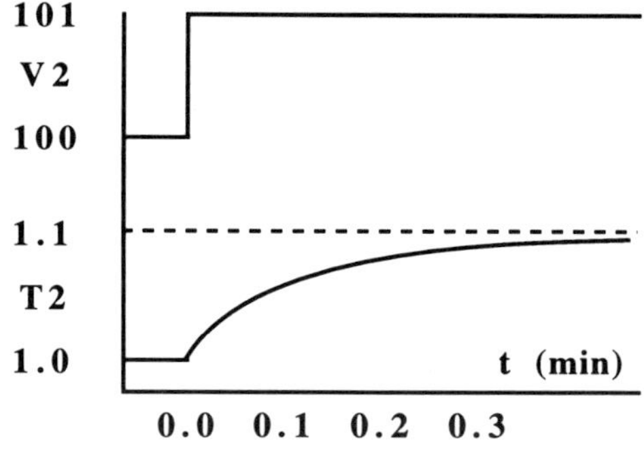

Catenary (Loop) Control

Some applications require such exceptionally **light tensions** that load cells or dancers may not have suitable resolution. If this material also had **limited stretch**, then draw control would also not be suitable. Fortunately, light tensions can be run using catenary control.

The shape of the sagging arc of a telephone line, clothesline (Figure 2.13), kite string or web between two rollers are all the same. The shape goes by the fancy mathematical term of catenary. We can use the catenary for light tension applications as well as for a poor man's tension measurement on machines not equipped with load cells or other tension sensors. The catenary curve as seen below, has a curve defined as:

$$(2.9) \qquad y = \frac{a}{2}\left(e^{x/a} - e^{-x/a}\right)$$

where x and y are the horizontal and vertical positions respectively and "a" is called the parameter of the catenary. Mathematically, "a" is the distance from the lowest point in the curve to the directrix. Physically, "a" relates weight and tension. For webs, a is calculated as:

$$(2.10) \qquad a = \frac{T_h}{w} \quad \text{where}$$

T_h = horiz'l component of tension (force/width)
w = basis weight (weight/area)

Both rollers on each end of a span must lie tangent to the curve defined above and shown in Figure 2.14 The application of the catenary to calculate tension from the web sag, span and basis weight as seen in Figure 2.15. This can be simplified for the specific case of a shallow horizontal curve as:

Figure 2.13
A Catenary Sag

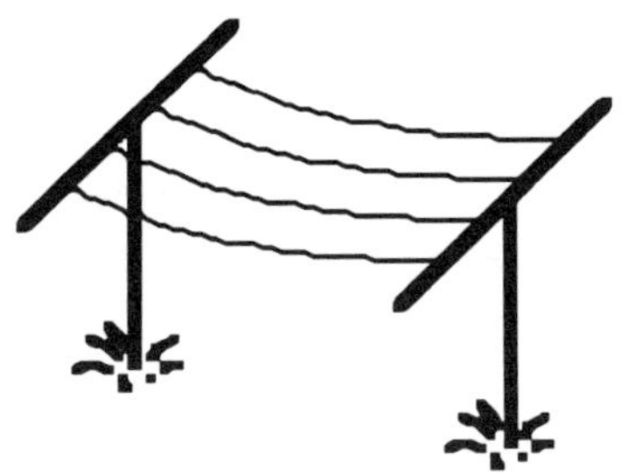

Figure 2.14
Rollers are Tangent to Catenary Sag

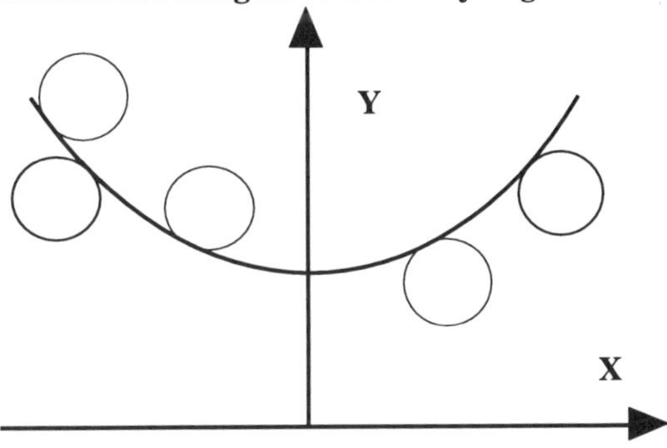

$$(2.11) \qquad T = \frac{L^2 w}{8 s} \quad \text{where}$$

L = span length between roller tangents
w = basis weight
s = sag

<u>Example Problem</u>
Given a 20# light paper grade sagging 0.25" on a horizontal span 50" long, what is the tension?

$$T = \frac{L^2 w}{8 s} = \frac{(50 \text{ in})^2 \left(\frac{20 \text{ lb}}{3000 \text{ ft}^2} \frac{\text{ft}^2}{144 \text{ in}^2}\right)}{8 (0.25 \text{ in})}$$

$$T = 0.0579 \text{ lb/in}$$

As an exercise, the reader can verify that this result when plugged into equations (2.9) and (2.10) will yield a legitimate catenary.

The technique works well on lightly stressed webs with long spans. Sag can be measured with a ruler, optically or ultrasonically.

Figure 2.15
Parameters of the Catenary

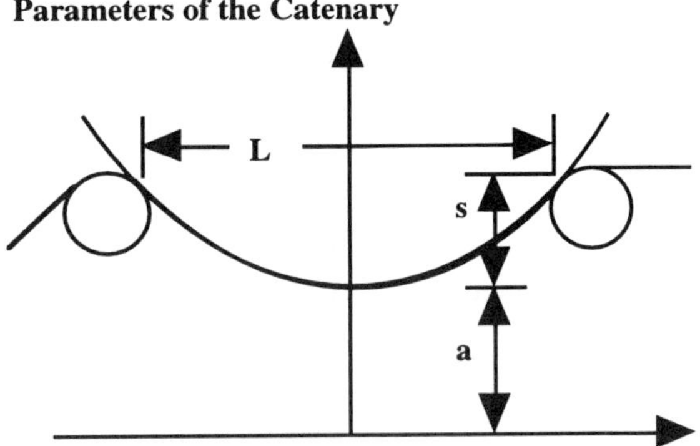

Tension from Motor Readings

Tension is a byproduct of motor torque. This knowledge enables us to approximate web tension from electric motor drive readings. It can be applied to unwinds, windup sections, and rollers. For unwinds and windups, the value will be for absolute tension. If applied to a roller, the calculation is for a tension differential across the roller. However, either measuring or controlling tension via amperage neglects inertial effects due to speed changes. The meter amperage used is the different between values of running with and without web tension.

Need to know:

- amperage from meter during run
- motor amp and power nameplate values
- web width
- speed

Efficiency can be assumed to be 90% for many systems (10% friction and other losses). However, friction can be high on some drive components and with some punky materials rollin on a drum.

English Units

(2.12)

$$P_R = P_{nameplate}\left(\text{Efficiency}\right)\left(\frac{\text{Amps}_{meter}}{\text{Amps}_{nameplate}}\right)$$

where
P_R = power during run (HP)
$P_{nameplate}$ = motor nameplate power (HP)
Efficiency = elect & mech efficiency ≈ 0.90
Amps_{meter} = meter reading during run (amp)
$\text{Amps}_{nameplate}$ = motor nameplate current (amp)

(2.13)

$$T = \frac{P_R \times 33{,}000}{W \times S}$$

where
T = tension (lb/in)
P_R = power during run (HP)
W = web width (in)
S = speed (ft/min)

Example for two-drum winder given:
front drum and rear drum motors
rated at 100 HP and 160A each,
front drum runs at 100A, rear drum at 60A
speed = 6000 fpm, width = 200 inch

$$P_R = (100+100)\left(0.90\right)\left(\frac{100+60}{160+160}\right) = 90 \text{ HP}$$

$$T = \frac{90 \times 33{,}000}{200 \times 6{,}000} = 2.48 \text{ lb/in}$$

Note: From (2.12) you can also size the tension power requirements (does not include accel/decel or losses).

Metric Units

(2.14)

$$P_R = P_{nameplate}\left(\text{Efficiency}\right)\left(\frac{\text{Amps}_{meter}}{\text{Amps}_{nameplate}}\right)$$

where
P_R = power during run (kW)
$P_{nameplate}$ = motor nameplate power (kW)
Efficiency = elect & mech efficiency ≈ 0.90
Amps_{meter} = meter reading during run (amp)
$\text{Amps}_{nameplate}$ = motor nameplate current (amp)

(2.15)

$$T = \frac{P_R \times 60}{W \times S}$$

where
T = tension (kN/m)
P_R = power during run (kW)
W = web width (m)
S = speed (m/min)

Example for an unwind given:
motor nameplate rating 150 kW
motor nameplate current 300 amp
motor running current 150 amp
speed = 2000 mpm, width = 5 m

$$P_R = (150)\left(0.90\right)\left(\frac{150}{300}\right) = 67.5 \text{ kW}$$

$$T = \frac{67.5 \times 60}{5 \times 2{,}000} = 0.41 \text{ kN/m}$$

Note: From (2.14) you can also size the tension power requirements (does not include accel/decel or losses).

Current Control

Since tension is a result of armature current in the ubiquitous DC motor, we can use it to crudely control web tension. All that need be done is to calculate from the previous section the required current to obtain a specified tension (unwind or windup) or tension differential (intermediate roller).

An example usage of this technique is on some large paper machinery unwinds. Here, the unwind tension is provided by large (50+HP) regenerative motors. Normally, the tension is controlled via load cell feedback. However, if the load cells should fail, which happens occasionally in that rugged environment, current control is the fallback to keep the process running. The operator merely switches to current and the former tension rheostat becomes a current controller which he sets to a value similar to what was normally observed.

Current control is open loop on the web, and as such results in compromised tension control. First and foremost is the large inertia component of tension during speed changes. In current control, the tension will rise during accel and drop during decel for unwinds and vice versa for intermediate rollers and winders. While some of this variation can be reduced by automatically adjusting current, inertia can vary many orders of magnitude on winding and unwinding rolls.

Other problems include drag, friction and other losses that are usually variable. Large losses will be seen on rolling nips, particularly if one of the elements is soft (covered roller or wound roll) and highly loaded. Other losses include bearing drag, sliding element drag (turnbars, stationary spreaders, etc.) and windage. Finally, there are many motor complexities that can also increase tension uncertainties when using the open loop current control scheme.

Torque Control

Pneumatic brakes, magnetic particle brakes and clutches, slipped core winding shafts and tendency drives are common converting components. These and similar devices are torque controlled actuators.

For example, the **air pressure** set by a regulator primarily determines the output torque of a pneumatic brake. Similarly, the magnetic particle device has an output torque approximately proportional to current. The slipped core winding shaft torque is primarily determined by axial load. Finally, the tendency drive torque is a reflection of bearing torque.

Unfortunately, none of these devices are very **repeatable**. For example, the pneumatic brake torque can also vary as a function of the friction coefficient of a particular pad, with rpm and with temperature (brakes fade at high temperature). Similar effects are found on the other torque devices as well.

Many plants operate these torque devices in **open loop** on the web. Thus, for example, an operator will set brake torque, and thus indirectly tension, through an air regulator. Unfortunately, the results will often be tension excursions exceeding standards of decency. The solution for poor tension control in this case is quite simply to use load cell (or dancer) feedback to run the devices instead.

Of even graver concern is the use of open loop torque devices on unwinds. The problem is compounded here because the torque must be reduced proportionally as the diameter of the unwind roll decreases. True, the operator can manually lower the setting as the roll is unwound, but what is the quality of the results? Also, there are diameter compensating devices that can reduce the error a bit, but why not simply install load cells instead?

<u>Good practice for web processes:</u>

No friction without feedback!

Multiple Drive Sections

Most web manufacturing and converting lines have more than one drive section. **A drive section is bounded on each end by a controllable drive**. If the section is in tension control, it must contain a dancer or load cell. A section has a nominally constant tension throughout, with minor changes in each span due to tension differential across undriven rollers, helper drive errors and so on.

The number of drive sections should be kept to a minimum to reduce cost, complexity and maintenance. Sometimes, machine builders to put in too many drive sections. For example, an "**isolation nip**" or pull roller is sometimes placed between an unwind and a critical process with the misconception that unwind tension fluctuations will be not be propagated downstream beyond that point. Unfortunately, in tension control the web bounce will pass through the isolation nip much like any similar sized undriven roller because the controller will not be able to react quickly enough. Similarly, the tension fluctuations will pass effortlessly across a draw zone because of the principle of strain transport.

Separate drive sections are required when there is a clear need for a different tension in one zone versus another, such as where the web material undergoes drastic physical changes (e.g., laminating, coating, drying and embossing). In these examples, the web's strength and consequently its appropriate running tensions may have changed considerably. Also, large process rollers often need to be driven which also creates another tension zone. However, the same tension settings should be considered upstream and downstream unless the web's strength has changed considerably. Indeed, tension sensitivity is an indicator of poor process health.

There is nothing in principle wrong with **mixing drive modes.** For example, a nonwoven unwind might have a dancer immediately following to set tension, with the remainder of the sections under draw control. The issue is that multiple drive sections are difficult as is, and mixing modes only increases complexity.

Figure 2.16
Drive Sections in Roll-to-Roll Converting

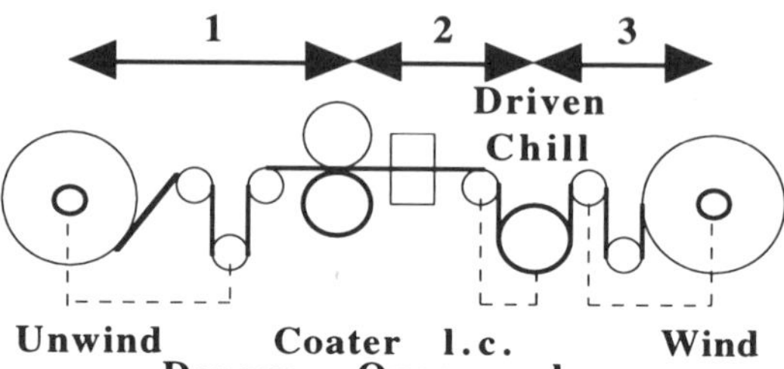

Figure 2.16 is an example of a 3 section drive roll-to-roll converting line. Here, the coater is chosen as the **speed reference** for the line. The speed reference sets and maintains the pace for the entire line. There must be one and only one speed reference in any line. The speed reference is usually chosen to be either end of the line or an important intermediate process.

In the first section, the dancer provides feedback to the unwind brake (or motor), and thus regulates the tension between the supply roll and the coater. The second section, between the coater and driven chill roller, is regulated by feedback from the load cell. However, if the material was intentionally yielded or drawn in the oven, the second section should be changed to draw control.

In this example, we've chosen to isolate the windup tension from the oven tension. This is commonly done to allow an automatic and programmed **taper tension** during the build of the windup roll while maintaining a constant tension throughout other critical upstream processes (drawing, slitting, etc.). This requires a load cell sensor, a roll diameter sensor, and a programmable controller or other means to taper the tension.

Note that the unwind and windup drives need to be **inertia compensated**. This automatically varies the gain as a function of current roll diameter to maintain an appropriate compromise between low gain sluggishness and high gain instability. Pneumatic or mechanical brakes attached to winding/unwinding rolls should also be inertia compensated for the same reasons.

Another example drive layout for a **laminator** is shown in Figure 2.17. In nearly all cases, the speed reference for laminating lines should be the laminator itself. In this example, there are three separate tension zones because there may be three distinct web strengths. The two unwinds are each load cell controlled to their appropriate tension, and the windup is load cell controlled to its appropriate laminate tension.

With lamination, however, there are additional constraints imposed on the tension settings of the ply constituents to avoid **curl**[17]. as given in Table 2.1.

Table 2.1
Tension Rules of Lamination

Web Handling

(2.16a) $$10\% < \frac{T_A}{T_{Ayield}} < 25\%$$

(2.16b) $$10\% < \frac{T_B}{T_{Byield}} < 25\%$$

MD Strain Matching to Avoid CD Curl
(difficult to meet with large mismatches in ply modulii)

(2.16c) $$\frac{T_A}{T_B} = \frac{c_A}{c_B}\frac{E_{1A}}{E_{1B}}$$

CD Strain Matching to Avoid MD Curl
(difficult to meet with mismatches in Poisson ratios)

(2.16d) $$\frac{T_A}{T_B} = \frac{\nu_{21B}}{\nu_{21A}}\frac{c_A}{c_B}\frac{E_{1A}}{E_{1B}}$$

where
T = lineal tension
c = caliper
E = modulus
υ = Poisson ratio
A = ply #1
B = ply #2
1 = machine direction
2 = cross direction

Note: The radius of curl resulting from strain mismatch can be calculated from the bimetallic strip procedure.

Figure 2.17
Laminator Drive Layout

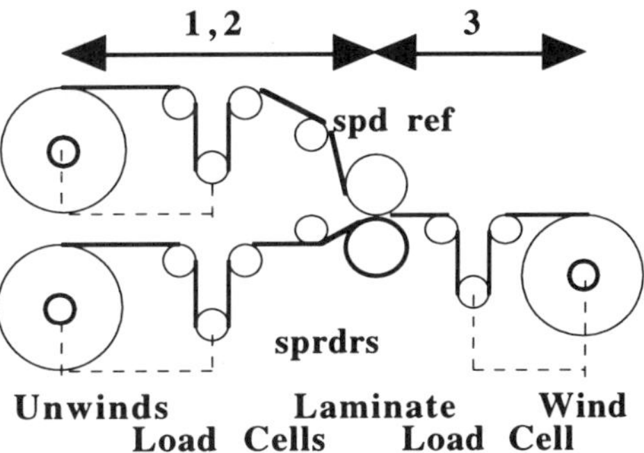

One complexity of multiple section drives is illustrated in Figure 2.18. Here, several calender stations are used to progressively decrease caliper, which also tends to elongate the web. This elongation must be taken up by running a progressively increasing speed as the web proceeds downstream. In this example, dancers provide tension feedback to the calender drive just downstream.

A problem occurs when, for example, a momentary slackness causes dancer D1 to call for a speedup of calender C1. This would put slack in the C1-C2 section which would then speed up C2, which puts slack into C2-C3 and so on. To avoid a domino tension disturbance, **feedforward** is used to also alert all downstream sections to speed up as well. Coordinating multiple drive sections, especially in draw, is a complex control problem.[18,19,20,21]

Figure 2.18
Feedforward on Multiple Drive Sections

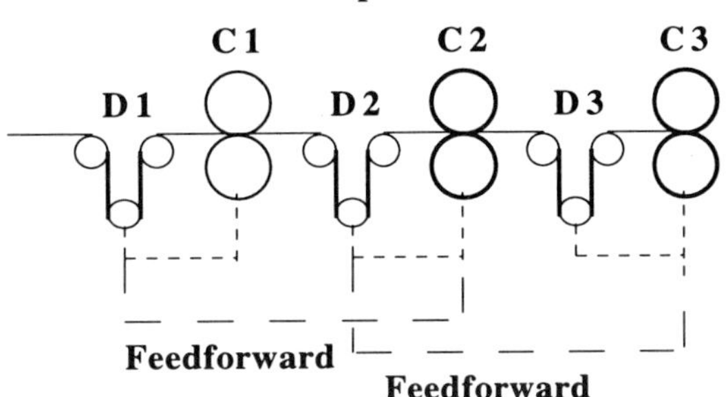

Down Web Tension Variations

Truly controlling web tension means maintaining tension at all points down a machine, across a machine and with time to be within some quality tolerance. In this section, we will discuss the variation of tension down a machine due to drag on undriven idler rollers and stationary members.

Figure 2.19 shows a simple unwind-rewind line with an intermediate process roller, such as a calender, which is also the speed reference. In this example, the two drive sections are load cell controlled; and we will assume that, since the web didn't change strength through the calender, both tensions will be set the same at 10 PLI. (lb/inch of width). Furthermore, we will let the several small rollers have 0.4 PLI of windage and **bearing friction**, but the **sliding drag** spreader has 1 PLI.

To construct the web tension we start with the load cell, which was set at 10 PLI. Since the load cell roller has 0.4 PLI of friction, the tension upstream and downstream must be 9.8 and 10.2 PLI respectively. Going downstream, we add the value of idler roller drag or bar friction and vice versa going upstream. Meeting at the calender we note that the drive is motoring instead of regenerating (braking) because the tension dropped.

In this example, we set steady-state downweb tension limits at ±10% of the setpoint, and that the spreader caused an out-of-spec tension excursion. This is not unusual for sliding elements or even some rotating elements such as undriven bowed spreader rollers.

The reader may argue that a 10% tension limit down a drive section is too tight. However, we must also allow for similar excursions due to cross web tension profiles from the machine (misalignment, etc.), cross web tension profiles from the web (baggy web), accel/decel torques on undriven rollers, tension/drive control errors, and variations in the webs caliper and/or strength. When all of these are added up, the actual web stress could vary more than 50% between locations/times. Hardly what we would call precision web handling control!

Figure 2.19
Tensions in a Load Cell Controlled Line

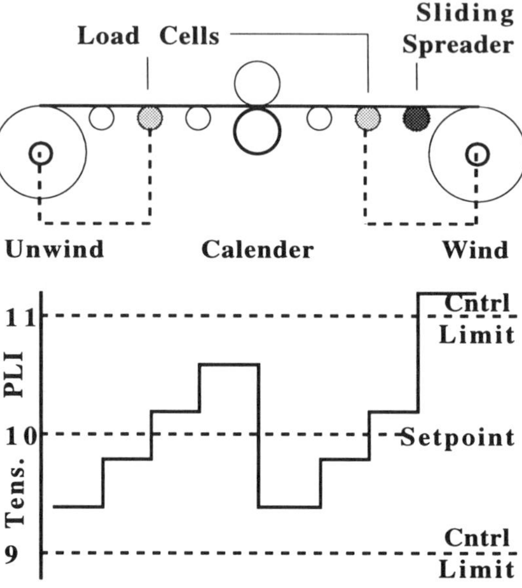

The principle of drive zone tension limits in effect determines how many (if any) undriven idler rollers or stationary elements can be put between driven rollers. In Chapter 4 we give the calculation for maximum idler roller count.

This example is one of steady state, where the machine is running at a constant speed. In this situation, the undriven elements provide a slight tension rise in the web as it passes over them. As will be seen in Chapter 4, maintaining drive zone tolerances becomes even more difficult when changing speeds.

When the web must accelerate a roller, the tension differential across it is the sum of some bearing drag plus a usually much larger acceleration tension component. This would make the stair steps of this example much steeper. So much so, that it is possible that the web might go slack between the calender and its following idler roller. When decelerating, the stair steps would be shallower, or more likely, descending going downstream. Inertial tension zone excursions can be caused equivalently by many small undriven rollers, or a single large undriven roller.

Cross Web Tension Profiles

So far, we've shown that controlling tension means holding it to within tolerance at the tension sensor (static and dynamic control), and that tensions must not rise or fall excessively across a tension zone (idler rollers). In this section, we will briefly introduce controlling MD tension across the web.

A major factor in creating nonuniform tension profiles is **web bagginess or camber.**[22] Briefly, a baggy web has a variation in natural length or tension across its width. Figure 2.20 shows an example web with a tight front quarterpoint and a baggy back edge. It is not uncommon for severely baggy webs to have zero tension in certain positions, while other positions may be stressed more than twice the nominal value determined by load cell control.

Another large factor in creating nonuniform tension profiles are **misaligned rollers**. Misalignment in the plane of the incoming web tips the tension profile just after the upstream roller. This in-plane web bending is shown in Figure 2.21. Misalignment out of the plane of the incoming web skews the tension into a parabola. This web twisting is shown in Figure 2.22.

Similar, but usually smaller, effects on web tension profile consistency are generated by roller deflection and roller diametral tolerances. Roller deflection steers the web much like a series of progressively misaligned segmented (partial width) rollers. It also causes the web path (on simply supported rollers) to be shorter through the center of the machine and longer at the edges. On **cantilevered machines**, the effect of roller bending and bearing play can cause severe apparent bagginess on the front side.

Roller diametral variations are significant only when they are very large, or the web is very stiff. For example, a concave roller has a parabolic tension profile much like Figure 2.22 on the incoming side, and opposite on the outgoing side. Diametral variations may be intended (crowning) or unintended, and are usually larger on covered rollers than on plain rollers.

Figure 2.20
Baggy Web Tension Profile Example

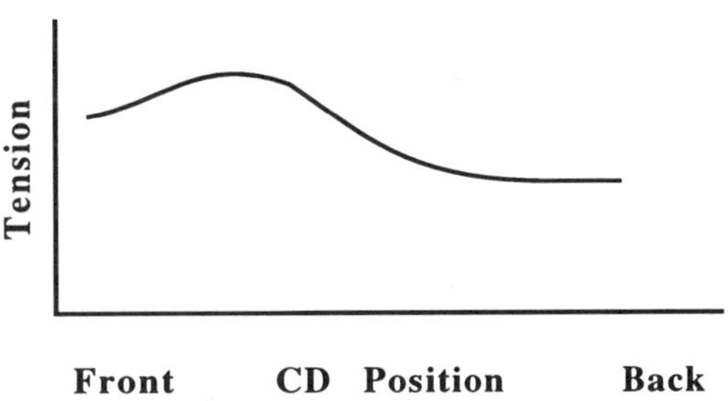

Figure 2.21
Misalignment (In-Plane) Tension Profile

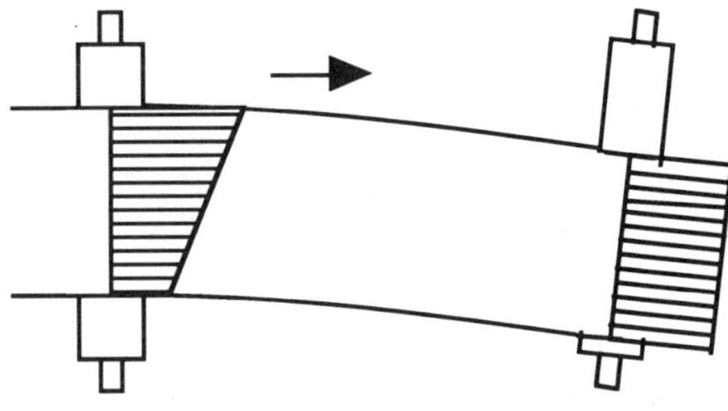

Figure 2.22
Misalignment (Out-of-Plane) Tension Profile

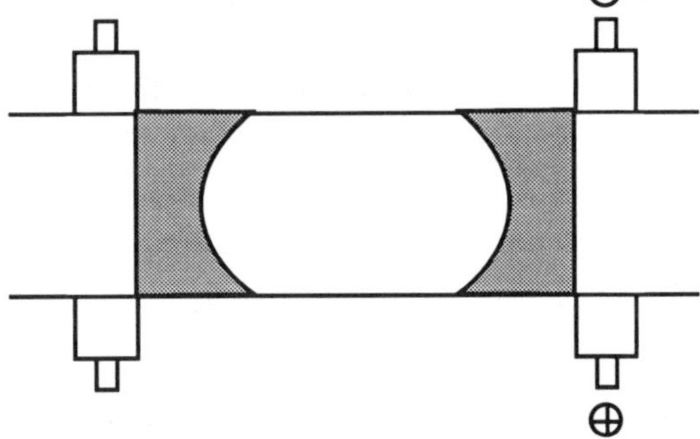

The combined effect of all these tension nonuniformities is one of the reasons that webs are seldom able to be sucessfully tensioned (as read by a load cell) at more than 25% of their strength. The load cell is only a snapshot of average tension across the width. It tells us nothing about variations down through the machine or across the width.

Bibliography

1. Roisum, David R. *Runnability of Paper.* Tappi J., vol 73, no 1, pp 97-101, January 1990, and no 2, pp 101-106, February 1990.

2. Roisum, David R. *What is the Best Way to Measure Length?* Converting Magazine, June 1995.

3. Roisum, David R. *How Can I Control Width Accurately?* Converting Magazine, June 1995.

4. Puckhaber, Charles F. *Print Registration Control in a Rotogravure Printing Press Combined with Extrusion Coating or Lamination Processes.* Tappi J., vol 78, no 3, pp 144-154, March 1995.

5. DuBois, G.B. *Selecting Proper Winding Tension for Various Web Substrates.* Tappi J., vol 76, no 5, pp 91-93, May 1993.

6. Jones, J.P. *Determining Tension Values.* Converting Mag., pp 64-72, November 1989.

7. Endersby, D. C. and Zucker, E. *Direct Tension Control by Strain Gauges and Disc Brakes for Rewinders.* Pulp and Paper Magazine of Canada, vol 67, no 9, pp 393-398, September 1966.

8. Viney, David. *Measurement and Control of Web Tension in Papermaking.* Paper, vol 198, no 1, pp 26, 28, 30, July 19, 1982.

9. Montalvo, William W. *Unraveling Web Tension Control.* Power Transmission Design, vol 27, no 1, pp 35-37, November 1985.

10. Bak, David J. *Dancer Arm Feeback Regulates Tension Control.* Design News, vol 43, no 6, pp 132-133, April 1987.

11. Weiss, Herbert L. *Getting Better Quality With Dancer Controlled Unwinders.* Package Printing, vol 28, no 12, pp 11-12, 103 and vol 29, no 1, pp 14, 50-52.

12. Martin, John R. *Tension Regulation Apparatus.* U.S. Patent 3,659,767 issued May 2, 1972.

13. Pfeffer, J.F. Jr. *Dancer Roll Assembly.* U.S. Patent, February 21, 1961.

14. Marhauer, H.H. *The Dynamics of Web Tension Measurement.* ISA Transactions, vol 5, no 3, pp 242-247, 1966.

15. Lin, P. and Lan, M. S. *Effects of PID Gains for Controller with Dancer Mechanism on Web Tension.* Proc. of the 2nd Int'l Conf. on Web Handling, Oklahoma State Univ., Stillwater, OK, June 6-9, 1993.

16. Reid, Karl N. and Lin, Ku-Chin. *Dynamic Behavior of Dancer Subsystems in Web Transport Systems.* Proc. of the 2nd Int'l Conf. on Web Handling, Oklahoma State Univ., Stillwater OK, June 6-9, 1993.

17. Roisum, David R. *How Can I Reduce Laminater Curl?* Converting Magazine, Web Works column, March 1995.

18. Brandenburg, G. *New Mathematical Models for Web Tension and Register Error.* Proc 3rd In'l IFAC Conf on Instrumentation and Automation in the Paper, Rubber and Plastics Industries, vol 1, Brussels, May 24-26, 1976.

19. Shelton, John J. *Dynamics of Web Tension Control with Velocity or Torque Control.* American Control Conf. Proc., Seattle, June 18-20, 1987.

20. Shin, Kee-Hyun. *Distributed Control of Tension n Multi-Span Web Transport Systems.* Ph.D. Thesis, Web Handling Research Center at Oklahoma State Univ., May 1991.

21. Reid, Karl N. and Lin, Ku-Chin. *Control of Longitudinal Tension in Multi-Span Web Transport Systems During Startup.* 2nd Int'l Conf. on Web Handling, Oklahoma State Univ, June 6-9 1993.

22. Roisum, David R. *What Causes a Baggy Web?* Converting Magazine, Web Works column, pp 22, April 1994.

Chapter 3

Deflection

In this chapter we show that roller deflection must be minimized to avoid web contraction and wrinkling, and to achieve uniform pressure profiles on nipped rollers. Design calculations for deflection are given and deflection standards are suggested. Finally, we show that deflection largely determines the required roller diameter.

Figure 3.1
Deflection Due to Gravity, Tension & Nip

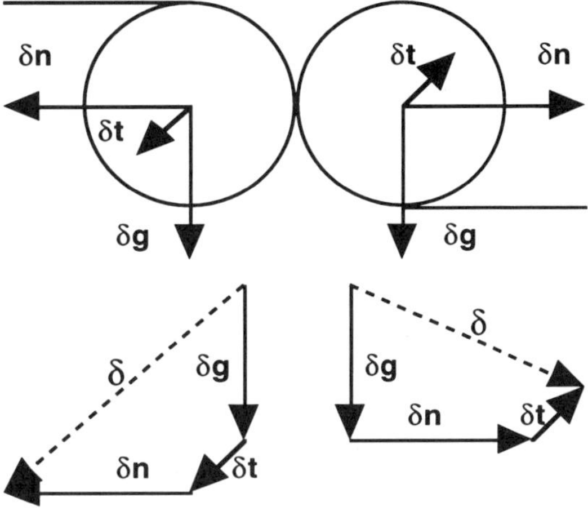

Why Deflection Must Be Minimized

There are three principle forces that contribute to the static deflection of a roller. These are sag due to roller **weight**, web **tension**, and **nip** as illustrated in the example Figure 3.1. While the details can vary considerably with the particular application, in general the nip (if present) will be the largest force except for very large diameter rollers, and tension will be the smallest except for very narrow rollers. The total deflection is the vector (tip-to-tail) sum of these components.

Figure 3.2
Nip Nonuniformity Due to Roller Deflection

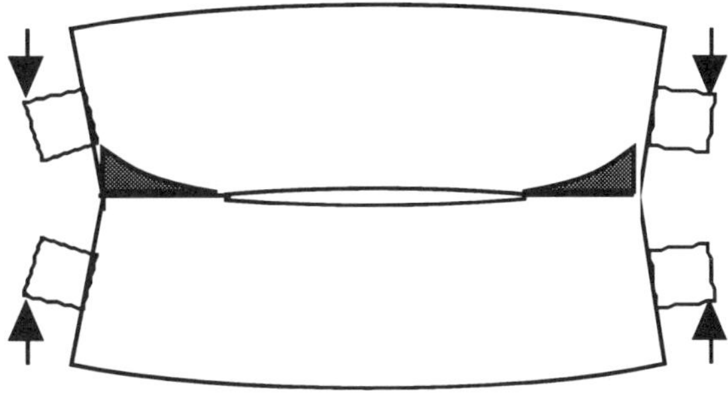

The problem with nipped roller deflection is illustrated in Figure 3.2. The end loaded rollers deflect away from each other, which causes a pinching on the ends.[1] For metal-metal nips, the profile may be so nonuniform that the nip pressure is concentrated only on 10-20% of each end and a gap results in the middle. This is why nipped rollers are often either **crowned** or **covered**. Nipped rollers are further described in Chapter 5, and covers are described in Chapter 9.

Figure 3.3
Web Contraction Due to Roller Deflection

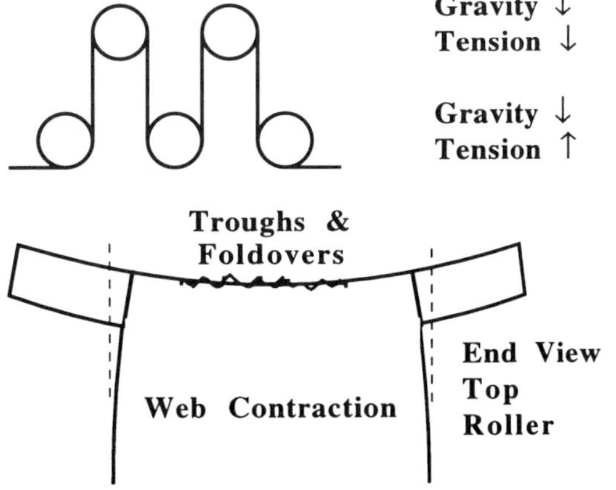

However, all rollers are subject to a more common deflection problem as shown in Figure 3.3. Schematically, the upper rollers in a machine are subject to gravity and tension which both act down. This causes a deflected shape of the roller that is the same as a bowed spreader roller pointing backward. This causes web contraction[2,3] that may result in troughs which can foldover and crease going over the roller. While we would not tolerate roller deflection as high as the bow on a spreader roller, the accumulative effect of several deflected rollers will require a **spreader** just to undo the deflection induced web contraction.

How Much Deflection Is Allowable

There can be no doubt that excessive deflection can cause severe web handling problems. Indeed, it is a common situation because many process engineers or even some machine designers do not recognize or understand the consequences of using slender rollers.

After the fact, excessive **nip** deflection can be diagnosed by a nonuniform nip profile as measured by nip impression paper[4,5] or other means.[6,7] Also, increasing product nonuniformity of ends versus center as nip loads are increased is strong evidence of a nip deflection problem.

After the fact, excessive deflection on **unnipped** rollers is even easier to diagnose. A symptom of this problem is an increased gathering or **troughing** of the web as it moves down through a machine. The fingerprint of roller deflection is when this troughing increases as **tension** is increased. This is because increase tension will often tend to reduce most other wrinkle conditions.[8]

The evolution of web machinery design expertise goes through three stages. The first, achieved in most cases, is a stress based design where the objective is to **avoid overstressing** machine components. The second stage, achieved by a slim majority primarily by experience and conservative seat-of-the-pants design rather than calculation, is to **minimize static deflection**. The last stage in the evolution is to **minimize dynamic deflection** such as resonance and excessive vibration. Static and dynamic deflection are far more demanding design criteria than stresses. Thus, a machine might not break and still perform unacceptably.

Thus, roller deflection standards must be set and observed. The question becomes how to establish these standards in light of the tremendous variety of web materials and machinery. The answer must come from modeling or experience.

Unfortunately, while modeling is the more precise means, it is also the more difficult. Modeling of nip induced deflection and the resulting nip uniformity must, in most cases, be performed using nonlinear finite element analysis.[9,10] Even nips involving linear materials are geometrically nonlinear because the contact footprint grows with load. Only rarely are nipped systems suitably described by the closed-form Hertzian contact equations.[11,12] Modeling of the contraction[13,14] and buckling tendency of webs going over deflected rollers[15,16] is even more difficult.

We assume that the reader will neither take the trouble to model the problem, nor will want to simply guess at a standard. Then, we can offer two much simpler alternatives.

The first is based on well established experiences of paper machine builders and contractors. The unpublished standards used by many of these large engineering departments are given in Table 3.1. These deflection standards are about 1/10 those of an appropriately sized bowed spreader roller, implying that a **spreader** might be needed after every tenth fully deflected roll just to undo the gathering caused by roller deflection. These deflection standards are also about 10X the commonly used **misalignment** standards. The reason is simple **economics**, while neither misalignment nor deflection is desirable, alignment is less costly than increasing roller sizes and thus can be made tighter.

The second deflection criteria is a very simple rule of thumb (which does not account for loading, wall thickness and material):

The slenderness (length/diameter) of most rollers should be between 10 and 20.

Table 3.1
Guidelines for Roller Deflection

<u>Class A - Deflection < 0.00008 x Face Width</u>
for *precision* or *nipped* applications
<u>Class B - Deflection < 0.00015 x Face Width</u>
for *most rollers*
<u>Class C - Deflection < 0.00030 x Face Width</u>
for *flexible materials* such as tissue, nonwovens, textiles and some machine forming elements such as aprons, belts, felts and wires, as well as for web *materials over 10 mils (0.25mm) thick.*
<u>Class D - Deflection < 0.00060+ x Face Width</u>
for conveyor belt rollers, spreaders, wind/unwind shafts and cores and similar

Deflection Definition and Forces

As far as the web knows, the only deflection of concern for an unnipped roller is the center with respect to the edges of the web as seen in Figure 3.4. Roller deflection outside of the web's deckle is not sensed by the web. (Although this does become an issue for critical speed, vibration and nipped rollers.) Deflection in the plane of the ingoing web run induces **contraction or spreading**. Deflection out of the plane of the ingoing web causes **twisting**, which is a minor concern except with very stiff and thick materials. Deflection in the direction of the wrap bisector causes a very slightly smaller path length on the center and thus **slackens the center** with respect to the ends. This is of only minor concern except with very stiff materials such as foils. The direction of largest concern with nipped rollers is away from each other.

At present, however, our failure criteria of roller deflection is not sufficiently detailed to warrant separate criteria for each mode. Thus, the deflection criteria for all rollers, applications and modes will be the same, that is, *the maximum deflection of the center of the roller with respect to the roller at the edges of the web.* Vector math must be used to find the resultant maximum as was illustrated in Figure 3.1.

The first step in calculating deflection is to determine the location and direction of all of the forces acting on a roller. The force resulting from **web tension** is distributed (nominally) uniformly across the width of the *web*, acts in the direction of the bisector of the wrap angle, and has a magnitude as given in Chapter 2. The force resulting from a **nip** is distributed (nominally) uniformly across the effective width of the *roller*, acts along the line connecting the roller axes, and has a magnitude as discussed in Chapter 5.

The forces from gravity are directed downward, but are not necessarily uniformly distributed. The constituent forces on a simply supported dead shaft roller (taken from Figure 1.1) are shown in Figure 3.5. The distributed tension acts across part of the face, while the distributed shell load (shell weight and nip) acts across the entire face. Finally, there are two point loads from the bearing supports.

Figure 3.4
Deflection of Web Edges Versus Center

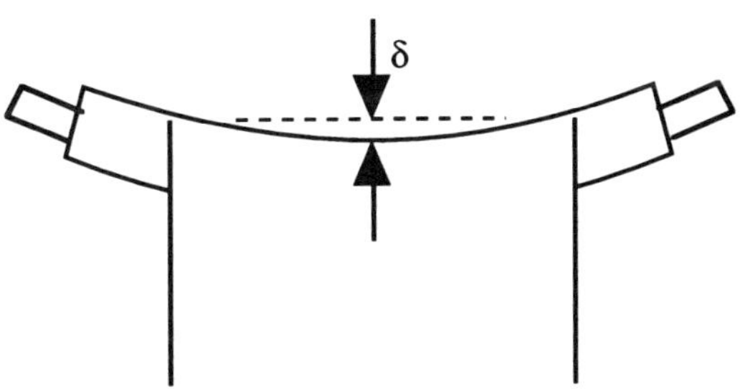

Figure 3.5
Forces Acting on a Dead Shaft Roller

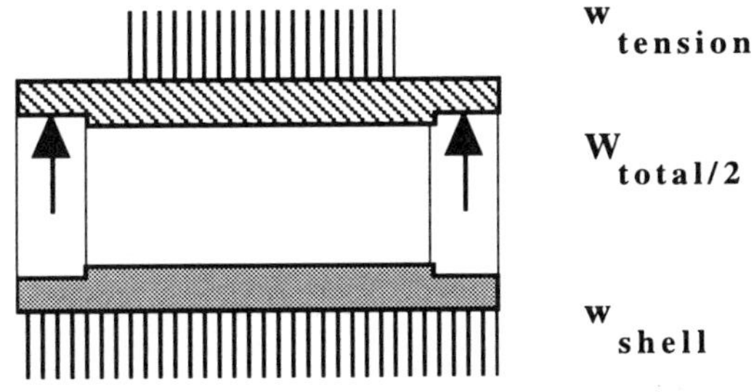

Figure 3.6
Forces Acting on a Live Shaft Roller

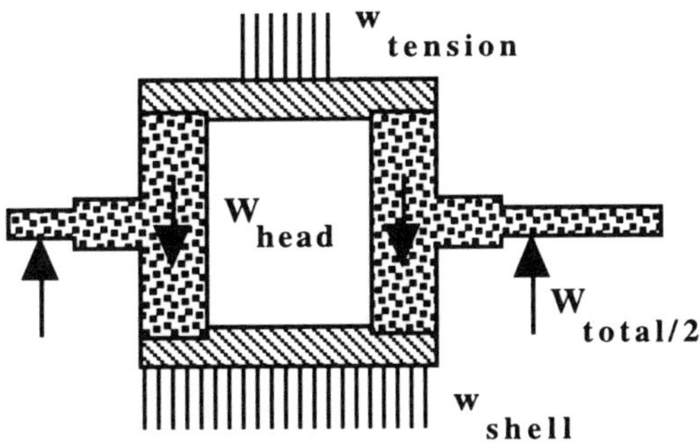

The forces acting on the live shaft roller (taken from Figure 1.2) are shown in Figure 3.6. They contain the same elements as does the dead shaft roller. However, there are two distinct differences. First, there are additional point loads due to the weight of the heads and journals. Second, the bearing support point loads are well outboard of the shell of the roll. This overhang, as we will see, causes a bending moment with further contributes to roller deflection.

Deflection Calculation Made Simple

For those who wish to calculate deflection rather than use the rough rule of thumb of a slenderness ratio (10 for nipped, 15 for unnipped), we can simplify the analysis using some approximations for simply supported rollers. In most cases this will be adequate for comparison against deflection standards such as was given in Table 3.1.

First, we will assume that the rollers are wrapped 180 degrees pulling downward. While slightly conservative, this would ensure that all rollers in a web line would be suitable for any existing or future web path, and eliminates the need for vector math. The distributed load is then

$$(3.1) \qquad w = 2t + n + s$$

where
w = total distributed down load (N/m) or (lb/in)
t = unit web tension (N/m) or (lb/in)
n = unit nip down load (N/m) or (lb/in)

$$s = \text{unit shell wgt (N/m) or (lb/in)} = \rho \frac{\pi}{4}\left(D^2 - d^2\right)$$

where
D = shell outer diameter (m) or (in)
d = shell inner diameter (m) or (in)
ρ = weight density in English units of lbf/in^3 and in metric units

$$\rho = \rho_m g = \rho_m \times 9.807 \frac{m}{sec^2}\left(\frac{N-sec^2}{kg-m}\right)$$

ρ_m = mass density in metric units of kg/m^3

In the case of a non-vertical nip, we will obviously need to resort to vector math or accept an even more conservative calculation with the greater possibility of oversizing the roller.

We will let the distributed tension act along the entire face of the roller instead of just over the web width. This again is conservative, but only slightly so because the web width is a very large portion of roller width in most cases. The only typical exception is the unwind/windup shafts used in some rewinders. Also, we are going to calculate the deflection with respect to the roller edges instead of the web's edges. This conservative approximation again simplifies the math. More accurate methods are given shortly.

The area moment of inertia of the shell is required and is calculated as

$$(3.2) \qquad I = \frac{\pi}{64}\left(D^4 - d^4\right)$$

Simply Supported Dead Shaft Roller Deflection

$$(3.3) \qquad \delta = \frac{5\, w\, F^4}{384\, E\, I}$$

Simply Supported Live Shaft Roller Deflection

$$(3.4) \qquad \delta = \frac{(12B - 7F)\, w\, F^3}{384\, E\, I}$$

where
δ = deflection of center (m) or (in)
F = face width of roller (m) or (in)
B = distance between bearing (m) or (in)
E = shell modulus (N/m^2) or (lb/in^2)

Note that the additional deflection terms in the numerator of the live shaft roller come from the bending moment on the journals. Table 3.2 gives properties for common shell materials. Figure 3.7 shows roller deflection geometry.

Table 3.2
Deflection Properties of Shell Materials

Aluminum - 6061-T6
ρ =density=2710 kg/m^3=0.098 lb/in^3
E = modulus = 69x10^9 N/m^2 =10x10^6 lb/in^2

Steel - ASTM-A36
ρ = density = 7860 kg/m^3 = 0.284 lb/in^3
E = modulus = 200x10^9 N/m^2 =29.0x10^6 lb/in^2

Composite - example carbon fiber epoxy
ρ = density = 1495 kg/m^3 = 0.054 lb/in^3
E = modulus = 165x10^9 N/m^2 = 24.0x10^6 lb/in^2

Figure 3.7
Deflection Geometry

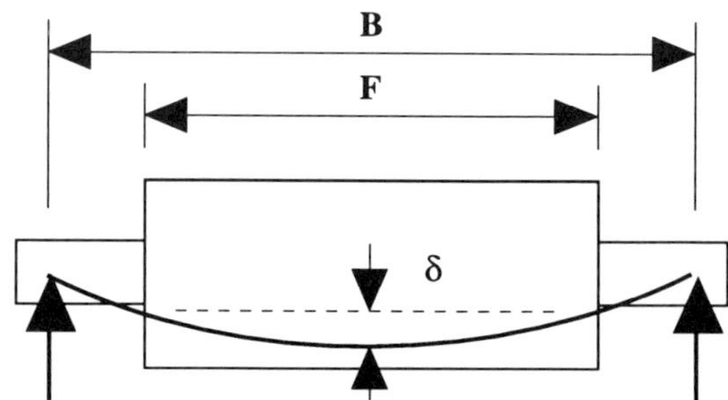

Deflection Calculation Example

English Version

Calculate the deflection for a 7" aluminum roller that was finished to a 6.88 "OD and has a 6" ID. The face of the live shaft roller is 80" and the bearing-to-bearing centerline distance is 88". This unnipped roller carries its own weight and a maximum of a 2 lb/in tension with a 180 degree wrap pulling down.

$$s = \rho \frac{\pi}{4} \left(D^2 - d^2\right)$$

$$s = 0.098 \frac{lb}{in^3} \frac{\pi}{4} \left(6.88^2 - 6^2\right) in^2 = 0.8723\ lb/in$$

$$w = 2t + n + s = (2)\,2 + 0 + 0.8723 = 4.872\ lb/in$$

$$I = \frac{\pi}{64} \left(D^4 - d^4\right) = \frac{\pi}{64} \left(6.88^4 - 6^4\right) = 46.37\ in^4$$

$$\delta = \frac{(12B - 7F)\, w\, F^3}{384\, E\, I}$$

$$\delta = \frac{((12)88 - (7)80)\,(4.872)\, 80^3}{(384)\,(10E6)\,(46.37)} = 0.006948\ inch$$

$$\frac{\delta}{F} = \frac{0.006948}{80} = 0.000087 < 0.000150$$

Thus, this roller does not deflect excessively. Indeed, we could drop down to a nominal 6" diameter roller (same wall thickness) and have a deflection of 0.01119", which gives a deflection ratio of 0.00014, which just meets our standards.

The reader should note that decreasing the diameter only 15% doubled the deflection. This is because the deflection decreases at approximately the 4th power of diameter, and increases with the 3rd power of length!

Metric Version

Calculate the deflection for a 180mm aluminum roller that was finished to a 175mm OD and has a 150mm ID. The face of the live shaft roller is 2.0m and the bearing-to-bearing centerline distance is 2.20m. This unnipped roller carries its own weight and a maximum of a 0.35 kN/m tension with a 180 degree wrap pulling down.

$$\rho = \rho_m\, g = 2710 \frac{kg}{m^3}\, 9.807 \frac{m}{sec^2} = 26{,}580 \frac{N}{m^3}$$

$$s = \rho \frac{\pi}{4} \left(D^2 - d^2\right)$$

$$s = 26580 \frac{N}{m^3} \frac{\pi}{4} \left(0.175^2 - 0.150^2\right) m^2 = 169.61 N/m$$

$$w = 2t + n + s = (2)\,350 + 0 + 169.61 = 869.6\ N/m$$

$$I = \frac{\pi}{64} \left(D^4 - d^4\right) = \frac{\pi}{64} \left(0.175^4 - 0.150^4\right)$$

$$I = 2.118E-5\ m^4$$

$$\delta = \frac{(12B - 7F)\, w\, F^3}{384\, E\, I}$$

$$\delta = \frac{((12)2.2 - (7)2.0)\,(869.6)\, 2.0^3}{(384)\,(69E9)\,(2.118E-5)} = 0.00015372m$$

$$\frac{\delta}{F} = \frac{0.00015372}{2.0} = 0.000077 < 0.000150$$

Thus, this roller does not deflect excessively. Just as with the English example, we could drop down a size and still have acceptable deflection.

The reader should note that decreasing the diameter only 15% doubles the deflection. This is because the deflection decreases at close to the 4th power of diameter, and increases with the 3rd power of length!

A Closer Deflection Calculation

A few simplifications were made in the previous deflection calculations. Here, we present a more accurate method.

Given the live shaft roller in the previous section, which has 8" long x 1" diameter steel journals, 2" thick heads, and a maximum web width of 75", what is the deflection?

The first step is to construct a shear and bending moment diagram for the load distribution shown in Figure 3.8. To do this, we need to know the weight of the journals, heads and shell.

$$Wgt_j = 0.284 \frac{lb}{in^3} \frac{\pi}{4} (1\ in)^2 (12) in = 2.676\ lb$$

$$Wgt_h = 0.284 \frac{lb}{in^3} \frac{\pi}{4} (6\ in)^2 (4) in = 32.11\ lb$$

$$Wgt_s = 0.098 \frac{lb}{in^3} \frac{\pi}{4} (6.88-6\ in)^2 (80) in = 69.79\ lb$$

The reaction at the bearings is then the weight of the roller plus the tension component

$$F_{bv} = \frac{1}{2}\left[(2.676+32.11+69.79)\ lb + (2)\,2\frac{lb}{in}\ 75\ in\right]$$

$$F_{bv} = 202.2\ lb$$

From engineering statics, the bending moment at the edge of the web can thus be calculated as 1371 lb-in. The deflection due to the bending moment can be calculated as

$$\delta_M = \frac{M L^2}{8 E I} = \frac{(1371\ lb-in)\ (75\ in)^2}{8\left(10E6\ \frac{lb}{in^2}\right) 46.37\ in^4}$$

$$\delta_M = 0.00208\ in$$

Similarly, we can calculate the deflection due to the distributed tension weights as

$$\delta_w = \frac{5 w L^4}{384 E I} = \frac{(5)\left(4.872\ \frac{lb}{in}\right)(75\ in)^4}{384\left(10E6\ \frac{lb}{in^2}\right) 46.37\ in^2}$$

$$\delta_w = 0.00433\ in$$

The total deflection of the middle of the roller with respect to the edges of the web is then the sum or superposition of the bending moment due to loads outside the web and the distributed load across the web due to roller shell weight and tension reactions as

$$\delta = \delta_M + \delta_w = 0.00208 + 0.00433 = 0.00641\ in$$

As expected of a good approximation, our previous simple calculation gave a very similar answer to the more accurate method shown here.[17] However, note how the more accurate method more clearly shows us the deflection penalty incurred due to journal stickout on live shaft rollers. **Journals** should be kept very short!!

Finally, for unusually complex roller systems, you may wish to use finite difference, finite element or other numerical techniques.

Figure 3.8
Roller Load Distribution Example

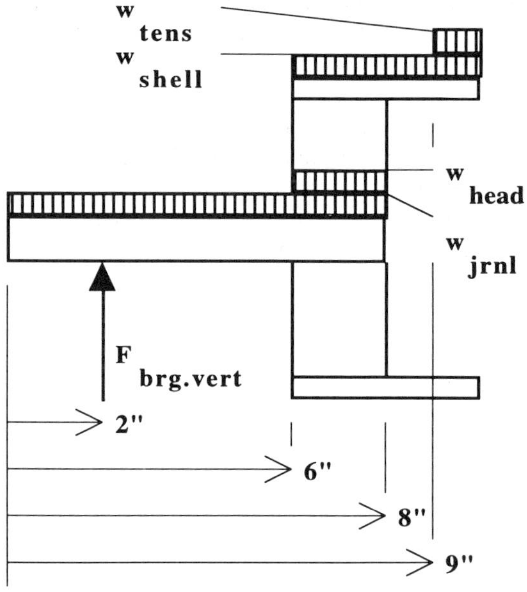

Cantilevered Machines

Cantilevered machines are sometimes used in narrow web converting (web widths <1m). They have two potential advantages over the far more common simply supported machines. First, they can be **threaded** easier because there is little if any framework interfering with access from the front side. Second, installing small supply wound rolls (which can be lifted by hand) onto a cantilevered chuck may be somewhat easier than chucking or shafting a simply supported unwind roll.

Cantilever machines must also conform to the same deflection standards as simply supported machines. However, cantilevers suffer deflection disadvantages for two reasons. First, the rollers are usually dead shaft rollers, whose resistance to bending is determined by the through shaft stiffness. This is in contrast to simply supported dead shaft rollers whose bending resistance is determined by shell stiffness. With the area moment of inertia going as the fourth power of diameter, the shaft will be much more flexible than the shell!

Figure 3.9 also illustrates another deflection penalty of cantilevered machines due support or framework compliance. Notice the nonzero slope at the front bearing, making it nonconservative to use the simple cantilever tip deflection formula for distributed loads as:

$$(3.5) \qquad \delta = \frac{w\,b^4}{8\,E\,I}$$

The figure illustrates just one type of support compliance, that is due to bending between discrete shaft supports. This can be calculated by superposition as illustrated in the example on the next page. More complex compliances are bending of the framework wall or shaft supports themselves. These will generally require finite element modeling using plate elements.

The deflection of cantilevers redistributes web tensions in a somewhat different manner than do simply supported systems. Simply supported rollers will bend in the center such that web path lengths through the machine are slightly less there than at the ends. This makes for a slightly **baggy center**.

Figure 3.9
Deflection of a Cantilevered Roller

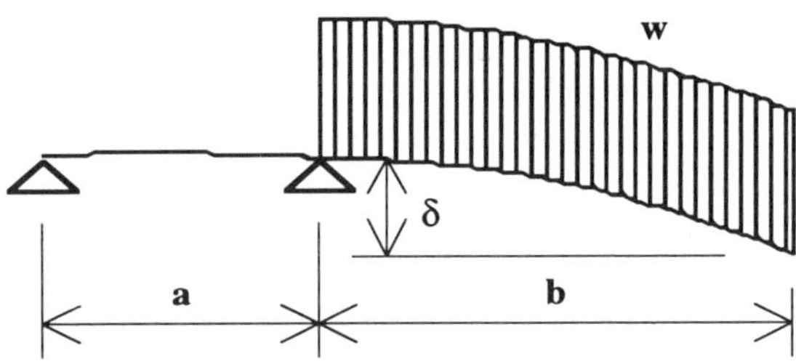

Cantilevers, on the other hand, will deflect at the front edge with respect to the back. This will make a **baggy front edge**. In extreme cases, but unfortunately not all that rare, the web will go completely slack on the front side. Figure 3.10 explains this **cross web tension profile** with a festoon segment of a machine. For illustration purposes, we will let the tension and roller weight cancel out so that the bottom rollers have no net deflection. Note how the path of the front side is much shorter because of top roller deflection. This trait is shared by all cantilevered machines.

The asymmetry of cantilever deflection, if excessive, also makes **alignment** much more difficult. On simply supported machines, the target is placed equidistant from each end of a roller so that it is nominally level and square under loads and deflections. However, to best align a cantilever, we should align it with a web threaded through the web run and tensioned at a nominal value.

Cantilever deflection is easy to measure. Put a dial indicator on the front side of a roller. Deflection is 1/2 the difference in readings when tensioned up versus tensioned down at 180 degrees of wrap.

Figure 3.10
Cantilevers Cause Front Side Slackness

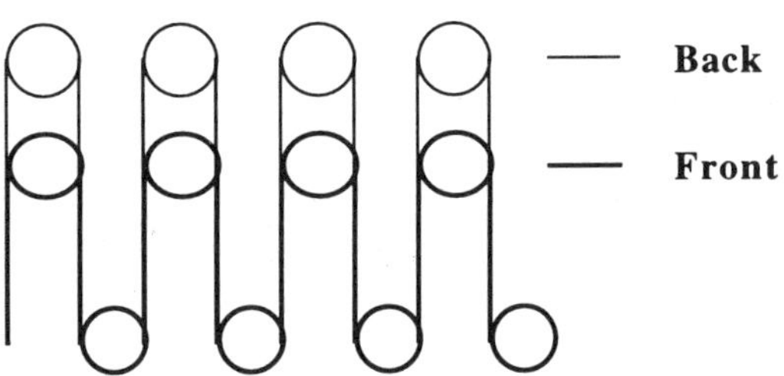

Cantilever Calculation Example

Given a cantilever machine with dead shaft rollers, calculate the end deflection. Assume a 12” bearing separation (a), and a 20” wide roller (b). The steel live shaft is 2” in diameter, the aluminum shell has a 4” OD and a 0.25” wall. The maximum web tension is 2 lb/in, and the worst wrap is 180 degrees pulling down. The mountings have 0.0005” radial clearance and the bearing/mount spring rate is 100,000 lb/in. We will solve for tip deflection using superposition of 4 different elements. Roark and Young is an excellent source of beam solutions.[18] The loads are shown in Figure 3.11.

Distributed Load & Area Moment

$$w_{st} = \frac{\pi}{4}(2\text{ in})^2\,0.284\,\frac{\text{lb}}{\text{in}^3} = 0.89\text{ lb/in}$$

$$w_{al} = \frac{\pi}{4}\left[(4\text{ in})^2 - (3.5\text{ in})^2\right]0.098\,\frac{\text{lb}}{\text{in}^3} = 0.29\text{ lb/in}$$

$$w_{tension} = (2\text{ lb/in})\,2\sin(180/2) = 4\text{ lb/in}$$

$$w_{total} = 0.89 + 0.29 + 4.00 = 5.18\text{ lb/in}$$

$$I = \frac{\pi}{64}\left((2\text{in})^4 - (0\text{in})^4\right) = 0.79\text{ in}^4$$

Mounting Clearance

$$\delta_{clearance} = \frac{20}{12}\,0.5''\text{ mils} = 0.8\text{ mils}$$

Mounting Compliance

$$\delta_c = F/k = \frac{(5.18\text{ lb/in})(20\text{in})\left(\frac{10}{12}\right)}{(100\text{ lb/mil})} = 0.9\text{ mils}$$

Cantilever Deflection

$$\delta_c = \frac{w b^4}{8EI} = \frac{(5.18)(20)^4}{8(29E6)(0.79)} = 4.5\text{ mils}$$

Figure 3.11
Cantilever Load Example

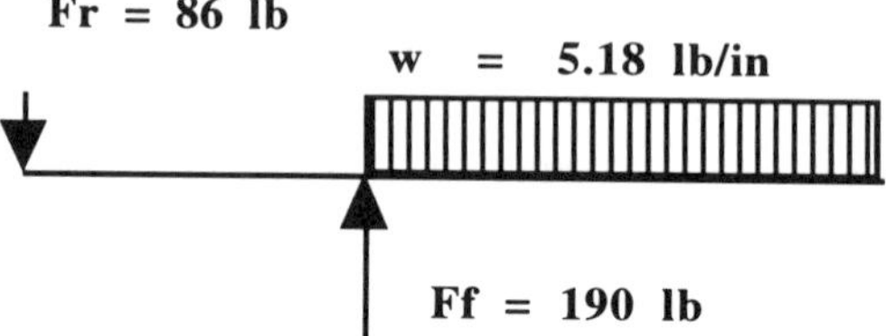

Slope Deflection

$$M = (5.18\text{ lb/in})(20\text{in})(10\text{in}) = 1{,}036\text{ in}-\text{lb}$$

$$\theta_f = \frac{ML}{3EI} = \frac{(1036)(12)}{3(29.5E6)(0.79)} = 1.78E-4\text{ radians}$$

$$\delta_s = b\theta = (20\text{in})(1000\text{mil/in})(1.78E-4\text{rad}) = 3.6\text{ mil}$$

Total Deflection

$$\delta_{total} = 0.8 + 0.9 + 4.5 + 3.6 = 9.8\text{ mils}$$

At first glance, this machine would appear quite sturdy with 4” diameter rollers and a tightly mounted central shaft with a slenderness ratio of only 10. Furthermore, this through **shaft is stressed** to a maximum of only 1300 psi. Yet, the total tip deflection is 0.00049 x Face, putting it into Class D case suitable for conveyors. This illustrates that deflection based criteria are usually much more demanding than stress based design criteria.

There are only a couple of strong deflection factors in this example problem. First is the width. While some claim that cantilevers are suitable for narrow web converting (<1m), it is difficult to achieve even half of that. Second is the shaft diameter, which could be increased. However, this would push bearing sizes larger and possibly even the roller shell diameter. Third, the 1 PLI tension is typical only for light webs. Note that making the rollers lighter with advanced materials would have a negligible effect in this case because the tension load is the largest component. Also note that in this example the front bearing was placed right at the shell. In most cases, there will be an inch or so of stickout here which would increase (slope) deflection even more.

Roller Deflection Summary

Rollers are sized first on deflection. The simplest approach is to simply keep the slenderness ratio (length/diameter) around 10 for nipped rollers and around 15 for unnipped roller. However, this can result in excessive deflection in some cases, and overdesign in others. Therefore a better approach is to calculate the deflection based on maximum static loadings. In most cases, the simple calculation (equations 3.3, 3.4 and 3.5) will be sufficient.

However, if the roller or geometry distributions are unusual (an 80" wide unwind roll on a 120" wide shaft for example), then the more accurate method should be used. This accurate method can be used whenever the distributed loads are uniform and shell deformation (ovalling or oil-canning) are negligible.

More complex systems will require numerical methods such as finite difference or finite element. Also, precision nipped rollers whose deflection *profile* (at all points across the width) is needed to grind a compensating crown may warrant full calculation efforts.

However, the weak link in sizing rollers is not calculating deflection. Rather it is determining an appropriate deflection standard. The guidelines given in Table 3.1 are typical of some reputable machine builders. Unfortunately, many builders have no explicit standards and will tend to undersize rollers to reduce the costs and eliminate the need for drives. However, as we have seen, excessive deflection can result in a number of web problems such as bagginess and wrinkling.

Another point of reference is to compare alignment standards (Chapter 12) against deflection standards. As seen in Figure 3.12, the deflection standards are much more liberal, especially if one compares slopes instead of position. One should not, however, compare roller tolerances (TIR etc.) which are typically better than 0.001" to deflection. This is because roller deflection is smooth and tends to cause fewer problems than abrupt cylindricity errors.

Figure 3.12
Alignment versus Deflection Standards for a 100" wide roller

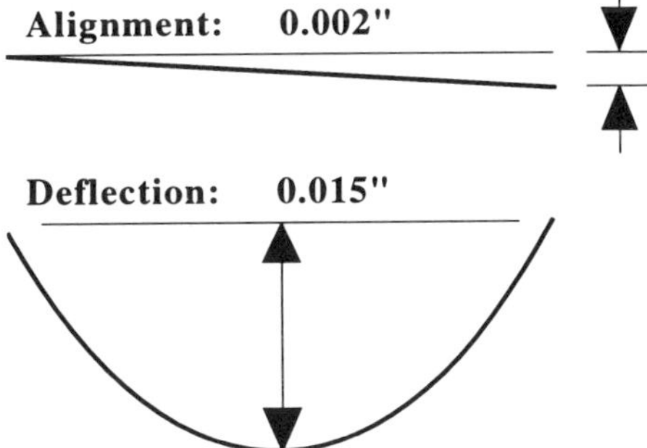

There is one additional criterion for roller sizing that should be mentioned at this time: critical speed. For high speed applications (thousands of MPM or FPM), **critical speed** is the more demanding criteria. This topic will be covered in much more detail in Chapter 11, and will include a discussion as to why most critical speed calculations are incorrect and can lead one to a false sense of security.[19]

After roller geometry (and material) has been determined using deflection standards, we must determine if an external drive is required or if the roller can be a simpler web driven idler. There are two questions that must be asked here. First, is there sufficient wrap and **traction** to accelerated/decelerate the roller and overcome bearing drag and windage without the web breaking loose from the roller? Second, is the total **downweb tension differential** across all rollers in a line due to acceleration/deceleration and drag small enough to hold tensions within some tolerance? These issues are explored in detail in Chapter 4.

Finally, what do you do with an existing machine that has excessive roller deflection? First, ruthlessly eliminate all non-essential rollers from the system. Second, consider replacing the remaining with larger diameter rollers. Third, trough wrinkling due to simply supported roller deflection can be mitigated with periodic spreader rollers. Fourth, front side looseness on cantilever machines can be minimized by aligning the rollers under tension/nip. Fifth, calculate the benefits of composite rollers.

Bibliography

1. Roisum, David R. *The Profile of a Nip.* Converting Magazine, Technical Report, pp 44-46, June 1994.

2. Roisum, David R. *The Mechanics of Web Spreading.* 2nd Int'l Conf. on Web Handling, Oklahoma State Univ., June 6-9, 1993.

3. Roisum, David R. *The Mechanics of Web Spreading - Part I.* Tappi J., vol 76, no 10, pp 63-70, October 1993.

4. Kennedy, W.B. *Nip Impressions.* Tappi J., vol 74, no 5, pp 277-280, May 1991.

5. Hamel, Jean et al. *Measurements of Nip Load Distribution in Calenders Using Uncalendered Paper.* Canadian Pulp and Paper Assoc, 77th Annual Mtg.

6. Brink, R. *Instrumentation for Roll Nip Studies.* TAGA Proc., pp 103-117, 1963.

7. Sussman, P.G. *Checking Nip Pressure Uniformity.* Paper Technology & Industry, October 1974.

8. Roisum, David R. *Why is My Web Wrinkled.* Converting Magazine, Web Works column, pp 16, February 1994.

9. Diehl, Ted and Stack, Kenneth D. and Benson, Richard C. *A Study of Three-Dimensional Nonlinear Nip Mechanics.* 2nd Int'l Conf. on Web Handling, Oklahoma State Univ., June 6-9 1993.

10. Rodal, Jose J. A. *Paper Deformation in a Calendering Nip.* Tappi J., vol 76, no 12, pp 63-74, December 1993.

11. Homer, E.M. et. al. *Elastic Theory of Nip Stresses and Hysteresis Effect.* Tappi J., vol 60, no 4, April 1977.

12. Deshpande, Narayan V. *Calculation of Nip Width Between Elastomeric Cylinders.* Tappi J., vol 61, no 10, 115-118, October 1978.

13. Delahoussaye, Ronald D. *Analysis of Deformations, Stresses, and Forces in Webs Encountering Spreading Rollers.* Ph.D. thesis, Web Handling Research Center at Oklahoma State Univ., December 1989.

14. Delahoussaye, Ronald D. and Good, J. Keith. *Analysis of Web Spreading Induced by the Curved Axis Roller.* 2nd Int'l Conf. on Web Handling, Oklahoma State Univ., June 6-9 1993.

15. Gehlbach, Lars S. and Good, J.K and Kedl, Douglas M.. *Web Wrinkling Effects of Web Modeling.* American Control Conf Proc, Minneapolis, June 1987.

16. Gehlbach, Lars S. and Good, J.K and Kedl, Douglas M.. *Prediction of Shear Wrinkles in Web Spans.* Tappi J., vol 72, no. 8, pp 129-134, August 1989.

17. Shelton, John J. *Deflection and Critical Velocity of Rollers.* 3rd Int'l Conf. on Web Handling, Oklahoma State Univ., June 18-21, 1995.

18. Roark, Raymond J. and Young, Warren C. *Formulas for Stress and Strain.* McGraw-Hill Book Co., 5th edition, 1975.

19. Roisum, David R. *Winder Vibration Can Reduce Operating Efficiency and Increase Maintenance.* Tappi J., vol 71, no 1, pp 87-96, January 1988.

Chapter 4

Traction, Torque and Power

In this chapter we show that web/roller traction is important so that web control is maintained. Design algorithms are given to determine if a roller can be safely driven by the web. If not, an external drive is required. Rollers must be driven if: traction is insufficient for the web to drive the roller, the tension differential across all undriven rollers is unacceptably large, or when separate tension zones are needed.

Figure 4.1
Modes of Web/Roller Interaction

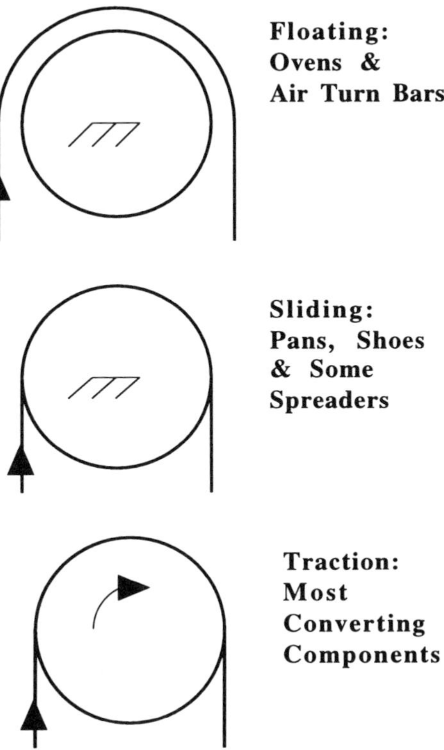

Why Traction Is Important

First a roller is nominally sized based on deflection (Chapter 3) or critical speed (Chapter 11). Then, we must determine if there is sufficient traction for the web to drive an idler roller, or for a driven roller to drive a web without slippage.

To understand why traction is important, we must first look at the three possible modes of web/roller interaction. As seen in Figure 4.1, the web can either float, slide or track with a web. **Floating** is used in many ovens so that a scratch prone web doesn't touch anything until it is dry, and on air turn bars. **Sliding** is used on some pans, shoes and spreaders. However, the vast majority of web handling components are intended to be in traction.

The important difference between the modes (for this discussion) is that the path of a web through a machine is different. As shown in Figure 4.2, the web will continue on a straight path through a machine in the presence of a misaligned floating element. Conversely, the web will bend in the plane of the web to meet a misaligned roller in traction at a right angle as is described shortly. The sliding is the intermediate case where the web will offset slightly, but not so much as to achieve perpendicular entry. Of course, the web will enter all other tracking rollers in a machine at right angles.

Figure 4.2
Path of a Web Through a Machine

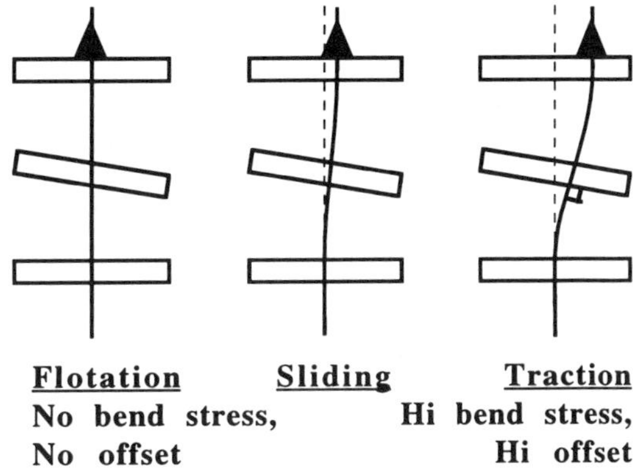

Thus, the **path** of the web through a machine is determined, in part, by the boundary conditions at each of the rollers. Also, the web will always take a smooth curve meeting those boundary conditions. However, these two constraints are not sufficient to describe the exact path through the machine because there are an infinite number of curves that will fit. To get a quantitative description of the web's edge (or center) path also requires the mechanic's equations of a (short) beam where the area moment of inertia (I = f(width, caliper)) and the web modulus are accounted for.

There is nothing inherently good or bad about any of the three web/roller modes of interaction, merely issues that must be kept in mind. For example, keeping a web from contacting an active or passive air **float** bar at all times and at all CD locations can be challenging, especially for baggy or heavy webs. Also, since the web is not effectively steered, edge position control is more difficult because the effective span ratio (draw length / web width) is greater. Air float bars are covered in more detail in Chapter 18.

Issues with **sliding** include drag induced tension differential, wear of the element and possible scratching of the web. Furthermore, if the web/shoe friction or web tension changes with either CD or MD position, edge position control and web flatness may be compromised.

Issues with **traction** include getting enough wrap, web/roller friction and tension to ensure the web doesn't slip during steady state. During accelerations, the demands also include minimal inertia. The traction mode is also the most prone to wrinkling due to roller misalignment and other causes.

All modes of web/roller interaction have a niche in web manufacturing and converting processes. While most lines are designed with all elements in traction, a few will require the occasional floating and sliding element. Thus, mixing modes within a machine is also acceptable as long as the choices are rational and the details are attended to. The problem arises only when a single element changes modes either with time (MD position) or CD position.

Changes with time can happen if the element is just marginally in a particular mode and the tension changes. For example, the air film can collapse briefly on a floating element during a tension surge. This would cause an even larger tension upset downstream as the web picks up drag going over the element. Another example is if a traction is briefly lost on an element due to a tension drop. This toggling between traction and sliding can exacerbate both tension and edge position problems. A common case here is an **edge position** offset occuring during speed changes. From web handling principles, we know that a tension change is occuring then that is causing a path change due to a mode shift on a misaligned roller.

Changes in web/roller interaction can also happen over a portion of the web's width. This more common situation can cause violent edge position deviations. For example, a floating web that just touches down on the back side of the machine will induce a large in-plane bending moment causing an abrupt change in the path of the web through the machine. The same thing happens to a smaller extent for changes between sliding and traction because the coefficients of static and kinetic friction are different.

An analogy to these issues is a car which is braking to slow down for a stop sign. If there is good traction, the car will continue in a path determined by the direction of steering. Even if the wheels lock up on an icy stretch, the car will still tend to follow a straight path. The problem happens when a single tire encounters a dry patch of road on the icy stretch. This could cause a violent turning of the vehicle. A similar situation happens if a front tire of a car suddenly deflates. The additional and unbalanced rolling resistance of the flat can pull the car in the ditch.

Thus, any web handling element should be designed to maintain its single mode of intended interaction at all times and positions. This means attention to details as given throughout this book, as well as maintaining close tension control. Finally, in the event that an element may toggle between modes, the effects may be minimized to an extent by precise machine alignment.

Which Mode Is in Effect

The easiest way to determine which mode is in effect on a particular element is to mark it with a colored pen or paint it with machinist blueing. In the case of flotation, the mark will not rub off because the web doesn't ever touch the element. In the case of traction, the mark will be retained for a long time because there is only a small amount of rubbing and microslip. However, the mark will be quickly worn away at any location where there is sliding.

After a period of time in a machine, a roller or other element can be examined to estimate the degree of slippage by examining the surface finish or by measuring changes in diameter or other geometrical features.

Most metals commonly used in machinery will tend to form an **oxide** or patina (exceptions include stainless steel and anodized aluminum). Aluminum oxide is a dull light gray, while oxides (sulfides, etc.) of steel may be gray, red, brown, or even a dark gray bordering on black. Also, some processes leave a trace of contaminants on the rollers which may change the surface color, finish and texture. Slipping on metal rollers will quickly remove the surface oxides and contaminants leaving a bright and shiny appearance. Also, the original **machining marks** will be removed and replaced by a random MD pattern.

Rubber (polymer) covered rollers may also give evidence of slippage. In this case, the surfaces may also change from a dull flat color to a shiny or reflective **glaze**. A common example is bowed spreader rollers which are often overbowed causing a traction loss and web sliding. In printing processes, slippage is easy to detect as a smearing of the ink. However, this slippage could either be due to gross web/roller slippage, or due to the microslip that accompanies many nipped systems because of nip strains of deformable soft covers.

The peaks of tungsten carbide (traction) coatings will tend to wear off during normal operation and especially if there is slippage. This causes a reduction in the **roughness** and tooth, which may require resurfacing after a period of a few years. Finally, after years of slippage, the roller will be worn to smaller diameters, or coverings and coatings to a smaller thickness, or grooving/holes to a shallower depth. This can be measured with commonly available tools such as micrometers and compared against ingoing measurements.

Flotation on air bars can also be easily checked in some applications with a long strip of paper. With a healthy amount of flotation, the strip can be entrained into the ingoing side and removed without breaking. However, a web with a small air film height will probably grab and break the strip.

Complete flotation on a roller is easy to detect because the roller will often come to a complete stop because **bearing drag** torque is greater than the torque produced by air shear. The propensity to sliding can also be estimated on some small idler rollers by attempting to rotate them by hand while the web is tensioned, preferably (for safety) while the machine is stopped.

Traction is the hardest mode to measure directly because very accurate speed measurements must be obtained. Even then, however, one can only ascertain that traction exists on at least part of the roller, but not necessarily across the entire width, and slippage always occurs on the exit of the wrap.

Accurate **speed measurements** of rollers can be done with live shaft mounted tachometers, or better yet optical encoders, or by metallic targets for proximity switch detection mounted on the head or shell of any roller. Speed or footage counters that nip against the roller shell or web, such as hand held tachs, seldom provide sufficient accuracy to detect slippage. Direct speed measurements of the web can be made, if there is sufficient optical variability, through laser doppler anemometry. Speed measurements of a web or roller can be improved if they have easily discernable registration or timing marks printed on them. The speed of the web is the distance between two photo-detectors divided by the time difference between sensor triggering. An example of a very accurate photo-detector is the Automatic Timing and Controls model 7062AFRN4X4NLX which has an incredibly fast switching time of 15 microseconds.

The Normal Entry Law

The Normal Entry Law is a building block for the description of the behavior of rollers, spreaders and guides[1] in traction as depicted in Figure 4.3.

Normal Entry Law

A web will seek a normal (perpendicular) entry to the local axis of a roll(er) in traction.

If the web is not currently at a right angle, it will move toward it at an exponentially decreasing rate.

The Normal Entry Law was described by J. David Pfeiffer, who was a very early web handling technologist.[2,3] He also proposed an early model of the corollary of how the web will seek a normal entry with time. This was more rigorously treated by Shelton[4,5,6] and others[7,8,9,10] to describe the dynamics of web guides.

However, the Normal Entry Law does not apply unless there is web/roller traction. Daly was one of the first to measure web/roller traction[11] using an endless loop web machine with test roller equipped with a torque measuring brake. From this testing, he was able to list the significant variables to increase paper web traction as given in Table 4.1. Similar work was performed more recently by Ducotey for other web materials.[12]

Three decades later, this list still holds well. However, we are now able to better predict traction given a few simple measurements. These include the static coefficient of web/roller friction and its reduction due to air (or other fluid) entrained at higher speeds. Friction is used in the band/brake equation to determine the maximum tension or torque transmissibility of any web/roller system. These subjects are covered in more detail in the next sections.

In all cases, two of the stronger traction factors are web tension and wrap angle. However, as was seen in Chapter 2, web tension is determined primarily by the strength of material. Thus, **wrap angle** is our most universal traction tool.

Figure 4.3
The Normal Entry Law for Rollers in Traction

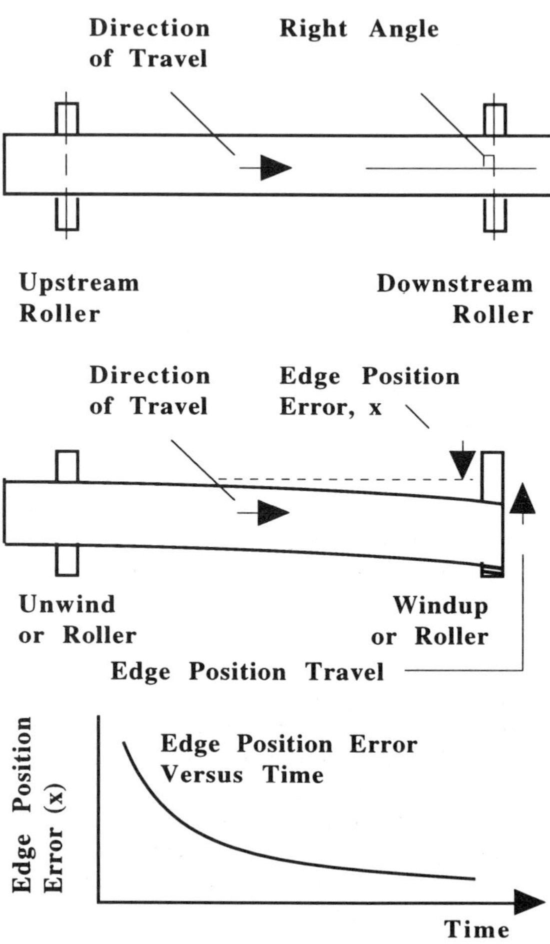

Unfortunately, many machine designers don't take the time to rearrange components as necessary to get a good wrap on all rollers. Obviously, it is much more difficult to move real rollers than it would be to move circles on a drawing. About the only penalty paid for high wrap angles is increased deflection, which is seldom an issue except for bowed spreader rollers or heavy webs.

Table 4.1
Paper Web/Roller Traction Factors

Web Tension - increase
Web Speed - decrease
Wrap Angle - increase
Roll Diameter - decrease
Web Porosity - increase
Web Moisture
Paper Grade

Static Coefficient of Friction

Coefficient of friction is a physical property of solids in contact.[13] The maximum available friction, F, is calculated from the coefficient of friction, μ, and the normal force, N, pressing the bodies together as

$$(4.1) \quad F \leq \mu N$$

This friction equation can be applied in both a static or a kinetic situation. The static case, which is the most important, is where the web just breaks loose from the roller. The kinetic case, which is the lower value of the two, is where the web is sliding over the roller.

The static coefficient of friction is most easily obtained from the inclined plane method[14] as shown in Figure 4.4. Here, the sled is surfaced on the bottom side with the same material as the roller shell, and the plane is surfaced with the web material oriented in the MD. The plane is lifted slowly and evenly until the sled breaks loose. From the vector freebody diagram of the forces on the sled, the angle of breakaway can be used to calculate static friction as:

$$(4.2) \quad \mu = \tan(\theta)$$

Another coefficient of friction measurement is the horizontal plane method[15] as shown in Figure 4.5. Here, a motorized winch pulls the sled across the plane and a load cell records the frictional resistance. The instrumentation for this method can either be a dedicated commercially available horizontal test bed, or a pulley and bed fixture that can be purchased for standard tensile testing machines. An extra output of the horizontal method is the kinetic coefficient of friction. Kinetic friction measurements have less variability because the raw signal is integrated and (rms) averaged, whereas the static measurements must capture a brief spike. Finally, web/roller friction can be reverse engineered from the band-brake equation discussed later.

Friction can be used for a variety of design purposes. For example, the web against web friction can be used to describe wound roll telescoping,[16] the stability of stacks or reams, and nip induced defects on paper.[17]

Figure 4.4
Friction - Inclined Plane Method

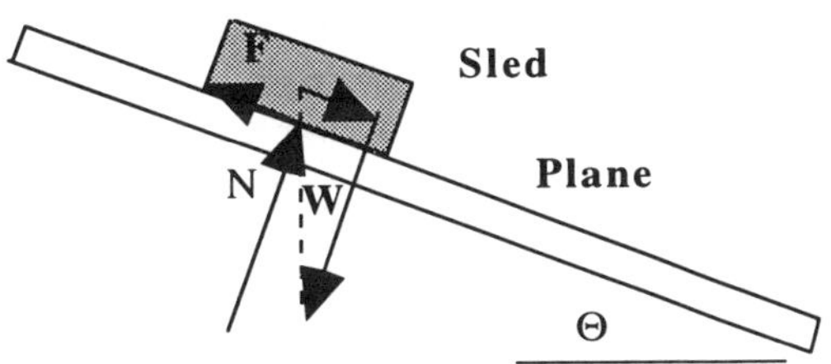

Figure 4.5
Friction - Horizontal Plane Method

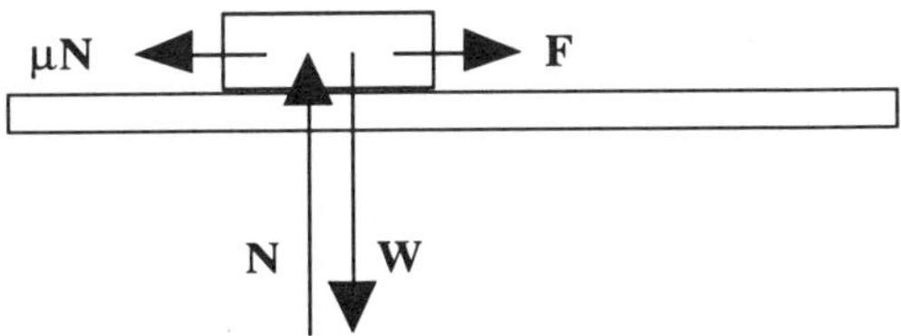

Table 4.2 gives a few examples of static friction of paper against itself and against a smooth aluminum and a slightly rougher steel surface. However, there is considerable variation among paper grades depending on their manufacture.[18] For example, newsprint against itself can be less than 0.25. Also, the web-web friction of grades like tissue and nonwovens and textiles[19] can occasionally go over 1.0.

Other extremes include film which can be much slipperier than paper, and adhesive coated products which have so much tack that the web may block (stick) to rollers or itself. The two factors which affect roller friction the most are surface **roughness** and surface **material** (covers versus plain metal or composites). Note that in the table below it was the surface roughness of the metal rather than its chemistry that accounted for the difference between the aluminum and steel.

Table 4.2
Static Friction of Paper Examples

Grade/Against	Itself	Aluminum	Steel
Bond	0.56	0.37	0.54
Kraft	0.63	0.36	0.54
Magazine	0.29	0.35	0.40
Newsprint	0.37	0.36	0.51

Friction Complications

There are many subtle parameters in the test measurement of friction. First and foremost is the difference between static and kinetic friction. While static is generally the higher of the two, kinetic friction is a more conservative value and has less measurement noise because it is an average over several seconds rather than a capture of a brief peak. Next, there are the inclined and horizontal classes, both of which deliver a static measurement.

There are several suppliers and models of instrumentation, whose analog and digital circuitry may process the raw load signals differently. Finally, there are perhaps more than a dozen friction measurement standards from organizations such as TAPPI, ASTM and so on. Thus, the inherently noisy friction coefficients are made further variable with the numerous test variations.

Most friction measurements use a weighted sled which either slides down an incline, or is pulled horizontally. These sled weights differ which affects the measure (beyond the obvious need to input the weight into horizontal instruments). Also, the sled area and thus unit pressure varies. While simple friction models taught in school say that friction is independent of weight and area, the realities are not so invariable. Finally, the compliance, shape and other properties of the common rubber sled backing can also affect the measurement.

Speed also affects friction measurements. On the inclined tester, one will often see the bed coast for a time after the sled has broken loose before the motor comes to a stop and the angle pointer is read. Speed affects the kinetic friction of horizontal methods because friction depends slightly on speed.

The mounting of the web and test material also affects the measurement. In particular, every test should begin with a virgin web sample because repeated testing can smooth many materials and thus reduce their friction. Care must be taken to check both sides of the web (because many are two sided), and to check both directions (+MD and -MD).

So far we have dealt only with the issues of friction measurement. However, web/roller friction during operation will be different than that predicted by lab testing. First and foremost is the reduction of effective traction with increasing line speed due to air entrainment as discussed in more detail in the next section.

Care must be taken that the test bed and roller surface condition are very similar. It is not unusual for rollers to be polished during operation so that their surface is much smoother. Similarly, grooves and textures tend to have their peaks rounded or worn down and their valleys filled in with dust or other contaminants.

Also, **static electricity** will increase traction if the web and roller carry opposite charges and vice versa. Static electricity can be generated due to microslippage between a web and roller or due to flexing of the web, and is a common occurrence with many films and with overdried paper. Static electricity in turn is affected by ambient humidity. Static is covered in more detail elsewhere in this book.

Since one of the ten commandments of roller design is to ensure that the web never slips, a conservative approach is indicated. Especially in light of the uncertainties of friction measurement and actual friction during operation. Perhaps, using the kinetic value with a safety factor of two might keep most systems out of the undesirable sliding condition.

In the next section, we will give a model for air entrainment and discuss its application. Later, we will discuss the band brake equation for maximum tension difference across a roller. Only then will we be able to calculate torque and power requirements for driven and undriven rollers.

Table 4.3 gives kinetic friction over rollers as pulled by a tensile test machine and reverse engineered from the band-brake equation.[20]

Table 4.3
Kinetic Friction over a Roller

clay coated paper	0.41
calendered kraft paper	0.30
polypropylene film	0.23

Air Entrainment

Air follows a moving web and roller because a boundary layer sticks to the surface. This air is pumped into the ingoing tangent between the web and roller, increasing the pressure above the ambient seen on the top side of the web. This is analogous to an airplane wing which obtains lift from a pressure differential between the top and bottom surfaces.

A **radial pressure** is developed on a wrapped roll equal to the tension (force/width) divided by the roller radius. This radial pressure is balanced in part by the normal force between the web and roller (which assists traction) and in part by the air pressure differential. If the forces were not balanced, the web would accelerate inward or outward. Thus, if the air pressure increases, the web/roller normal force must decrease correspondingly. Eventually, with enough speed, the air pressure will just equal the radial pressure and all normal force and thus traction will be lost.

These principles are best illustrated on a very high speed converting winder which accelerates from a stop to a running speed and then later decelerates. At low speeds, there is little air pumped in between the web and roller so that traction is maintained. However, at increasingly higher speeds the air lifts the web off of the roller. If this is an idler roller, it will coast down to a stop (or to a very slow speed) because there is no traction to overcome bearing drag, much less accelerate the roller. This liftoff may be innocuous as the tension surge is a (hopefully) small one caused merely by the removal of roller acceleration and bearing drag forces.

However, as the winder eventually slows down, the air film will collapse until the web contacts the stopped roller. Then a very large tension spike is seen as the web slides and attempts to bring the roller up to speed. This tension spike can cause web breaks, necking (width), guiding (edge position) and marking problems.

As mentioned earlier, floating, sliding or tracking can all be acceptable web/roller modes of interaction. However, the transition from one mode to another on any element is not.

All nonporous webs entrain some amount of air (or other fluid) into the ingoing tangent. The thickness of the air film layer increases almost proportionally to roller diameter, speed, fluid viscosity and inversely to tension. The calculation for this is given in the next section.

The dependence on viscosity is why it is extremely difficult to obtain good web/roller contact where they are immersed in a fluid bath. Indeed, some coater rollers are intentionally run at different speeds than the web with little undue tension differential or power requirements. Thus while air entrainment is only rarely a problem below 100 ft/min or even 1000 ft/min, nonporous webs in baths can lift off a roller at just a couple of ft/min.

Another large factor in air entrainment is the porosity of the web. For example, the entrained air pressure is easily bled off through a nonwoven, apertured film or other porous web. However, even porous papers (as measured by standard test equipment) can act relatively nonporous with respect to air entrainment. Thus, the reason paper tends to track well is because of typically higher surface roughness which protrudes through the air film to retain some contact with the roller.

Another factor in air entrainment is the surface roughness of the roller. Indeed, we can crudely size the required surface cross section by using the mathematical model in the next section. Grooves and texturing allow the air bleed through channels on the roller surface rather than allowing it to pressurize the underside of the web. Rollers can be texturized for increased traction at both low and high speeds by grooving, knurling or spray coatings.

If one wants to intentionally float the web, the design tools would be low tensions, large roller diameters with smooth surfaces. Additional assistance can be obtained by pressurizing the roller (opposite of a vacuum roller) and to overdrive the roller to much higher speeds than the web.

Air Entrainment Model & Application

Air follows a moving web and roller. Some of this air will be pumped into the inlet between them[21] as seen in Figure 4.6. The height of this air boundary layer between a nonporous web and a stationary bar or a rotating roller can be calculated from foil bearing theory as[22,23,24]

(4.3)

$$H = 0.643\, r \left(\frac{6\,\mu V}{T}\right)^{2/3}$$

where

$$V = V_R + V_W$$

$$T = T_0 - B\,V_W^2$$

and where

H = air film layer thickness

r = radius of roller or bar

μ = absolute viscosity

V_R = speed of roller (or for bar = 0)

V_W = speed of web

T_0 = tension

B = basis weight (mass per area)

Be careful of units, dimensional consistency and conversion factors, especially for viscosity. Note that the second term (centrifugal reduction) in the tension equation is negligible for most cases.

The height calculated here is for the constant height region. This is reduced to about 72% at its minimum value close to the exit region. The calculation is also for the center of a wide web. As seen in Figure 4.7, the height decays (quite possibly to zero) at the ends, and end effects are not considered in this model.

There are a couple of cautions when applying this model. The first and foremost is that the web is assumed to be uniformly tensioned across the width rather than baggy[25]. The loose or baggy lanes can bleed air around the roller rather than pressurizing the web, while the tight bands will tend to collapse the film because of greater tension. The second caution is that the model does not account for any through web air flow due to porosity. The extent of this effect will be estimated later.

Figure 4.6
Anatomy of Air Entrainment

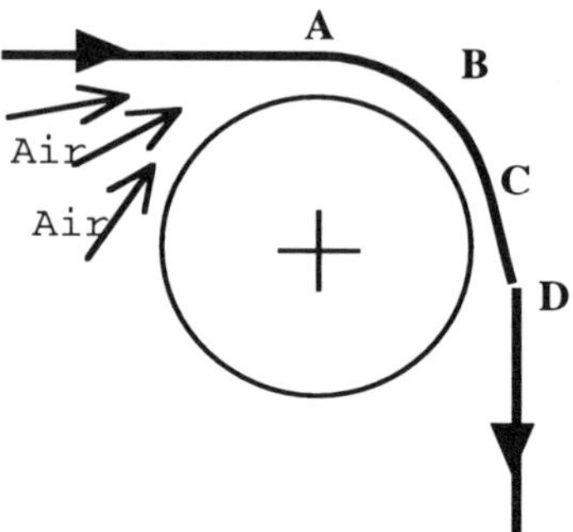

A - Inlet Transition
B - Constant Height Region
C - Minimum Height
D - Exit Transition

Figure 4.7
End View of Air Flotation

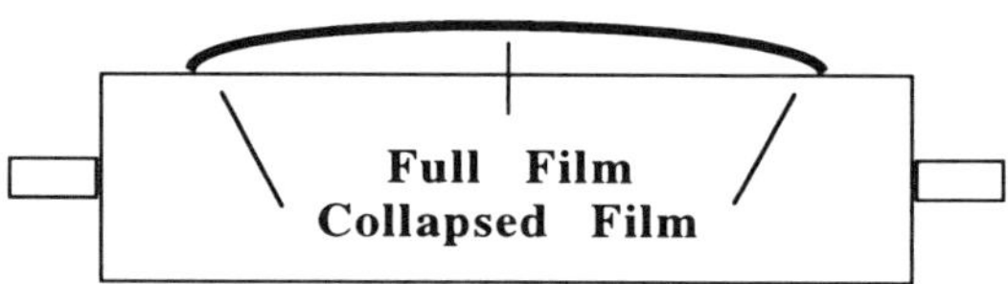

Figure 4.8
Nip Rollers Reduce Air Entrainment

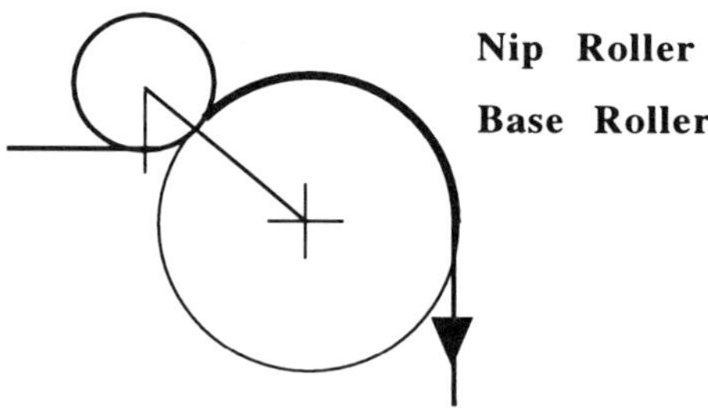

Air film height is greatly reduced with the addition of a wrapped nipping roller as seen in Figure 4.8. There are three considerations for nipping rollers. The first is that the nip roller must be wrapped because a nip after the ingoing tangent greatly reduces the effectiveness of air exclusion. Second, the effect of nip loading is very large compared with the factors in equation 4.3. Third, it might be desirable for one of the rollers to be covered so that the surfaces conform despite geometrical imperfections.

Air Entrainment Example Calculation

English Units

Given a 30#/3000ft^2 newsprint web running at 5000 ft/min under a tension of 1 lb/in over a 24" diameter roller. The absolute viscosity of air at STP is about 0.018cp.

What is the air film thickness in the constant height region?

$$V = (5000+5000)\frac{ft}{min}\,12\frac{in}{ft}\,\frac{min}{60\ sec} = 2000\ in/sec$$

$$T = 1\frac{lbf}{in} - \left(\frac{30lbm}{3000ft^2}\right)\left(\frac{5000ft}{60sec}\right)^2\left[\frac{lbf-sec^2}{32.2lbm-ft}\right]\left[\frac{ft}{12in}\right] = 0.820lb/in$$

$$H = 0.643\ x\ 12in\left\{6\left[0.018cp\frac{2.09x10^{-5}lbf\cdot sec}{cp\cdot ft^2}\frac{ft^2}{144in^2}\right]2000\frac{in}{sec}\ /\ 0.820\frac{lbf}{in}\right\}^{2/3}$$

$$H = 0.00876in$$

This example shows the web is floating only about 0.009" (about 3 x web thickness) above the roller surface, even at high speeds and light tensions. Thus, air film thicknesses are quite small. However, significant reductions in web traction start at air film heights of 1/4 of the surface roughness of the web/roller, and may be effectively floating at 4X the surface roughness. This, unfortunately, is a generalization of a very complex topic and surface roughnesses may be difficult to define or measure.

Since the surface roughness of this example web may be a few ten thousandths of an inch (micrometers in metric units), our web will have little traction if not completely floating. Thus, grooving or texturing is indicated. The design philosophy for either case will be the same: provide a path for the air with a cross sectional area equal or greater than the air film height (times roller width) as described in the next section.

Metric Units

Given a 50g/m^2 newsprint web running at 1500 m/min under a tension of 0.2 kN/m over a 60cm diameter roller. The absolute viscosity of air at STP is about 0.018cp (1 cp = 0.01 g/cm/sec)

What is the air film thickness in the constant height region?

$$V = (1500+1500)\frac{m}{min}\,\frac{min}{60sec} = 50\ m/sec$$

$$T = 200\frac{N}{m} - \frac{50g}{m^2}\left(\frac{1500m}{60sec}\right)^2\left[\frac{N-sec^2}{kg-m}\right]\left[\frac{kg}{1000g}\right]$$

$$T = 168.8\ N/m$$

$$H = 0.643\ x\ 0.3m\left\{6\,\frac{\left(0.00018\frac{g}{cm\cdot sec}\,\frac{kg}{1000\ g}\,\frac{100\ cm}{m}\right)50\frac{m}{sec}}{168.8\frac{N}{m}}\,\frac{N\ sec^2}{kg\cdot m}\right\}^{2/3}$$

$$H = 0.0001945m = 0.1945mm$$

The reader will note that some air (or fluid) is always pumped between the web and roller; it is only a matter of how much and how this affects traction. This is also true of nipped rolls as well, although the calculation has yet to be derived or published. Once brought in, some fluid will leak out the edges, some will pass through the web if it is porous, and the remainder will pass circumferentially around the roller.

The amount of air brought under the web is approximately proportional to roller diameter and speed, and inversely to web tension. This model is also valid for liquids, which have viscosities that are several orders of magnitude greater than gases. Thus, while high speeds are required to float a web on air, flotation inside a fluid bath (such as a treater or coater) can occur at speeds of only a few feet per minute.

Grooving

The purpose and function of grooving is to allow air to pass around the roller in the groove channels rather than pressurize and float the web. It does not, as sometimes believed, provide any spreading function nor does it substantially change the coefficient of web/roller friction. We can use the air entrainment calculation to help size grooving using a simple assumption, that is, to provide a unit groove cross sectional area equal to the air film height.

We will illustrate this concept by checking the common annular (or spiral) grooving pattern used in the paper industry shown in Figure 4.9. The effective area of a groove pattern converted to a thickness would be

(4.4) Effective Groove Thickness = wd/p

where
w = width of a groove
d = depth of a groove
p = pitch of the grooving

For the grooving shown,

EGT = 0.060in x 0.100 in / 0.25 in = 0. 024"

which is 3 times the film height in the example in the previous section, and more than adequate to allow a passage around the roller. The only improvement might be to shorten the pitch so that the air can reach the grooves quickly enough. Currently, the air must travel (0.250 - 0.060)/2 = 0.095" or 12 x the nominal film thickness axially to reach a groove from the center of the land. In reality, however, it will be much greater as the film thickness with grooving should be quite small.

If we reduce the pitch, we can also reduce either the width or depth of the grooving and still maintain our EGT. However, narrow grooves tend to fill or foul with contaminants. Conversely, wide grooving can cause MD troughing and/or CD contraction on thin webs.[26,27] How then can we optimize grooving?

Figure 4.9
A Common Grooving Geometry

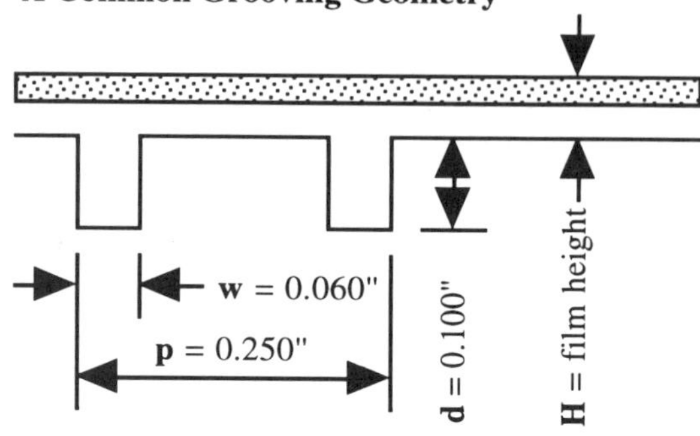

First, let us set the groove width at 10X minimum web caliper as a compromise between narrow easily fouled grooves and excessively wide grooving that can cause trough wrinkles. Then for our example of the previous section, w = 10x0.003=0.030" which we will round up to the nearest conventional size of 1/32" or 0.031".

Next, we must select a w/p. If w/p is small, machining costs will increase because it is quicker to cut a few deep grooves than many shallow grooves. If w/p is large, however, then the air will see a long axial path to get to a groove. Let us propose a w/p of 0.5 as a compromise between cost and performance. This would give us equal groove and land widths. Then for our example of the previous section, p = w / 0.5 = 0.031"/0.5 = 0.0625" which already is a common size.

Next, we need to decide a safety factor for the EGT compared to the air film thickness, H. Let us use 4. Then for our example, EGT = 4 x H = 4 x 0.00876" = 0.0350".

Finally, we calculate groove depth as d = EGT x p / w = 0.0350 x 0.0625 / 0.03125" = 0.070" or about 1/16". Thus, our 'optimized' grooving is 1/32" wide x 1/16" deep on a 1/16" pitch.

While readers should select their own parameters, this section illustrates how modeling can give us a better starting point.

Texturizing

Texturizing a roller, much like grooving, allows the air to flow around passages in the roll shell rather than pressurize and float the web. Unlike grooving, however, texturizing provides an increase in web/roller friction because it has tooth with respect to the MD. Common methods of texturizing include knurling of soft metals such as aluminum, thermal spray coatings such as tungsten carbide, and shot blasting of metal shells.

In order to use a similar approach to sizing as in the last section, we are going to need to make a few simplifying assumptions about the topography of the surface. While surfaces might be shaped more like a mountain range, we are going to simplify that random pattern to pyramids aligned in rows. A cross-section view in the MD would then give a sawtooth appearance as shown in Figure 4.10. With this simplification

(4.5) Effective Texture Thickness = 0.5h

where
h = pk-pk surface roughness
which would be of a similar order as RMS roughness.

For our example problem with a safety factor of 4, we would want

h = (4) x H / 0.5 = 4x 0.00876"/0.5 = 0.070"

This roughness would be much greater than provided by tungsten carbide coatings or shot peening. Thus, one might select knurling, tooling, casting or other method capable of large textures. However, the example was for very high speed equipment. Rough spray coatings will be suitable for most converting equipment.

Figure 4.10
Texturized Surface Approximation

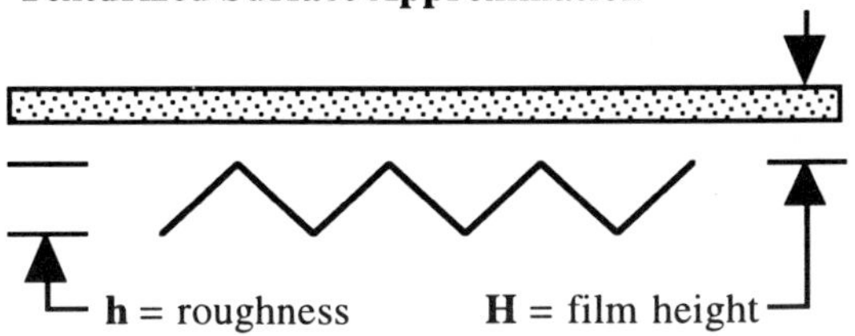

Web and Roller Porosity

In addition to passing the entrained air/fluid around a roller or out the sides, it can pass through a web or roller given sufficient porosity. Examples of porous webs include most nonwovens, tissues and apertured films. While flat paper grades are considered porous as measured by lab instruments, the porosity is usually too small to greatly reduce film heights and thus improve traction. Examples of porous rollers include porous PVC, sintered metals, and drilled shell rollers.

The mechanics of air passing through the web is a bit different than passing around a roller. In particular, the dwell time comes into play as the entrained air escapes through the web due to a pressure defined nominally by web tension divided by roller radius. In order for the web to act porous (reduced film height and improved traction), a significant amount of the air must pass through in the time it takes the web to pass from the ingoing and outgoing tangents.

Example with Permeability
Given a paper web with a permeability (calculated from porosity measurements) of 2 $in^3/sec/in^2/lb/in^2$ running at 5000 ft/min under a tension of 1 lb/in over a 24" diameter roller or into a 24" diameter wound roll, what is air float wrap angle? The pressure under the web is (note the 18% centrifugal component considered on pages 46 and 47)

(4.6)
$$P = \frac{T}{r} = \frac{0.820\ \mathrm{lb/in}}{12\ \mathrm{in}} = 0.0683\ \mathrm{lb/in}^2$$

The time for the air film height thickness of 0.00768" (calculated from the previous example) to pass through one square inch of area is

(4.7)
$$t = \frac{Q}{kPA} = \frac{0.00876\ \mathrm{in}^3}{\left(2\frac{\mathrm{in}^3}{\mathrm{lb-sec}}\right)\left(0.0683\frac{\mathrm{lb}}{\mathrm{in}^2}\right)\left(1\mathrm{in}^2\right)} = 0.064\ \mathrm{sec}$$

Since the 24" diameter roller turns 1 revolution in 0.075 second, it would take 308 degrees of wrap for the air to pass through the web, and before the web was in full contact with the roll.

Measuring Traction

Traction is defined as the maximum tension difference that can be sustained across a driven or nondriven roller before slippage occurs. It is a fundamental parameter when deciding whether a roller needs to be driven (by an electric motor or other means) to keep the web in traction, and how much torque the drive can impart on the roller. This is a very important design and operating criteria because loss of traction usually implies loss of web control in both the CD and MD.

Unfortunately, it is often just too difficult to accurately predict web/roller traction even at low speeds based on static coefficient of friction tests, and especially so at higher speeds where air entrainment becomes a significant factor. Thus, web/roller traction is best measured as a function of web, web tension, wrap angle, roller diameter, roller surface and speed. Speed, as it turns out, is the predominant factor for nonporous webs running over smooth (ungrooved or texture) rollers for processes operating faster than a crawl.

Traction should be measured by any department which has significant web machine design responsibilities such as engineering departments of large web companies, as well as roller and machine builders. This can be done on almost any good pilot machine with a few simple modifications.

As seen in Figure 4.11, the pilot machine need only have a load cell tension control just upstream, downstream or both with respect to a roller test position. Additionally, an accurate (encoder based) speed measurement must be available at the test roller and an adjacent roller. These criteria can be met by many web makers, rewinders and endless loop machines.[28]

The test roller is a live shaft roller mounted on very low friction bearings, attached to an encoder speed readout, precision brake (or drive) and torque meter if not equipped with load cells both upstream and downstream of the test position.

Figure 4.11
Traction Test Apparatus Example

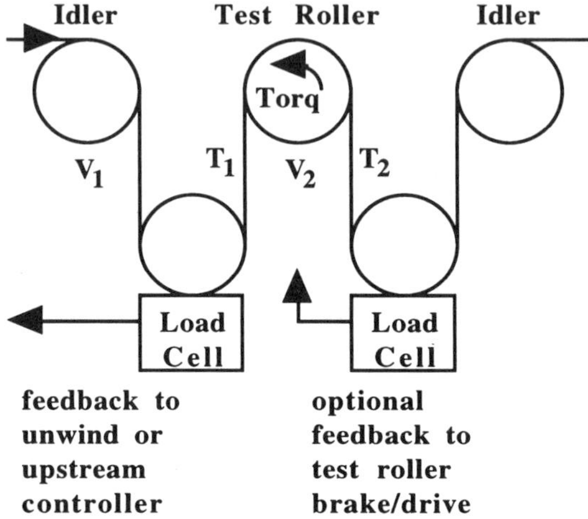

The test procedure is straightforward. The brake torque on the test roller is increased slowly and steadily until a statistically discernible difference in speed is measured between the test roller and an adjacent idler roller. Then, the effective coefficient of traction can be reverse engineered from the band-brake (derived in a later section) equation

$$(4.8) \qquad \mu_E = \frac{1}{\beta} \ln\left(\frac{T_2}{T_1}\right)$$

where
μ_E = effective coefficient of web/roller friction
β = wrap angle (radians)
T_1 = upstream tension (lb/in or kN/m)
T_2 = downstream tension (lb/in or kN/m)

Note that the one of the load cells is redundant (but provides a good verification) if the test roller is equipped with a torque meter because

$$(4.9) \qquad Torq = rw\,(T_2 - T_1)$$

where
Torq = test roller brake torque (lb-in or kN-m)
r = roller radius (inch or m)
w = web width (inch or m)

Note that bearing torque on the test roller must be calculated and added to brake torque as shown in a later section.

Traction Factors

The rationale choices of a dependent variable include coefficient of friction (unitless) or traction (tension differential in lb/in or kN/m). From a design point of view, friction is perhaps the better choice because it can be inserted into the band-brake equation. However, in this section we will use traction for illustration purposes. The choices of an independent variable could include web, web tension, wrap angle, roller diameter, roller surface and speed. In this illustration we will use speed because it is often the second largest factor. (The largest is wrap angle which is already accounted for in the band-brake equation.)

Thus, Figure 4.12 shows a generic plot of maximum tension difference that can be sustained across a roller without slippage as a function of web line speed. As seen here, low speed behavior is dominated by wrap angle, tension and web/roller friction. However, at higher speeds several additional factors will shift the curve as listed.

To limit the number of variables of testing, one notes that the wrap angle is already accounted for in the band-brake equation and needn't be explicitly varied and tested. Similarly, incoming tension is accounted for at low speeds, but must be varied to get the effects due to air entrainment. Indeed, it is the air entrainment variables that will effectively determine the amount of testing required. However, it is very important to cover the entire range of intended applications as testing is far less costly than violating one of the 10 commandments of web handling that states "Thou shalt not allow a roller to slip."

Experience has shown that friction is a very noisy mechanical parameter in itself, to say nothing about uncertainties in the test apparatus. Thus, don't be surprised if there is a bit of scatter. Curve fitting and statistics will help put order and certainty to the data sets.

Figure 4.12
Traction Factors

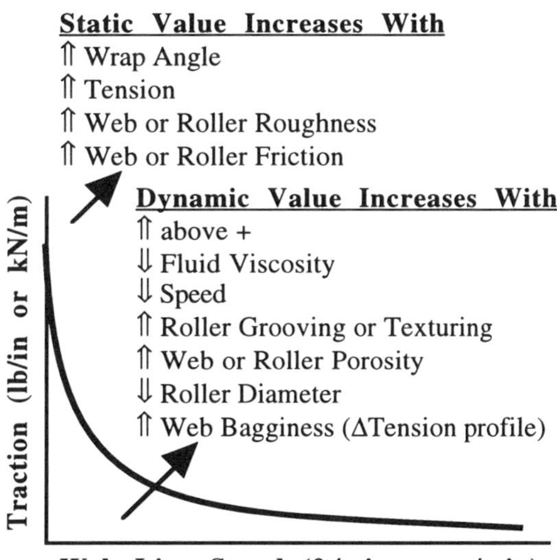

After the coefficient of friction has been fitted as a function of important variables, one is almost ready to use the results for design. However, first we must select a safety factor to allow for measurement and application uncertainties. While statistics might provide a more thorough answer, we probably will do just fine by using a safety factor of 2.

Whether we are involved in design or responsible for web processes, we must make sure that no roller will slip during operation. This means verify that there is sufficient traction to accelerate an undriven roller and overcome bearing drag and windage. For every roller type, we will calculate the minimum wrap angle needed to maintain traction at the minimum web design tension. Similar checks will be made for rollers that are motor driven to intentionally step tension up or down between tension zone boundaries. Helper driven rollers, which are not intended to step tension but rather are used to merely overcome bearing drag and windage, need not usually be checked.

The Band-Brake Equation

The band-brake equation was derived more than a century ago to calculate the driving capability of belts or the braking capability of band or strap brakes.[29] Since then, it has become part of the curriculum of almost every mechanical engineering design course.[30]

The forces acting on a web segment over a roller are shown in Figure 4.13. Summing the forces in the y-direction gives

(4.10)
$$-(T+dT)\sin\left(\frac{d\theta}{2}\right) - T\sin\left(\frac{d\theta}{2}\right) + dN + r\omega^2 dm = 0$$

In this section, we will omit the acceleration component which is generally small, and assume the normal force comes strictly from tension (as opposed to a suction roll vacuum). These terms will be retained in a following section. Thus, equation 4.10 simplifies to

(4.11)
$$-(T+dT)\sin\left(\frac{d\theta}{2}\right) - T\sin\left(\frac{d\theta}{2}\right) + dN = 0$$

Summing the forces in the x-direction gives

(4.12)
$$(T+dT)\cos\left(\frac{d\theta}{2}\right) - T\cos\left(\frac{d\theta}{2}\right) - F = 0$$

where F is the friction force on the segment

$$(4.13)\quad F \le \mu_k dN$$

Eliminating dN by substituting 4.12 into 4.11 gives

(4.14)
$$dT\cos\left(\frac{d\theta}{2}\right) - 2\mu_k T\sin\left(\frac{d\theta}{2}\right) - \mu_k dT\sin\left(\frac{d\theta}{2}\right) = 0$$

Inserting the small angle approximations

$$(4.15a)\quad \cos\left(\frac{d\theta}{2}\right) \rightarrow 1 \text{ when } d\theta \rightarrow 0$$

$$(4.15b)\quad \sin\left(\frac{d\theta}{2}\right) \rightarrow \frac{d\theta}{2} \text{ when } d\theta \rightarrow 0$$

Figure 4.13
Forces Acting on a Web Segment

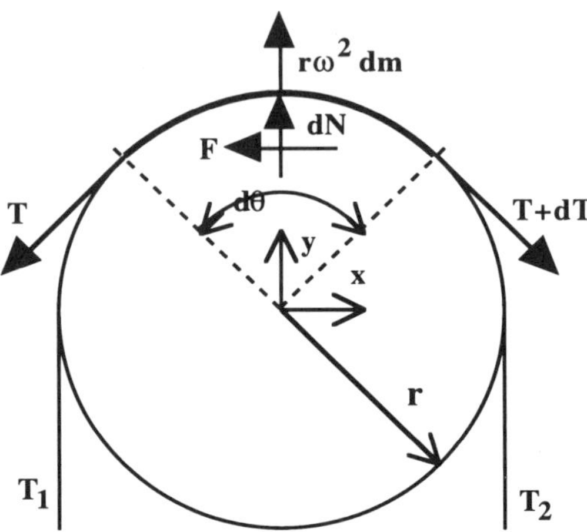

into 4.14 and neglecting the small cross product term $dTd\theta$ gives

$$(4.16)\quad \int_{T_1}^{T_2} \frac{1}{T}\, dT = \mu_k \int_{\theta=0}^{\theta=\beta} d\theta$$

where $dT = \mu_k T\, d\theta$ and β is the wrap angle. This equation can be solved to give the band-brake equation as

$$(4.17)\quad \frac{T_2}{T_1} \le e^{\beta\mu_k}$$

where T_2 is the high tension side and T_1 is the low tension side and thus can be used in either motoring and breaking directions.

Note that as either the wrap angle or low tension decreases to zero, no tension differential can be pulled across the roller. As a side observation, this equation is not limited to 180 degree wraps (the maximum usually obtainable) and explains why a rope wound several times around a stationary spindle (**capstan**) can hold very large forces with only a small pull.

Finally, note that we have shown the equation in its proper form as an <u>inequality</u>, instead of the usual equality. In other words, this describes the maximum tension difference, not the actual which is based on roller torques.

Band Brake Application

The band brake equation tells us how much tension difference can exist across a tracking roller before macro slippage occurs if we use the static friction value. One typical design application in the case of a nondriven idler roller would be to first determine the required tension difference to overcome bearing drag, windage and accelerate the roll, which is given in an upcoming section. Then, we would determine the minimum wrap angle necessary to ensure traction.

However, we can place an upper bound on the required tension difference. For example, the total tension difference from all **idler rollers** in a tension zone should not exceed 10% of the nominal minimum tension. For the **driven roller** which isolates two tension zones without drawing or strength change, we should need less than 15% of the breaking strength of the material across the zones because the good running tensions for almost all applications is 10-25% of strength as described in Chapter 2.

Nondriven idler roller example

Given a machine running products which vary from 5-10 PLI strengths, determine the minimum wrap angle for the 10 nondriven rollers which have a measured minimum static coefficient of friction under operating conditions of 0.2.

The minimum running tension would be 0.10x5=0.5 PLI. Thus, we would expect no more than a 10% total downweb tension rise, or equivalently 1% for each of the rollers to accelerate it and overcome bearing drag and windage. Thus, (T2-T1)max = 0.01x0.5=0.005 PLI, or T2=0.5+0.005=0.505 PLI. Solving the band brake equation 4.17 for wrap angle gives

$$\beta \geq \frac{1}{\mu_k} \ln\left(\frac{T_2}{T_1}\right) \tag{4.18}$$

A safety factor of 2 on the friction then gives

$$\beta \geq \frac{1}{0.1} \ln\left(\frac{0.505}{0.500}\right) = 0.100 \text{ rad} \; \frac{360^\circ}{2\pi \text{ rad}} = 5.7^\circ$$

Driven roller example

Given the previous web and machine, determine the minimum wrap angle for a driven roller separating the machine into two tension zones. The friction coefficient on this roller 0.4 because it has a textured rubber cover.

The design tension range for the machine would be 10-25% of the breaking strength of the web. Taking the light weight grade again, the minimum tension on the first section would be T1min=0.1x5=0.5 PLI and the maximum tension on the second section would be T2max=0.25x5=1.25PLI. Thus from equation 4.18 with a safety factor of 2 on friction,

$$\beta \geq \frac{1}{0.2} \ln\left(\frac{1.25}{0.5}\right) = 4.581 \text{ rad} \; \frac{360^\circ}{2\pi \text{ rad}} = 262^\circ$$

Notice how much wrap is needed for the driven roller even with its exceptionally high friction coefficient. (The maximum physical wrap angle for equal diameter rollers is 240°.) However, in these examples we have bounded the maximum required tension differential based on web handling principles. The actual process may require less. If we did require this difference, however, we may need to consider nipped drive rollers or suction rollers to enhance traction.

Also note the characteristic difference in required wrap between driven and nondriven (or helper driven) rollers. In general, driven rollers will have difficulties getting enough wrap and friction because a noticeable portion of the web's strength must be pulled. Conversely, as we shall see, nondriven rollers have difficulties getting the inertia or acceleration down far enough to avoid tension disturbances which are but a fraction of the web's strength. In other words, wide machines (which require large diameter rollers for deflection) and/or fast accelerating machines (such as most rewinders) will require a drive even if there is sufficient traction across rollers.

Band Brake and Macroslippage

The band brake equation is also valid for a grossly slipping or sliding web/roller conditions. Indeed, as we will see in the next section, the web is microslipping on rollers in traction anyway. The only difference is that the differential speed on grossly slipping rollers or bars might be high enough to change the kinetic friction (which is always the one used for design) noticeably (up or down) beyond that measured on slow slip speed lab testing.

Macroslippage is a web/roller condition common with pans, shoes, some spreaders, and air bars that are not properly sized to ensure continuous flotation.

Example

Given a D-bar or bent pipe spreader that is lightly wrapped at 15 degrees, what is the parasitic **sliding drag** tension? The tension just downstream is load cell controlled to a T2 of 1.0 PLI, and the kinetic friction at high speeds on the highly polished bar is only 0.1.

From the band brake equation 4.17,

$$\frac{T_2}{T_1} = e^{\left(15^\circ \frac{2\pi \text{ rad}}{360^\circ}\right)0.1} = 1.027$$

Therefore, T1=1.000/1.027=0.974PLI and

$\Delta T = 1.000 - 0.974 = 0.027$ PLI

which is

100%(0.026/1.000) = 2.6%

of the nominal line tension.

Obviously, we can have only a few sliding members in a web line, and all would have to be low wrap and low friction in order to keep overall process tension in control.

Band Brake and Microslippage

The band brake equation gives us a very useful insight into the MD tensions, strains and slippage of a web going over a tracking roller. Somehow, the tension must step up (or down) going over every roller (unless the helper drive is precisely matched to load). The ingoing tension is at T1 right up until contact with the roller because a sustained (non-dynamic) tension difference can't exist in a free web span. Similarly, the outgoing tension is held at T2 from the exit point of the roller to the next roller.

From this, we can conclude that the tension must change within the wrapped zone. If tensions changes, then strains must also as given by Hooke's law. Since the roller is relatively rigid, there must be slippage between the web and roller for changing strains. The band brake equation tells us the shape of the tension and strains which charge or decay exponentially as seen in Figure 4.14. Initially, the web enters locked onto the roller, but will traverse into the band-brake **slippage zone** where microslippage adds or relieves tension to match the downstream tension given by the exponential. This microslippage zone can cause web marking as well as roller polishing and **wear**. As differential tensions increase, the size of the slippage zone and magnitude of slippage increase. In no case, however, is MD tension or CD guiding lost with at least some portion of the wrap tracking.

Figure 4.14
Tensions Across a Roller

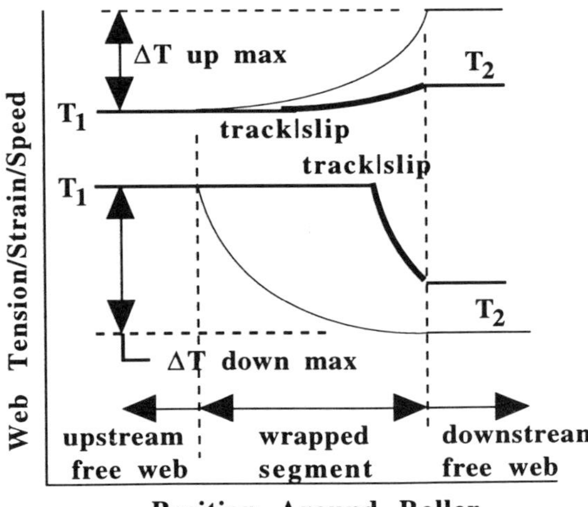

Nipped Pull Rollers

Many **process** rollers are nipped such as for calendering, coating, embossing, laminating, printing and so on. This nip also increases the traction capacity of that drive point. However, coatings or other liquids will often reduce effective friction because they act as a lubricant, and embossing increases the effective friction tremendously as the material is literally swaged into the pattern.

Some **transport** rollers are nipped to provide additional traction for a drive point. In some cases, this is a poor design decision because there are only a few situations where an additional tension zone must be created. One example is to isolate a taper tensioned winder from an upstream slitting station. Another is to bring the tension back down after picking up a significant tension increase from bearing drag or inertia on an inordinate number of idler rollers. Finally, a nip is required for large tension differentials, such as during threading where the tail is under near zero tensions.

The problem with any nip is that it is an intolerant system. Imperfect nips can ruin a thin material that is perfect because the material will draw unevenly[31]. Conversely, a perfect nip can ruin an imperfect material by turning troughs or wrinkles into creases. Every nip roller set must be considered to be precision components.

An example of a nipped roller system is shown in Figure 4.15. Often, the bottom roller will be loaded against a fixed top roller which provides for a quick drop feature to increase safety and reduce the potential damage to the rollers if they become wrapped. The best loading systems are often stiff pivoting arms as they can have rigidity without friction. Slide loading systems are often a compromise between looseness and binding.

In many cases, a smooth metal roll is the wrapped element. This has the advantage of wrapping the more geometrically precise element. Unfortunately, that configuration will not give the best traction if the nip is opened up because the process or traction is no longer needed. The metal roll should be the one driven if **speed control** is used on that section.

Figure 4.15
A Nipped Roller System

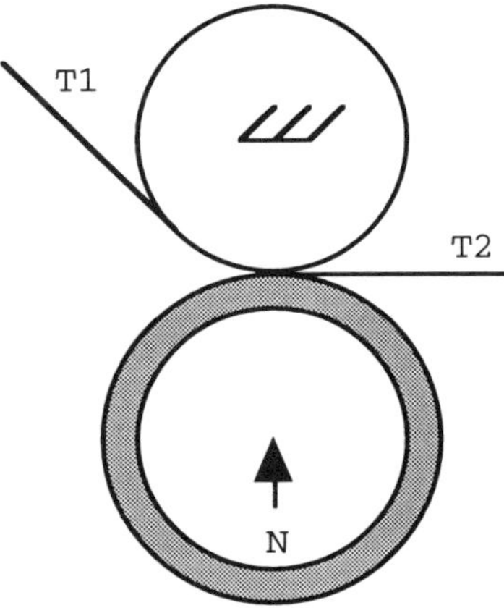

The slip criteria for the above system can be derived from elemental traction equations[32], plus an additional concept. That is, the web will be tensioned going through the nip zone as much as[33]

(4.19) $\Delta T_{nip} = \mu N$

where N is the nip load and is the coefficient of friction of the web against the slipperiest roller.

The procedure begins by defining four tensions: incoming onto a roller, ingoing into the nip, outgoing from the nip and outgoing from a roller. Band belt equations are written for the ingoing and outgoing wraps, and the tensioning from the undriven nip roller is defined above.

As a practical matter, however, the tension difference across the entire pull roller section is determined conservatively and primarily by the above equation because nips are usually an order of magnitude or more than typical web tensions.

If a maximum performance pull station is required, both rollers can be driven. Never, however, run both nipped rollers in speed control, such as by **belting or gearing** them together. Inevitable mismatches in surface speed will ensure a constant slippage on one of the interfaces. This will reduce pull roller effectiveness as well as induce unnecessary roller wear and potential web marking.

S or Bridle Wrapped Drive Rollers

An alternative to the nipped pull roller station is S-wrapped or bridle rollers as seen in Figure 4.16. This drive point arrangement has the advantage of good traction by virtue of the high effective wrap angles, without the material damage potential of nipped rollers. While the figure shows only two rollers, the principle is readily extended to stacks of multiple unnipped rollers. This principle of operation is to increase the effective wrap angle beyond the 180 degrees or so possible with a single roller, and has nothing to do with increased surface area. No matter how many rollers, however, the band brake equation tells us that a minimum tension must exist on both the inlet and outlet in order to drive the web.

There are a couple of issues with S-wrapped rollers. First, the short draw between the rollers demands very high **alignment** tolerances to avoid wrinkling. Second, the desirable surface speed of each of the rollers must be determined. In most cases the surface speeds should be matched for maximum performance.

Figure 4.17 shows an example MD web tension variation through a two roll drive system, where the set is braking, hence causing the tension to rise from T1 to T2. The inter-roll tension, T12 can be calculated from the band-brake equation, provided that T12>T1. If not, then T12=T1 and one of the rollers is redundant because there is sufficient traction across the other. Note how roller #2 is in a state of constant slip, and will operate acceptably as long as V2≤V1. If not, the ability to generate the tension differential, T2-T1, would be compromised. However, it would be desirable to keep V2=V1 to reduce the relative speed difference between the web and roller #2 to reduce roller wear and web scratching. Note how roller #1 is in traction on the first part of the wrap, and then transitions to slippage on the latter. Again, the slippage zone can be calculated as T1 was given and we just calculated T12.

In a **drawing** (material yielding) application, the web tension must be brought up from a line tension, to a yield tension, and back down to a line tension. The ingoing tension, T1 will be floating unless there is an upstream tension

Figure 4.16
S-Wrapped Drive Rollers

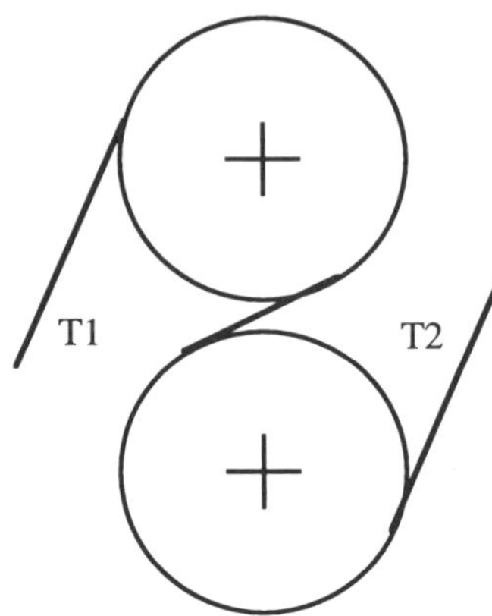

control zone. The yield tension, T2, is difficult to determine and should not be tension controlled. Rather, it should result from the material properties as determined by the strain of the web in that zone. Thus, roller #3 (not shown) will run faster and will be under **speed control** with respect to the S wrap set. Furthermore, the tension will in many cases need to be stepped back down to a running line tension after drawing is complete.

With a drawing process, the step up of tension from line (10-25% of yield) to yield is a 4:1-10:1 tension ratio. This is usually much more than can be provided by a single unnipped roller. Thus, multiple rollers will be almost inevitable if the nip is to avoided. However, most other process tension changes in a line may be modest enough to be achieved with only a single unnipped drive roller. In any case, gross slippage will result from insufficient traction and control over the tension differential will be lost.

Figure 4.17
Tension Through an S Wrap Drive

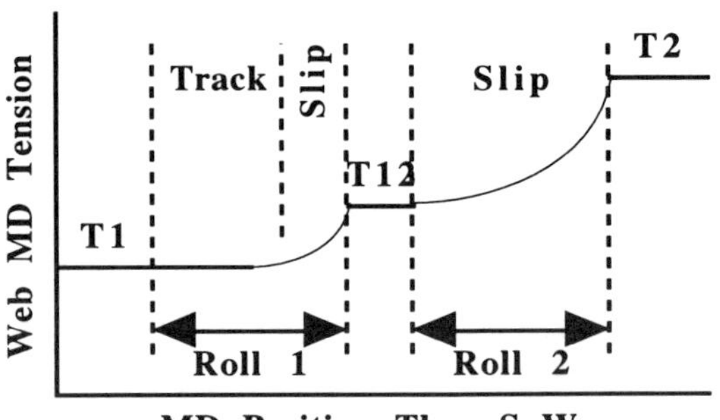

Vacuum, Pressure and Centrifugal Effects

The band-brake equation is readily extended to vacuum rollers which enhance traction, or air bars/rollers which reduce traction. Centrifugal effects are analogously derived in many machine design handbooks[34], but are seldom significant except at very high speeds. Centrifugal effects do not change tension but do reduce the normal pressure between the web and a bar or roller as was given in the air entrainment section

The band-brake equation including these higher order terms is written as

$$(4.20) \quad \frac{T_2 - T_p}{T_1 - T_p} \le e^{\beta \mu_k}$$

where all of the terms have been previously defined except the reduced tension Tp.

Vacuum Rollers

$$(4.21a) \quad T_p = -V r$$

where V = vacuum in force/area and has a positive sign, and where r is the radius of the roller. Note that even a modest vacuum, such as 0.1 psi, can be quite effective with a large radius to bring the left-hand side of (4.20) closer to unity, thus requiring a smaller wrap angle.

Air Bars

$$(4.21b) \quad T_p = P r$$

where P is the presssure in force/area and has a positive sign, and where r is the radius of the roller or bar. Note that even a modest pressure, such as 0.1 psi, can become quite effective in bringing the denominator of (4.20) closer to zero and thus making the left-hand side go to infinity as well as the required wrap angle. Also note that the equation breaks down when Tp approaches the lessor of T1 or T2 as the web is essentially **floating** by then.

Centrifugal Force

$$(4.21c) \quad T_p = \frac{B s^2}{g}$$

where B is the basis weight (mass/area) of the web and s is the speed. In the English system, basis weight is often in units of force/area and would have to be divided by the acceleration of gravity.

Also note that when Tp approaches the lower tension, T1, that the traction is essentially zero and the required wrap angle goes to infinity.

Centrifugal Example Problem

Given a 30#/3000ft^2 newsprint tensioned at 1 lb/in, find the velocity where the web will leave a 6" diameter roller due to centrifugal effects.

Setting (4.21c) equal to tension T1 = 1 PLI and solving for speed

$$S = \sqrt{\frac{T g}{B}}$$

$$S = \sqrt{\frac{\left(1 \frac{\text{lb}}{\text{in}}\right)\left(386 \frac{\text{in}}{\text{sec}^2}\right)}{\frac{30 \text{ lb}}{3000 \text{ ft}^2}\left(\frac{\text{ft}^2}{144 \text{ in}^2}\right)}} = 2357 \text{ in/sec}$$

$$S = 3757 \frac{\text{in}}{\text{sec}}\left(\frac{\text{ft}}{12 \text{ in}}\right)\left(\frac{60 \text{ sec}}{\text{min}}\right) = 11{,}800 \text{ ft/min}$$

There are three things to note about this problem. First, and most importantly, there is a **speed limit** of sorts to web handling. At or above this speed, the web will depart from rollers unless equipped with a vacuum or nip. Second, this speed limit is only dependent on the web tension divided by basis weight. This is related to a material property called the **breaking length** (strength ÷ basis weight), which is commonly used in the paper industry to define the structural efficiency of webs. The only difference is that centrifugal effects use web tension, which is typically 10-25% of strength.

Roller Drag

While the band-brake equation tells us the maximum tension difference can exist across a bar or roller, the actual tension difference can be less than or equal to this amount when the web is in traction with a roller. When in traction, the tension difference across rollers is composed of three terms: drag, inertia and drive.

Roller drag acts like friction in that it always opposes motion. This tends to make tensions climb as the web progresses downstream over multiple rollers. This also, though perhaps not intuitively, makes the downstream rollers tend to turn faster than upstream rollers.

This drag itself is composed of several elements. First, all rollers are supported by bearings which have some turning resistance. The magnitude of **bearing drag** is very difficult to predict as it depends on the type of bearing, seals and lubrication. In general, decreasing friction, improved bearing life and increased cost are ordered as: continuous lube, oil sump, followed by grease. Furthermore, bearing drag will change somewhat with time due to wear, temperature which affects lubrication viscosity, speed and many other factors.

Second, all rollers (except those running in a vacuum, move air (or other fluids). While they are not very powerful fans, you can detect the moving air with a light tissue streamer or smoke. This **air resistance** acts like a weak brake, except if the fluid is a liquid when the resistance can be quite high.

Third, rollers in nip will see a high amount of rolling resistance hysterisis in the cover, wound roll and/or web as it turns cyclic stresses into heat. **Nip rolling resistance** can be quite high for soft wound rolls and soft covered rollers.

Finally, driven rollers will have mechanical **drive transmission losses** at belts, couplings, gearboxes, and bearings.

Measuring Drag

Values for bearing drag are seldom published, and are best only approximate when they are. Moreover, bearing drag is only one of the retarding forces on rollers, and the others are even harder to predict. Fortunately, we can measure the composite drag torque with an easy spin down test. This makes use of the rotational equivalent of Newton's law:

$$(4.22) \quad \text{Torque} = I\,\alpha$$

The mass moment of inertia, I, is calculated from roller dimensions and material as given on a detailed drawing. In the case of a nipped roll(er) set, one of the rollers must be picked as the reference. The total inertia is the sum of the reference roller plus the velocity corrected inertia of the other roll(er). The angular acceleration rate, α, is calculated from the linear acceleration rate, and thus the time it takes to coast down from a known speed as:

$$(4.23) \quad a = r\,\alpha = \frac{\Delta V}{\Delta t}$$

An improvement in the accuracy of the drag measurement at speed is to use a change in velocity from just above normal operating speed to just under. However, this requires a continuous speed readout. Using the easier range of operating speed to zero speed will underestimate drag slightly because the windage increases strongly (square of) with speed, and other drag components are slightly speed dependent.

This drag measurement technique is easy because it only requires a speed readout and stopwatch. However, it will only give the composite or total drag, with no information as to the strength of the individual constituents. The following examples show a real life drag calculation.

Drag Calculation Example - English Units

An aluminum idler roller coasts from 7500 FPM to 5000 FPM in 55 secs; what is the drag torque in this speed range? How many rollers could we allow in a line. The 40" long roller has a 6.88" OD and a 6.00" ID, and the head and journal inertias are negligible.

(4.24) $$WR^2 = [0.000682]\,\rho L \left(D_O^4 - D_I^4\right)$$

where
WR^2 is the mass moment of inertia in lb-ft^2
ρ is the shell density in lb/in^3 and
the D's are diameters in inches

$$WR^2 = [0.000682](0.10)(40)\left(6.88^4 - 6^4\right) = 2.58$$

Torque is calculated as

(4.25) $$\text{Torq} = \frac{WR^2\,\Delta V}{[80.5]\,D_O\,t}$$

where
Torq is torque in ft-lb
ΔV is the change in speed in FPM and
t is time in seconds.

$$\text{Torq} = \frac{2.58\,(7500-5000)}{[80.5](6.88)(55)} = 0.212\ \text{ft}-\text{lb}$$

From equation 4.9 we can calculate the drag induced tension differential for a 36" wide sheet going over that roller as

$$\Delta T_D = \frac{0.212\ \text{ft-lb}\ \dfrac{12\ \text{in}}{\text{ft}}}{\left(\dfrac{6.88}{2}\ \text{in}\right)36\ \text{in}} = 0.0205\ \text{lb/in}$$

If the minimum web tension were 1 PLI and we wanted no more than a 0.1 PLI tension increase across a tension zone at steady state speed (no inertial tension as given in the next section), the maximum number of rollers we could have is

$$N = \frac{0.1}{0.0205} = 4\ \text{rollers}$$

Drag Calculation Example - Metric Units

An aluminum idler roller coasts from 2300 MPM to 1500 MPM in 55 secs, what is the drag torque in this speed range? How many rollers could we allow in a line? The100 cm long roller has a 17.5 cm OD and a 15 cm ID, and the head and journal inertias are negligible.

The mass moment of inertia of the roller is

(4.26) $$I = \rho L \frac{\pi}{32}\left(D_O^4 - D_I^4\right)$$

$$I = 2710\frac{\text{kg}}{\text{m}^3}\,1.0\text{m}\,\frac{\pi}{32}\left(0.175^4 - 0.15^4\right)\text{m}^4 = 0.115\ \text{kg}-\text{m}$$

The angular acceleration rate is

(4.27) $$\alpha = \frac{a}{r} = \frac{2\,\Delta V}{D\,t}$$

$$\alpha = \frac{2\,(2300-1500)\dfrac{\text{m}}{\text{min}}\,\dfrac{\text{min}}{60\text{sec}}}{(0.175\text{m})\,(55\ \text{sec})} = 2.77\ \text{rad}/\text{sec}^2$$

The torque is

(4.28) $$\text{Torq} = I\,\alpha$$

$$\text{Torq} = (0.115\text{kg}-\text{m})\left(2.77\frac{1}{\text{sec}^2}\right)\left[\frac{\text{N}-\text{sec}^2}{\text{kg}-\text{m}}\right] = 0.318\text{N}-\text{m}$$

From equation 4.9 we can calculate the drag induced tension differential for a 0.9m wide sheet as

$$\Delta T_D = \frac{0.318\text{N}-\text{m}}{\left(\dfrac{0.175}{2}\,\text{m}\right)0.9\text{m}} = 4.04\ \text{N/m}$$

If the minimum web tension were 0.2 kN/m and we wanted no more than a 0.02 kN/m tension increase across a tension zone at steady state speed (no inertial tension as given in the next section), the maximum number of rollers we could have is

$$N = \frac{0.02}{0.00404} = 5\ \text{rollers}$$

Inertial Tension

When a machine changes speed, there will be an additional **tension difference** across each roller. It is this imbalance in tension that supplies the necessary force, and thus torque, to accelerate and decelerate a roller. When accelerating, the tension will increase going over a roller even more than provided by drag. When decelerating, the inertial tension will subtract from drag causing a lessor increase, or in many cases a net tension drop going over a roller.

One way to conceptualize inertial tension is to replace the rotating roller with an equivalent mass bonded to the web. As seen in Figure 18, this yields a schematic similar to a railroad train where the right end is the pull provided by the engine (drive roller), and the intermediate positions are cars (idler rollers). During acceleration, the caboose coupling at the left end (tension sensor) has to only pull the drag and accelerate only that last car (ΔT). However, the couplings closer to the engine must pull more cars. Finally, if the engine slows down, the tension on all couplings will be reduced and perhaps go into compression.

The magnitude of the inertial tension component depends on only two factors as can be obtained from equation 4.9 as

$$\text{(4.29)} \quad \text{Torq} = I\alpha = \frac{2\,\Delta T_I\, w}{D_O}$$

where
T_I is the inertial tension
w is the machine width
D_O is the roller's OD
I is the mass moment of inertia and
α is the rotational acceleration rate

The first factor is mass moment of inertia of the roller as was given in the last section. Since, as was described in Chapter 3, the minimum roller size is often determined by deflection, diameter will increase approximately proportional to machine width. However, inertia increases by the fourth power of diameter as was given in the last section. Thus, the mass moment of inertia of a roller is primarily determined by machine width. Roller material, wall thickness and other variables are quite minor by comparison.

Figure 4.18
Tension Buildup to Accelerate Idler Rollers

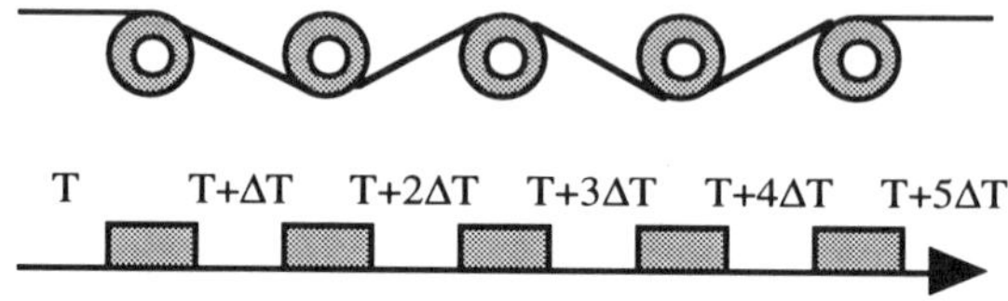

The second factor in inertial tension is the **acceleration rate** which might vary from as little as 1 ft/min/sec (0.3 m/min/sec) to as high as 300 ft/min/sec (100 m/min/sec). As a practical matter, however, acceleration rates tend to fall into two categories. The first is continuous operations such as web manufacturing lines which nominally run at a constant speed except during grade changes or during the occasional startup/shutdown event. Continuous operations will tend to only accelerate/declerate at around 10 ft/min/sec (3 m/min/sec). This slow accel rate reduces the required drive torques and horsepowers, as well as allowing ample time for drives to respond. The only penalty paid for this slow accel is a reduction in saleable production for perhaps a minute or so.

However, discrete processes such as roll-roll converting, slitting, and rewinding may spend a significant portion of their time accelerating and decelerating.[35,36] Thus, acceleration and deceleration rates are a compromise between net throughput and the cost and complexity of fast speed changes. A common acceleration and deceleration rate for a rewinder is 100 ft/min/sec (30 m/min/sec).

The **unwind** roll is a very demanding drive application for many reasons. First, the inertia changes by many orders of magnitude from the beginning of an unwinding cycle to the end. This will require **inertia compensation** to set the appropriate drive gain changes. Second, the torque range often exceeds the range of many motors and brakes. Thus, unwind tension may climb uncontrollably on light webs if the unwind diameter and accel rate are large, even if the brake is turned off. The solution is a drive motor or combination drive motor and brake. Similarly, tension may drop to zero if there is insufficient brake due to a high inertial tension reduction. These subjects are covered in more detail in other chapters.

Inertia Tension Example - English Units

How many undriven idler rollers can be used in a tension zone at acceleration rates of 10 and 100 ft/min/sec without the tension climbing more than 25%? The other parameters are as described in the previous example.

From (4.25), the acceleration torque at 10 ft/min/sec is

$$Torq_I = \frac{2.58\,(10)}{[80.5](6.88)(1)} = 0.0466\ \text{ft}-\text{lb}$$

From (4.9), the resulting inertial tension increase will be

$$\Delta T_I = \frac{0.0466\ \text{ft}-\text{lb}\ \frac{12\ \text{in}}{\text{ft}}}{\left(\frac{6.88}{2}\ \text{in}\right)36\ \text{in}} = 0.00452\ \text{lb}/\text{in}$$

at 10 FPM/sec and 0.0452 PLI at 100 FPM/sec.

The net tension increase across a single roller, for acceleration, is the sum of drag and inertial tension or

$$\Delta T_{A10} = 0.0205 + 0.00452 = 0.0250\ \text{PLI}$$

$$\Delta T_{A100} = 0.0205 + 0.0452 = 0.0657\ \text{PLI}$$

The maximum number of undriven idlers for a 25% tension increase (above a minimum 1 PLI) across a tension zone during acceleration is

$$N_{10} = \frac{0.25}{0.0250} = 10\ \text{rollers}$$

$$N_{100} = \frac{0.25}{0.0657} = 3\ \text{rollers}$$

Thus, in general terms, the maximum number of idler rollers is primarily determined by the machine width, minimum strength of the product, acceleration rate and your standard for an allowable downweb tension differential through a tension zone. Some relief can be obtain with low inertia or composite rollers. If more is required, some rollers will need to be driven.

Inertia Tension Example - Metric Units

How many undriven idler rollers can be used in a tension zone at acceleration rates of 3 and 30 m/min/sec without the tension climbing more than 25%? The other parameters are as described in the previous example.

From (4.27), the angular acceleration rate at 3 MPM/sec is

$$\alpha = \frac{2(3)\frac{\text{m}}{\text{min}}\ \frac{\text{min}}{60\text{sec}}}{(0.175\text{m})\ (1\ \text{sec})} = 0.571\ \text{rad}/\text{sec}^2$$

From (4.28), the acceleration torque is

$$Torq = (0.115)(0.571) = 0.0657\ \text{N}-\text{m}$$

From (4.9), the resulting inertial tension increase will be

$$\Delta T_I = \frac{0.0657\ \text{N}-\text{m}}{\left(\frac{0.175}{2}\ \text{m}\right)0.9\ \text{m}} = 0.834\ \text{N}/\text{m}$$

at 3 MPM/sec and 8.34 N/m at 30 MPM/sec

The net tension increase across a single roller, for acceleration, is the sum of drag and inertial tension or

$$\Delta T_{A10} = 4.04 + 0.834 = 4.87\ \text{N/m}$$

$$\Delta T_{A100} = 4.04 + 8.34 = 12.38\ \text{N/m}$$

The maximum number of undriven idlers for a 25% tension increase (above a minimum 0.2 kN/m) across a tension zone during acceleration is

$$N_{10} = \frac{0.25\text{x}200}{4.87} = 10\ \text{rollers}$$

$$N_{10} = \frac{0.25\text{x}200}{12.38} = 4\ \text{rollers}$$

Again, the maximum number of undriven idlers is limited !!

Wrap Angle

One of the common web handling questions I get is "How much wrap angle is required?" The answer to this for almost all applications is "The wrap angle must be sufficient to avoid web/roller **slippage**." The traction mode of web/roller interaction should be maintained because slippage incurs the very real risks and dangers of web marking, roller wear and general loss of web control. The next question that inevitably follows is "Can you give me a rule of thumb for wrap angle." Unfortunately, there is no rule of thumb for any web or application that would be safe but not overly conservative. However, we can offer solid design recipes.

The case of the **unipped driven roller** was given in the earlier section entitled Band Brake Application. There, the wrap needed to be sufficient to create the maximum anticipated tension difference between the zones separated by the drive roller. For **nipped drive rollers**, the total traction or tension difference capability is the sum of the wrap and nip contributions as given in the section on Nipped Pull Rollers. However, in many cases the nip is dominant and sufficient in itself so that no wrap is required at all. Thus, calendering, embossing, laminating and printing stations often have little or no wrap on either roller.

We now also understand how the band brake equation shows that a nip (or similar mechanism) is required in the event that there is extremely low or zero tension on one side of a driven roller. This situation, though not common, is seen on some **threading** applications, where the drive bounds a section in **loop** or catenary control, and where the web is **sheeted** or cut off. Here again, the traction may be provided by the nip so that no wrap is required.

We also touched on the minimum required wrap angle for an **idler roller** in the Band Brake Application section. However, here the constraint was imposed that the tension should not rise unacceptably (more than 10-25% of its setpoint) in a tension zone due to idler rollers. There, the tension rise though caused by drag and inertia was limited by good web handling principles rather than by explicitly calculating drag and inertia.

On the next page, we will use already developed techniques to calculate the minimum wrap angle on any **idler roller** to avoid slippage by explicitly calculating drag (see Measuring Drag) and inertial (see Inertial Tension) factors. The only other measurement we need is the coefficient of web/roller traction. A simple but crude method (see Static Coefficient of Friction) and accurate but complex method (see Measuring Traction) were given earlier. However, traction can also be measured adequately (in non air-entrainment cases) using a weighted strip of web, a force scale and the procedure given in Figure 4.19.

The actual wrap angle should be a little more than calculated to allow for a safety factor. However, one may use minimal wrap (and use a large diameter roller) if one wants to avoid wrinkles crossing rollers in a specific and strategic position. Many spreaders require more than a minimum wrap to allow for spreading forces. Heavier wrap angles are seldom a problem except where deflection might be a consideration, such as on bowed rollers.

Also note that the wrap angle has no influence on **bending stresses of a thick web** as is sometimes believed. The bending strains are determined merely by the ratio of web thickness to roller diameter and are completely independent of angle. Thus, it is no more damaging to the web to wrap 180 degrees than it is to wrap 10 degrees.

Figure 4.19
Measuring Web/Roller Traction

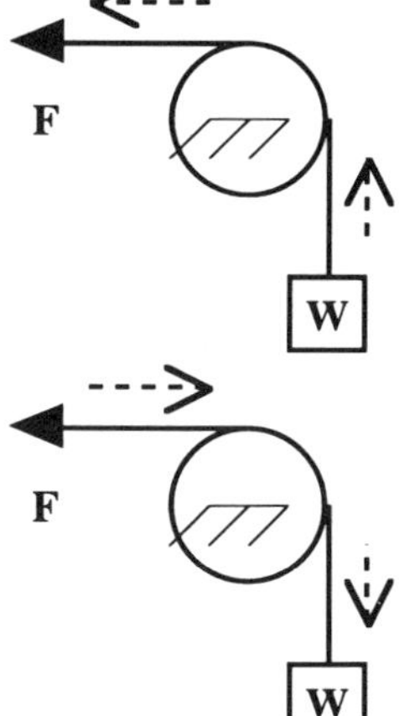

Pull Weight Up for 90 degree wrap

$$\mu = \left(\frac{2}{\pi}\right) \ln\left(\frac{F}{W}\right)$$

Let Weight Down for 90 degree wrap

$$\mu = \left(\frac{2}{\pi}\right) \ln\left(\frac{W}{F}\right)$$

Wrap Angle Calculation - English Units

Given a 1 PLI weak web, a modest 10 ft/min/sec acceleration rate and a 0.75 in-lb bearing drag, what is the minimum wrap angle to accelerate the idlers. The 4" OD x 3.5" ID x 60" long aluminum idler rollers have a typical 0.3 coefficient of traction with the 56" wide web. Equations, definitions and units have been given in earlier sections.

Inertia is calculated from equation (4.24)

$$WR^2 = [0.000682](0.10)(60)\left(4^4 - 3.5^4\right) = 0.433 \text{ lb} - \text{ft}^2$$

Inertial torque is calculated from (4.25)

$$Torq_I = \frac{0.433\ (10)}{[80.5](4)(1)} = 0.01345 \text{ ft} - \text{lb}$$

thus giving an acceleration tension from (4.9) of

$$\Delta T_{accel} = \frac{0.01345 \text{ ft-lb}}{\left[\frac{\text{ft}}{12 \text{ in}}\right](2 \text{ in radius})(56 \text{ in wide})} = 0.00144 \text{ lb / in}$$

The drag tension is

$$\Delta T_{drag} = \frac{0.75 \text{ in-lb}}{(2 \text{ in radius})(56 \text{ in wide})} = 0.00670 \text{ lb / in}$$

The tension upstream of the idler roller, T_1 may be as little as 10% of the web strength, or as little as 0.1 PLI.

The tension downstream of the idler roller is then

$$T_2 = 0.1000 + 0.00144 + 0.00670 = 0.10814 \text{ lb / in}$$

From equation (4.18) we can calculate the minimum wrap angle to avoid slippage during acceleration as

$$\beta \geq \frac{1}{0.3} \ln\left(\frac{0.10814}{0.1000}\right) \text{rad} \frac{360°}{2\pi \text{ rad}} = 14.9°$$

While this is a modest minimum wrap (no safety factor is yet applied), the reader should not get too comfortable. Only modest increases in roller diameter, acceleration rate, or bearing drag might quickly turn this application into impractical wrap angles.

Wrap Angle Calculation - Metric Units

Given a 0.175 kN/m weak web, a modest 3 ft/min/sec acceleration rate and a 0.083 N-m bearing drag, what is the minimum wrap angle to accelerate the idlers. The 10 cm OD x 8.89 cm ID x 1.5 m long aluminum idler rollers have a typical 0.3 coefficient of traction with the 1.4 m wide web. Equations, definitions and units have been given in earlier sections.

Inertia is calculated from equation (4.26)

$$I = 2710 \frac{\text{kg}}{\text{m}^3} 1.5\text{m} \frac{\pi}{32} \left(0.10^4 - 0.0889^4\right) \text{m}^4 = 0.01498 \text{ kg-m}$$

Angular acceleration is calculated from (4.27)

$$\alpha = \frac{2(3)\frac{\text{m}}{\text{min}}\frac{\text{min}}{60\text{sec}}}{(0.10\text{m})(1 \text{ sec})} = 1.000 \text{ rad / sec}^2$$

Inertial torque is calculated from (4.28)

$$Torq = (0.01496\text{kg-m})\left(1.\frac{\text{rad}}{\text{sec}^2}\right)\left[\frac{\text{N-sec}^2}{\text{kg-m}}\right] = 0.01496 \text{ N} - \text{m}$$

thus giving an acceleration tension of

$$\Delta T_{accel} = \frac{0.01496 \text{ N-m}}{(0.05\text{m radius})(1.4 \text{ m wide})} = 0.214 \text{ N / m}$$

The drag tension from (4.9) is

$$\Delta T_{drag} = \frac{0.083 \text{ N-m}}{(0.05\text{m radius})(1.4\text{m wide})} = 1.186 \text{ N / m}$$

The tension upstream of the idler roller, T_1 may be as little as 10% of the web strength, or as little as 17.5 N/m. The tension downstream of the idler roller is then

$$T_2 = 17.5 + 0.214 + 1.186 = 18.9 \text{ N / m}$$

From equation (4.18) we can calculate the minimum wrap angle to avoid slippage during acceleration as

$$\beta \geq \frac{1}{0.3} \ln\left(\frac{18.9}{17.5}\right) \text{rad} \frac{360°}{2\pi \text{ rad}} = 14.7°$$

Drive Tension

Often, a roll or roller is connected to a drive element. This may be a mechanical drive, such as detailed in Chapter 13, or an electrical drive motor, such as described in Chapter 14. These drives may be governed by tension control schemes as described in Chapter 2.

The torque provided by the drive component will alter the tension differential across a roller above and beyond that determined by the drag and inertial components described earlier. If the drive component is braking (absorbing power), the tension will rise going over a roller. If the drive component is motoring (providing power), the tension will tend to drop going over a roller.

The torque provided by a drive component on the ends of a machine (such as unwind or windup) will establish the tension at a steady state speed. Keep in mind, however, that the drive will need to provide (usually quite large) additional torques to change speed. The torque provided by a drive component not located on the ends of a machine will alter the tension, unless it is established by feedback from a load cell or calibrated dancer.

Of the many types of drive actuators, most will fall into two categories. The first are mechanical friction drives such as tendency drives, pneumatic brakes, as well as non-motor electric brakes and clutches. The magnitude of the torque (and thus tension) produced must be obtained from charts provided by the component manufacturer. However, these charts are only approximate because the coefficient of friction and the force inducing element will vary considerably. Thus, it is best not to use friction-based drives in open loop control.

The other common category of drive element is the DC drive motor. The tension produced by the motor can be calculated from armature current as displayed on an ammeter. A simple example of how to do this is given in the next section. The only complication is to account for mechanical losses (due to gearboxes, nips, etc.) by using the difference in amps at steady state with and without a tensioned web, and accounting for the accel/decel terms which can be calculated out.

To Drive or Not to Drive

The typical cases to require a drive point are given in Table 4.4. The common theme is to justifiably control the web's tension as it proceeds down through a line. The nominal tension is determined by the strength of the web at each point in the line. The allowable (downweb) tension excursion is a quality control or cost-benefit decision. Tight product specs, such as close geometrical tolerances, requires tight tension control.

In all cases, the inclusion of a drive point is not a judgment call. Rather, it is a design recipe that falls out from the mechanical design of the machine, the web's properties, and a precision standard.

Table 4.4
Guidelines for Including a Drive Point

1. To intentionally **step a tension up or down** as required by large changes in the material's strain (drawing, hygrothermal expansion) or strength (coating, embossing, drying, laminating) upstream and downstream of the drive point. The minimum stress-strain change which would justify the complexity of another drive point is very application dependent.

2. Reduce an unacceptably large unintentional tension differential across a roller(s) due to **drag** (nip rolling resistance, fluid drag, bowed spreader or other high drag components) or **inertia** (light webs, wide machines, high accel/decel rates). The maximum total downweb tension change through a tension zone might be limited to between 10% and 25% of the lightest design tension.

3. To maintain **traction** on a (light, nonporous) web traveling at high speeds over a roller which may be lightly wrapped or slippery

4. Establish a tension zone endpoint at an **unwind**, **rewind** or **sheeter**.

5. Isolate a **taper tensioned** (center) **winder**, or in rare cases an unwind, from the rest of a line.

Tension from DC Motor Readings

This section shows how to calculate web tension from DC motor drive readings. It can be applied to unwinds, windup sections, and rollers. On winders, unwinds or other endpoints, it is the **absolute tension**, while on intermediate rollers it is the **tension differential**. This example only shows the drive component of tension, drag losses and inertial components must also be accounted for as given earlier.

Need to know:

- amperage from meter during run
- motor amp and power nameplate values
- web width
- speed

Efficiency can be assumed to be 90% for many systems (10% friction and other losses). However, friction can be high on some drive components and with some punky materials rolling on a drum.

English Units

(4.30)

$$P_R = P_{nameplate}\,(\text{Efficiency})\left(\frac{\text{Amps}_{meter}}{\text{Amps}_{nameplate}}\right)$$

where

P_R = power during run (HP)

$P_{nameplate}$ = motor nameplate power (HP)

Efficiency = elect & mech efficiency $\approx$ 0.90

Amps_{meter} = meter reading during run (amp)

$\text{Amps}_{nameplate}$ = motor nameplate current (amp)

(4.31)

$$T = \frac{P_R \times 33{,}000}{W \times S}$$

where

T = tension (lb/in)

P_R = power during run (HP)

W = web width (in)

S = speed (ft/min)

Example for two-drum winder given:
front drum and rear drum motors
rated at 100 HP and 160A each,
front drum run at 100A, rear drum at 60A
speed = 6000 fpm, width = 200 inch

$$P_R = (100{+}100)\,(0.90)\left(\frac{100 + 60}{160 + 160}\right) = 90 \text{ HP}$$

$$T = \frac{90 \times 33{,}000}{200 \times 6{,}000} = 2.48 \text{ lb/in}$$

Note: you can also size the tension power requirements of a drive from equation 4.31 (does not include acceleration/deceleration or losses):

$$P_R = T \times W \times S \,/\, 33{,}000$$

Metric Units

(4.32)

$$P_R = P_{nameplate}\,(\text{Efficiency})\left(\frac{\text{Amps}_{meter}}{\text{Amps}_{nameplate}}\right)$$

where

P_R = power during run (kW)

$P_{nameplate}$ = motor nameplate power (kW)

Efficiency = elect & mech efficiency $\approx$ 0.90

Amps_{meter} = meter reading during run (amp)

$\text{Amps}_{nameplate}$ = motor nameplate current (amp)

(4.33)

$$T = \frac{P_R \times 60}{W \times S}$$

where

T = tension (kN/m)

P_R = power during run (kW)

W = web width (m)

S = speed (m/min)

Example for an unwind given:
motor nameplate rating 150 kW
motor nameplate current 300 amp
motor running current 150 amp
speed = 2000 mpm, width = 5 m

$$P_R = (150)\,(0.90)\left(\frac{150}{300}\right) = 67.5 \text{ kW}$$

$$T = \frac{67.5 \times 60}{5 \times 2{,}000} = 0.41 \text{ kN/m}$$

Note: you can also size the tension power requirements of a drive from equation 4.33 (does not include acceleration/deceleration or losses):

$$P_R = T \times W \times S \,/\, 60$$

Summary

Designing or evaluating a line always begins with the properties of the web such as it's stress-strain curve which determines nominal tensions. However, other properties such as the coefficient of web/roller and web/web friction are also needed for roller and drive sizing. Some applications will require knowledge of more arcane properties such as stress-strain at elevated temperatures, hygrothermal expansion rates, creep and stress relaxation constants and so on.

Then we calculate minimum roller sizes based on deflection (Chapter 3), critical speed or other limitation. Most transport rollers can be identical in all respects throughout the machine, while process rollers will need to be larger because of nips, higher tolerances, or to obtain sufficient dwell time (such as on dryer cans).

After transport and process rollers are sized, the components can be arranged to make a manufacturing line. The design goal is to minimize transport roller counts, make sure all rollers are sufficiently wrapped, and that ingoing and outgoing spans are appropriate.

Next, drive points must selected (Table 4.4), sized and layed in. Usually, process rollers will need to be driven, and will require a load cell roller to be added between drive points. The torque requirements of the drive are given by

(4.34)

$$Torq_{Drive} = Torq_{Drag} \pm I\alpha - \frac{D_0 w}{2}\left(T_2 - T_1\right)_{max}$$

Also, the ratio of the tensions can not exceed the available traction (plus a safety margin) as given in the band-brake equation. If so, the drive roller will need to have a grippier surface, a higher wrap angle, or as a last resort, a nipping or suction roller.

Next, undriven rollers also must be checked for potential slippage. The worst case would be at the lightest wrapped and tensioned roller with the slipperiest surface. If it would slip, it needs to be driven.

Finally, we must determine how much the tension will be allowed to change down through a tension zone. As seen in Figure 4.20, the tension rises in steps going over undriven idler rollers. The size of the steps during run is bearing drag. The size of the steps during accel and decel is bearing drag plus and minus inertial tension respectively. The total tension climb will be worsened during acceleration, but may be improved during deceleration.

If the tension change is too much, the following steps should be taken in the given order. First, look again to see if any more transport rollers can be removed. Second, if the tension change is modest and is largely due to inertia, low inertia composite rollers or a reduced acceleration rate may be enough. Third, if the tension change is due to bearing drag, different seals or lubrication may bring the system into spec. Fourth, if the above changes are insufficient, the roller must be driven. In some cases, this can be a (open loop) helper drive as opposed to a (closed loop) tension controlled feedback drive.

The tension control strategy, as well as a choice of speed reference can be made later in the design cycle. Indeed, modern drive controls can accommodate a wide variety of strategies, such as changing from tension to draw control, merely by configuring the software.

Figure 4.20
Tension Changes over Idler Rollers

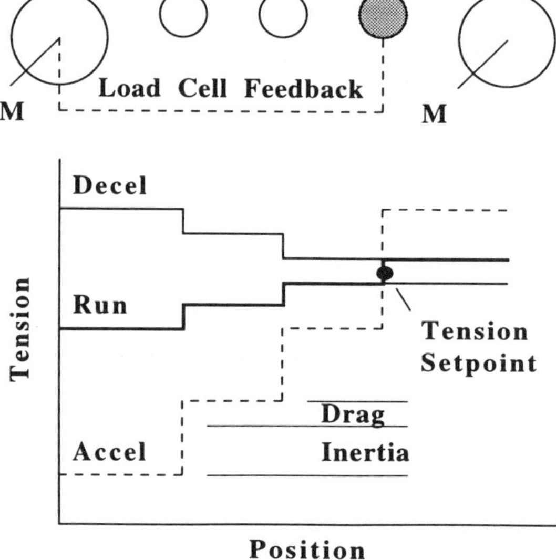

Bibliography

1. Gallahue, William M. *Web Guiding and Control in the Paper Industry.* Paper Trade J., pp 38-41, March 1965.

2. Pfeiffer, J. David. *Web Guidance Concepts and Applications.* TAPPI Finishing and Converting Conf. Proc., pp 23-31, October 1977.

3. Pfeiffer, J. David. *Web Guidance Concepts and Applications.* Tappi J., vol 60, no 12, pp 53-58, December 1977.

4. Shelton, John J. *Lateral Dynamics of a Moving Web.* Ph.D. thesis, Oklahoma State Univ., July 1968.

5. Shelton, John J. and Reid, Karl N. *Lateral Dynamics of a Real Moving Web.* Transactions of the ASME, September 1971.

6. Shelton, John J. and Reid, Karl N. *Lateral Dynamics of an Idealized Moving Web.* J. of Dynamic Systems Measurement and Control,Transactions of the ASME, September 1971.

7. Fang, Bin. *Lateral Web Behavior of Interactive Web Spans.* Ph.D. thesis, Oklahoma State Univ., Web Handling Research Center, December 1990.

8. Kardamilas, C.E. *Stochastic Modeling and Control of Lateral Web Dynamics.* Ph.D. thesis, Oklahoma State Univ., Web Handling Research Center, May 1990.

9. Kardamilas, C.E. and Young, Gary. *Stochastic Modeling and Control of Lateral Web Dynamics.* American Control Conference Proc., San Diego, May 23-25, 1990.

10. Eriksson, Leif G. *Measurement and Control of the Tension Distribution Across the Web in a Newspaper Printing Press.* 1st Int'l Conf. on Web Handling, Oklahoma State Univ., pp 167-195, May19-22, 1991.

11. Daly, David A. *Factors Controlling Traction Between Webs and Their Carrying Rolls.* Tappi J., vol 48, no 9, pp 88-90, September 1965.

12. Ducotey, Keith S. *Dynamic Coefficient of Friction Including the Effects of Air Entrainment.* MS thesis, Oklahoma State Univ., Web Handling Research Center, December 1987.

13. Szeri, A.Z. *Tribology, Friction, Lubrication and Wear.* McGraw-Hill, 1980.

14. Anon. *TAPPI Test Methods* (T503, T542, T548, T815). TAPPI PRESS, 1991.

15. Anon. *TAPPI Test Methods* (T549, T816). TAPPI PRESS, 1991.

16. Roisum, David R. *The Mechanics of Winding.* TAPPI PRESS, 1994.

17. Roisum, David R. *Nip Induced Defects of Wound Rolls.* TAPPI Finishing and Converting Conf. Proc., October 1994.

18. Roisum, David R. *The Mechanics of Roll Winding.* Internal publication of the Beloit Corp., February 1986.

19. Howell, H.G. *The Laws of Static Friction.* Textile Research J., pp 589-591, August 1953.

20. Spielbauer, Thomas M. and Walker, Timothy J. *Theory and Application of Draw Control for Elastic Webs with Nipped Pull Rollers.* 2nd. Int'l Conf. on Web Handling Proc., Web Handling Research Center, Oklahoma State Univ., June 6-9, 1993.

21. Daly, David A. *Factors Controlling Traction Between Webs and their Carrying Rolls.* Tappi J., vol 48, no 9, pp 88A-90A, September 1965.

22. Knox, K. L and Sweeney, T. L. *Fluid Effects Associated with Web Handling.* Ind. En. Chem Process, vol 10, pp 201-205, 1971.

23. Tajuddin. *Mathematical Modeling of Air Entrainment in Web Handling Applications.* MS Thesis, Web Handling Research Center at Oklahoma State Univ., December 1987.

24. King, S. L. and Func, B. A and Chambers. Frank W. *Air Films Between a Moving Tensioned Web and a Stationary Support Cylinder.* 2nd Int'l Conf. on Web Handling, Oklahoma State Univ., June 6-9, 1993.

25. Roisum, David R. *What Causes a Baggy Web?* Converting Magazine, Web Works column, pp 22, April 1994.

26. Roisum, David R. *What Causes Machine Direction Trough Wrinkles.* Converting Magazine, Web Works column, pp 22, March 1994.

27. Roisum, David R. *The Mechanics of Web Spreading - Part II.* Tappi J., vol 76, no 10, pp 75-86, December 1993.

28. Ducotey, Keith S. *Dynamic Coefficient of Friction Including the Effects of Air Entrainment.* M.S. thesis (December 1987) and Ph.D. thesis (1991) at the Web Handling Research Center at Oklahoma State Univ.

29. Reynolds, Osborne. *On the Efficiency of Belts or Straps as Communicators of Work.* The Engineer, pp 396, November 1874.

30. Deutschman, Aaron D. and Michels, Walter J. and Wilson, Charles E. *Machine Design - Theory and Practice,* pp 697-698. Macmillan Publishing Co., Inc, NY, 1975.

31. Roisum, David R. *The Profile of a Nip.* Converting Magazine, Technical Report, pp 44-46, June 1994.

32. Spielbauer, Thomas M. and Walker, Timothy J. *Theory and Application of Draw Control for Elastic Webs with Nipped Pull Rollers.* 2nd Int'l Conf. on Web Handling, Web Handling Research Center at Oklahoma State Univ., Stillwater, OK, June 6-9, 1993.

33. Good, J. Keith and Fikes, M. *Predicting Internal Stresses in Center-Wound Rolls with an Undriven Nip Roller.* Tappi J., vol 74, no 6, pp 101-109, June 1991.

34. Lingaiah, K. *Machine Design Data Handbook.* McGraw-Hill Inc., 1994.

35. Roisum, David R. *How can I Improve Productivity.* Converting Magazine, Web Works column, pp 22, July 1994.

36. Roisum, David R. *The Mechanics of Winding.* Handbook published by TAPPI PRESS, Atlanta, GA, 1994.

Chapter 5

Nipped Rollers

This chapter describes the stresses in a nip and how they vary with loading, modulus, and geometry. The objective is to keep these stresses uniform so that calendering, laminating, printing and other nipped roll processing is uniform.

Nipped Rollers for Web Transport

A nip, as described by the dictionary, is a pinching force between two surfaces. However, in converting the term refers more specifically to rollers that are forced together. Nips are ubiquitous in web manufacturing and converting. So much so, that they are often overlooked until problems arise. Nips, just as rollers, can be classified by usage as either strictly for transport, or to process the web.

A transport nip is not intended to permanently modify (deform) a web. A common application for a transport nip is to step the web strain or tension up or down between sections of a process. As was discussed in the last chapter, the nip is only desirable to avoid slippage where the web/roller friction and wrap angle is very low or for positions requiring large tension differences across the roller. Low tractions might be found in applications where air or fluid entrainment is significant. Low wrap angles should be avoided at the design stage. Large tension differences may be seen during threading, drawing or sheeting operations.

This transport nip goes by a number of names such as pull rollers, draw rollers, pinch rolls and so on. As seen in Figure 5.1, the nip commonly consists of an undriven metal roller nipped against a driven covered roller. However, the hard roller should be the one driven if the drive point is in draw/speed control, even though it will have less traction. Also, it is very risky to drive both rollers simultaneously in draw or speed control (gearing, belting or servo). The reason is to maintain a known surface speed, which as we will see later, increases on a soft roller with increasing nip load.

Figure 5.1
Nipped Pull Roller

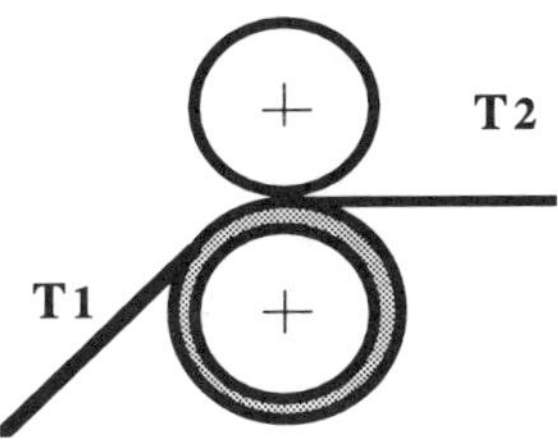

Nips are also used in surface winding as seen in Figure 5.2. Here, the nip provides additional winding tightness or hardness than could be achieved by web tension alone as shown by Pfeiffer[1,2,3,4] and Good.[5,6,7] Also, the nip helps exclude air from the wound roll when winding nonporous materials.[8,9,10,11,12] If too much air is brought into a wound roll, interlayer traction may be lost and the roll can telescope.[13] Unfortunately, the winding nip can also cause several types of web and roll defects.[14,15] Nip induced defects of wound rolls are exacerbated by a nonuniform nip as is commonly caused by caliper profile variations.

A nip certainly improves traction and greatly reduces air/fluid entrainment. However, nips can cause severe problems. First, an imperfect but saleable web can be damaged (wrinkled) by a perfect nip such that it must be rejected as waste. Alternatively, an imperfect nip can ruin a perfect material. Thus, the choice to include a nip is never to be taken likely. While process nips are required, transport nips can often be avoided by careful design. If included, transport nips must be regarded as precision systems with similar tolerances as process nips.

Figure 5.2
Nip Rollers in Winding

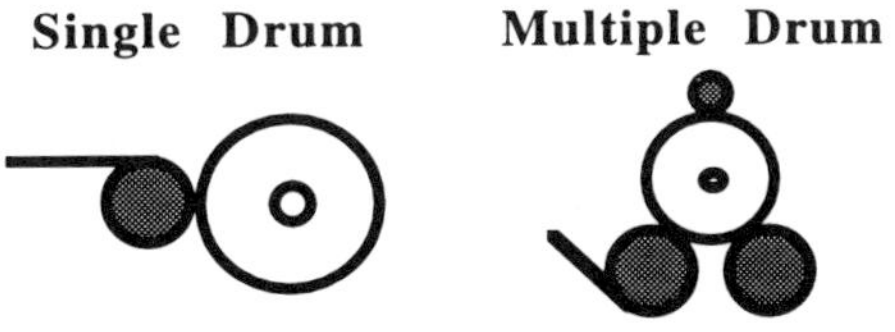

Single Drum **Multiple Drum**

Nipped Rollers for Web Processing

Transport nips, such as pull rollers and surface winding, are not intended to modify the web. Process nips, however, use the nip to alter a web toward its desired final form. Table 5.1 lists some of the more common web process nips.

In a **calendering** operation, the pressure between two smooth rollers will densify and consolidate porous materials. Calendering is used on magazine and related paper grades to increase the gloss and decrease the roughness to make it better suited for printing. This process is often supplemented with elevated temperatures and moistures to reduce the yield point of the material. Heated calender rollers also are used to bond nonwoven fibers together to form a mat. The lineal nip loading, which is nip force divided by width, can be as much as 400 kN/m (2300 lb/in) in calendering. High nip loads can cause cover life issues, which will be exacerbated at high speeds and elevated temperatures.

Rolling, which is a specific application of calender rollers, will decrease the caliper or thickness of the web in response to the Z direction pressure in a nip. However, here the materials will retain much of their ingoing density, so that the decrease in caliper is accompanied by a nearly proportional increase in length as well as a slight increase in width. The distinction between rolling and calendering is more of material properties than of equipment. Rolled materials, which include food products, plastics and metals, have a very high bulk modulus. Thus, rolling is used in the manufacturing of plastic materials where they are first formed most easily in thicknesses greater than their desired end use caliper.

Laminating, which is another specific application of calender rollers, also uses the nip pressure to bond two or more webs together. Since few materials will bond strongly with pressure alone, the laminating rollers may be heated or an adhesive may be added between the laminate plies. Adhesive labels, flexible packaging and many other web products make extensive use of laminating calenders. Laminating tensions should conform to the rules given in Chapter 2 to avoid end product curl.

Table 5.1
Process Nips in Converting

Calendering
Coating
Corrugating
Die-Cutting
Embossing
Laminating
Pressing
Rolling
Slitting

Embossing rollers differ from calendering rollers only in that the surface of one or both of the rollers have a strong texture which is imparted into the web during the process. Embossing can also create a weak bond between materials, though its primary purpose is most often to enhance product aesthetics. Common examples of embossed products include tissue, toweling, nonwovens as well as some building products such as wallpaper and flooring.

Rotary **die-cutting** and score **slitting** could be thought of as an extreme in roller texture. Here, there is a complete penetration of one of the textured rollers into the anvil backing roller.

Nipped rollers are also used to add a **coating** to the web, or the control the thickness or application rate of the coating. Coating machines are quite varied in their configurations.[16,17]

Finally, **printing** can be thought of as a coating operation. Here, however, the ink coating is intended to be placed only in specific locations, but nonetheless uniform in those locations. Machine and cross-machine registrations are important web handling considerations for printed material, especially those with multiple colors or where detail is high such as reproducing photos. Printing is the most mature of the nipped roller web processes, so that a wealth of literature can be found.[18]

This section is intended to merely list the wide variety of process nips. Next, we will show that all nipped roll systems have a common quality requirement, namely, control of the stresses in the nip.

Nipped Roller Issues

The nominal nip load must lie within a specific range for transport roller nips. If it is too low, traction can be lost on rollers, wound rolls may telescope because too much air is brought in, and so on. If the nip load is too high, however, the web, roll, or roller could be damaged due to Z direction overstress. In most applications this criteria is easily met as the acceptable range is quite wide.

A separate and more difficult constraint is to keep the nip uniform with respect to the machine direction position or time, and especially with respect to cross direction position. If the linear nip load varies with the CD, the web may not be drawn into the nip evenly. This is the major cause of wrinkling and other web damage in the nipped roller or the nipped wound roll systems, especially for thin materials.

The important point here is that while the nip load variation might be contained within the acceptable range, it could vary abruptly enough with CD position to cause problems. Figure 5.2 illustrates these nip load quality criterian. These criterian, as we will see, are related to absolute and station-to-station diametral tolerances.

Process rollers also share the same nip uniformity considerations as do transport rollers. However, not only must the web be drawn into the nip evenly, it must be processed evenly as well. Variations in process roller nips can cause product variations, even if the web is transported successfully. The demand for a uniform product greatly closes up the limit range for process rollers, often to less than ±10% of the setpoint.

Process nip variations with respect to CD will result in a streaky product. If it is a calender, the web will have CD varying caliper or density. The nip roller set is used for printing or coating, the application weight may mirror the nip profile.

If the nip varies regularly with respect to the MD, product barring may result. The distance between the bars is useful to distinguish between the two primary causes. If the bars result from mechanical vibration, such as a resonance, they will be spaced one wavelength apart. If the bars

Figure 5.3
Nip Uniformity Across the Width

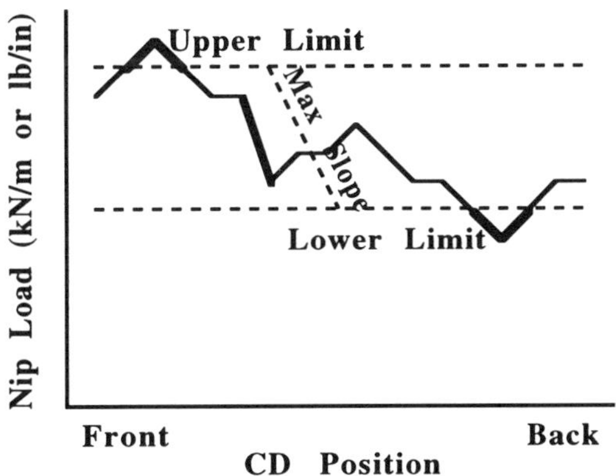

result from rollers that are not round, the distance between bar repeats will be an integral multiple of one of the roller circumferences.

Process nip loads can also vary irregularly with MD position or time. Sometimes this is due to random variations in the incoming web. However, high mechanical friction in the nip loading mechanism will cause nip fluctuations to vary unpredictably within the hysteresis band.

The nip can also vary irregularly simultaneously in the CD and MD, resulting in a mottled product. Most often, this is due to incoming web variations.

One powerful tool to sort out the elemental constituents of nip variations is to do a 1D or 2D FFT (Fast Fourier Transform) on a measurable product attribute. This will often show that seemingly random patterns have an order that corresponds to several machine elements either at or upstream of the process nip.

If a uniform product is desired, the pressure or stresses imposed by the nip rollers onto the web must be controlled in both the CD and MD to some tolerances. It is these tolerances that define acceptable loading systems, roller geometrical precisions, roller stiffnesses and material variations.

Web Stresses in a Nip

The web is highly stressed inside a nip, perhaps higher than at any time during its manufacture or end use. Web stresses inside a nip are multi-axial, meaning that they are imposed from several directions simultaneously. The six potential stress components are listed in Table 5.2. Of these, three are normal stresses which tend to change the web's length, width and thickness. The other three are shear stresses which tend to change the web's shape such that, for example, a square section is forced into a parallelogram. The diligent readers may wish to consult their mechanics of materials textbook to review these concepts.

Of the normal stresses, the Z-direction is the largest and most important by far. This should not be surprising as this is the direction of the applied load. The Z-direction is the working direction that provides the enhanced traction in the case of transport nips. It is the direction that densifies calendered paper, bonds nonwovens and laminates, transfers printing ink and so on. It is this direction, more than any other, that we need to know and control, particularly through nip load and cover hardness.

The shear stresses, on the other hand, can be viewed as an inevitable byproduct of nips. In other words, shear is not often a desired stress, but it is unavoidable to a certain degree. Of these, the in-plane shear is often the most destructive and avoidable. In-plane shear can result when material is drawn unevenly into the nip. In most cases, this is because the web or roller(s) has a profile that varies across the width. In the case of the web, it would be a caliper or basis weight variation. In the case of rollers, it is primarily a diametral variation across the width. The end result of either can be wrinkles on light weight webs. If the nip variation has a high gradient (rate of change) somewhere across the width, the result may be corrugations (also known as ropes or chain wrinkles) that are permanently set into the material. If the variation is still high but the gradient more modest, the web won't lay flat on the ingoing or outgoing side. The important note is that the nip profile or consistency across the width is an important quality control parameter.

Table 5.2
Nip Stress Components

Normal Stresses

Symbol	Direction	Source
$\sigma 11$	MD	web tension etc.
$\sigma 22$	CD	poisson expansion
$\sigma 33$	ZD	nip load

Shear Stresses

Symbol	Direction	Source
$\tau 12$	in-plane	profile problem
$\tau 13$	out-of-plane	various
$\tau 23$	out-of-plane	various

Nip stresses are exceedingly complicated, especially when the system has covered rollers, thick webs, complex web properties, plastic deformation, or MD/CD variation. Nonetheless, the basic mechanics principles derived a century ago still provide useful insight even if they can't always be used for quantitative analysis. Afterwards, we will discuss more involved mechanics or FEM (Finite Element Modeling) techniques that will provide closer answers to real world complexities.

Alternatively, nip stresses can also be measured or inferred. For research purposes, there are a variety of pressure sensors that can be imbedded into a roller or thin sensors that can be drawn into a nip much like a web. For maintenance and troubleshooting, nip impression papers can be used to check profile consistency and in some cases nip stresses.

After discussing nip stress modeling and measurement, we will cover the design of nip loading systems. Finally, we will discuss the role of geometry such as deflection and roller diametral tolerances.

Hertzian Contact

Nip contact stresses were first studied by Hertz in 1882, and later modeled mathematically. The first practical motivation was the spalling failure of cast iron railroad wheels (not unlike roller cover failures). Since then, many diverse applications were found for this type of analysis including the design of bearings, pins, rollers, wheels and other curved surfaces in contact.

The design equations for two parallel, elastic, isotropic cylinders in frictionless contact begin with the calculation of a material and a geometric parameter

(5.1)

$$c = \frac{1-\mu_1^2}{E_1} + \frac{1-\mu_2^2}{E_2}$$

$$k = \frac{D_1 D_2}{D_1 + D_2}$$

where

E = Young's modulus

μ = Poisson ratio

D's = roller diameters

subscripts 1 and 2 refer to the roll(er)s

Then, the maximum surface contact stress and nip width, as seen in Figure 5.4 are calculated as

(5.2) $$\sigma_c = \sqrt{\frac{2N}{\pi c k}}$$

(5.3) $$b = 1.60\sqrt{N c k}$$

where N is the nip load (force/width).

A couple of words of caution are appropriate at this point. First, the stress calculated here is the peak ZD stress on the web and the roller surfaces that is located at the midpoint of the nip zone. The ZD stress distribution is nearly parabolic about this midpoint and tapers to zero at the endpoints of the **nip footprint**. Also, there are stresses inside the rollers that are covered in a later section. Most importantly, however, the Hertz derivation is not general enough to be applied directly to calculate stresses for most web handling applications. It does not account for webs, covers, anisotropy, rolling friction and so on.

Figure 5.4
Contact Stresses at the Nip's Surface

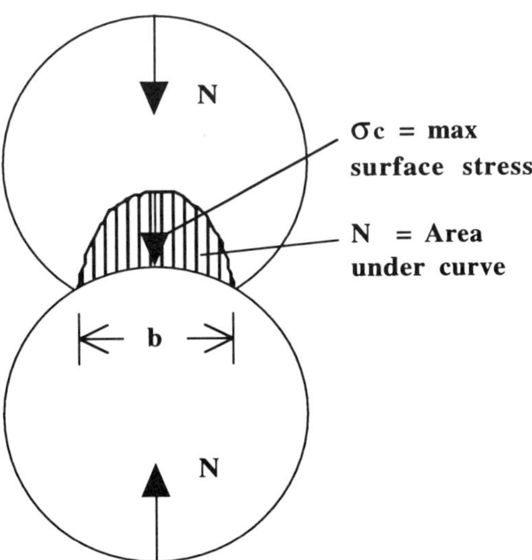

Thus, this derivation is useful only to approximate the trends in real systems as opposed to providing close numerical results.

From the material parameter given in (5.1), we can see that the peak stress is largely determined by the softest element. Thus, the stresses on a covered roller system are determined by the cover, and insensitive to whether the mating roller is aluminum or steel for example. Similarly, the diameter of the smaller of the rollers is a key geometrical factor.

The peak ZD stress goes as the square root of the nip load, and the inverse square root of the material compliance and geometrical factors. What this means is that quadrupling the externally applied nip load will only double the ZD stresses.

The system of equations also confirms the well known observations that the peak stress will increase and the nip width will decrease with increasing cover hardness and decreasing roller diameter. What may not be apparent, however, is that the area under the Z pressure profile will remain the same and is determined solely by the magnitude of the externally applied nip.

Determining Peak ZD Nip Stress

The development of the Hertzian contact equations are not general enough for most real nip systems as mentioned earlier. Nonetheless, the results can be used to accurately estimate peak ZD stress. All that is needed is knowledge of the nip load, nip width, and a simple model.

The lineal nip load has units of force per unit width (kN/m or lb/in). In other words, it is the resultant force from the nip loading system and the roller weight, divided by roller width. Calculating and calibrating the nip load, as a function of cylinder pressure for example, is usually very simple and straightforward. This is covered in a later section. Also, a calibrated and known nip load is so fundamental to process understanding, that there is absolutely no excuse for its omission. Thus, we would expect that calibrated nip load is readily available for this or any other process purpose.

The next thing that is needed is the width of the nip zone. This is the MD distance along the contacting length of the nip, and is known more commonly as the footprint. The nip width varies from as little as a few mils (0.001") for lightly loaded metal/metal systems, to many inches for soft wound rolls (tissue or nonwovens) nipped against drums. While nip width could be calculated, accurate models are exceedingly complicated and defeats the utility of the this simple method. Thus, we will resort to measurement.

Wide nips, say greater than 1 cm, can be measured conveniently with a simple tape measure or roller, provided that access to the end can be obtained. Intermediate nip widths, say between 1 mm and 1 cm, are most conveniently measured with a static nip impression test covered later. Briefly, it require loading the nip against a nip impression paper, which is much like a carbon paper. From the width of the mark, the nip width can be estimated. Caution must be used, however, in that the actual width is somewhat greater than the thickness of the mark. Unfortunately, there is no simple way to measure the width of narrow nips, such as those less than 1 mm.

The last requisite for a simple peak ZD nip stress determination is a model. This is obtained by simultaneously solving equations (5.2) and (5.3) to eliminate the material and geometrical parameters.

$$\sigma_c = 1.277 \frac{N}{b} \tag{5.4}$$

where

N = lineal nip load (N/m or lb/in)
b = nip width (m or in)

The careful reader may wonder why this result can be applied when the basic derivation itself was simplistic. The answer is that equation (5.4) merely makes the assumption that all nip load distributions are nearly parabolic in shape as defined by Hertzian contact which gives the leading coefficient. Even outrageous pressure distributions, such as a square (coefficient = 1.0) or triangle (coefficient = 2.0), would only affect the answer moderately.

Procedure

1. Determine the lineal nip load.

2. Measure the nip width using a tape measure or nip impression paper.

3. Calculate the peak pressure from (5.4).

The peak pressure can be compared, for example, to a bond activation pressure on laminations, or an ink transfer pressure for printing. Alternatively, it might be compared to some failure criterion to avoid material damage.

Another important concept for rolling nips is that the stresses are kinematically and dynamically similar to stamping. The rolling and sliding aspects are quite minor in comparison. Thus, the astute product developer can often prototype nipped systems quicker on a benchtop press than on pilot or production machines.

Interior Contact Stresses

Surface stresses as described in the last section are not the whole story. What about stresses under the surface of the rollers? Plots of the three depth varying principal stresses are given in Figure 5.5. These principal stresses are in the Z-direction, machine direction and the shear stress. (Consult your strength of materials textbooks for an explanation for the mechanics term "principal stress." Basic contact mechanics are covered in many advanced mechanics of materials textbooks.[19,20])

The Z-direction stresses (radial direction of the roller) are everywhere compressive and highest at the surface, and then taper asymptotically to zero deep in the interior.

The MD stresses have a similar pattern as the ZD, but of a slightly smaller magnitude. They are maximum at the surface and taper toward the interior.

The shear stress distribution is both surprising and most important. The maximum shear stress is located at about one nip width deep into the roll(er) surface. This shear stress is a primary cause of the pitting and **delamination of roller covers**. This is much like the spalling of railroad wheels that first drove contact analysis, as well as a common mode of ball/roller bearing failure. Here, the reversing shear stress starts a crack parallel to the surface, which then may propagate to the surface at an angle. Also, these shear stresses are a primary ingredient for **interlayer slippage** of wound rolls, which is the driving force for a number of wound roll defects.

The stresses plotted here are for a location corresponding the tangent connecting roll centerlines, in other words, the center of contact. As one moves outward from the center, all of the compressive stresses will be lower. Indeed, one may see a tensile ZD stress as a bubble on ingoing side of the winding nip. This also allows the outer layers of a wound roll to break loose from themselves.

Figure 5.5
Contact Stresses Interior of the Nip

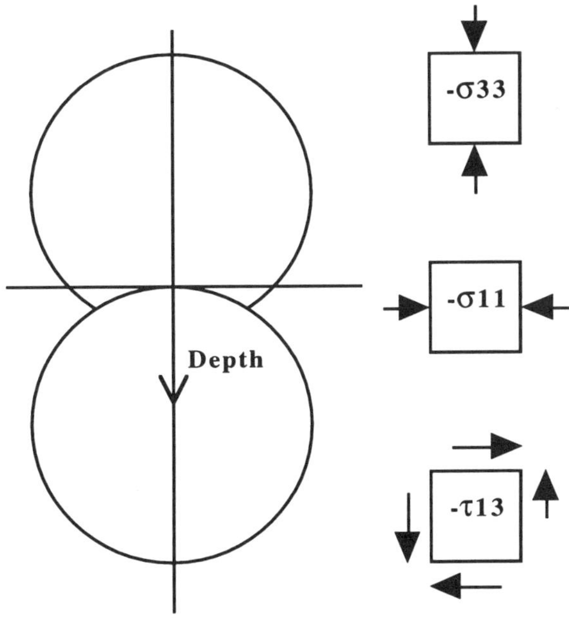

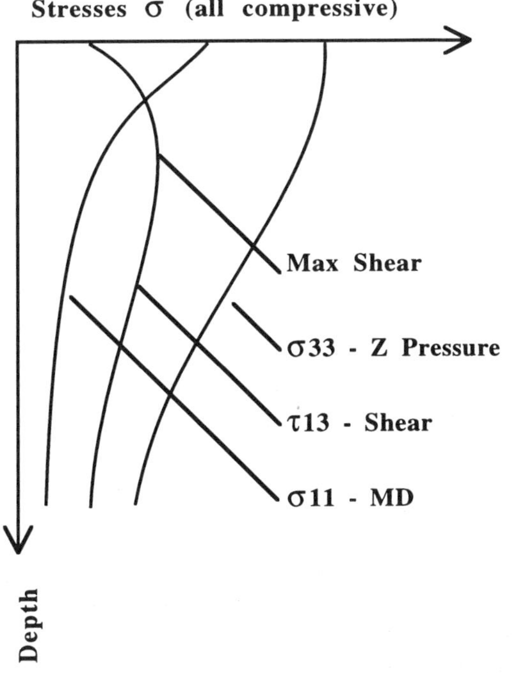

Complex Nip Models

The simple Hertzian contact model is derived for two linear, elastic, isotropic and homogeneous cylinders loaded together in frictionless contact. While the trends and tendencies will hold, the simple model is too restrictive to accurately predict stresses for most real nip systems. Figure 5.6 schematically illustrates some of the non-Hertzian elements.

First, the complex nip system always includes a web and sometimes a cover. Even a thin web will spread out the pressure distribution, as well as reduce its peak, on hard (i.e., metal-metal) nips by perhaps an order of magnitude. However, a cover will spread out the pressure distribution by perhaps two orders of magnitude depending on the hardness and thickness of the cover. Mathematical solutions for elastomeric covered rollers have been derived,[21,22,23,24] and a web could be modeled as a cover.

However, most covers and webs do not have simple linear stress-strain curves. Rather, as seen in Figure 5.7, they unload to the right of the load curve. Covers will be elastic under normal operation because they retain their shape when unloaded. However, the area between the load and unload curves represents an energy loss due to hysteresis. This is major factor for rolling resistance and heat generation.

In addition to hysteresis, inevitable microslip in the nip also adds to rolling friction. This tangential force component is superposed onto nominal contact stresses. Though complicated, rolling friction has been modeled.[24,25,26,27] The results of rolling friction are to skew the stress distributions and increases shear. Rolling friction can be quite high on soft covered rollers or soft wound rolls.

Lastly, webs which are permanently deformed, such as shown in Figure 5.8, are highly nonlinear problems. Nip systems that alter webs include pressing, supercalendering of paper,[28,29] rolling of plastics and metals and so on. These types of problems have just begun to be modeled with finite element computer programs.[30,31,32] Usually, however, most research is still done experimentally[33,34,35] on pilot equipment.[36]

Figure 5.6
Nip Components

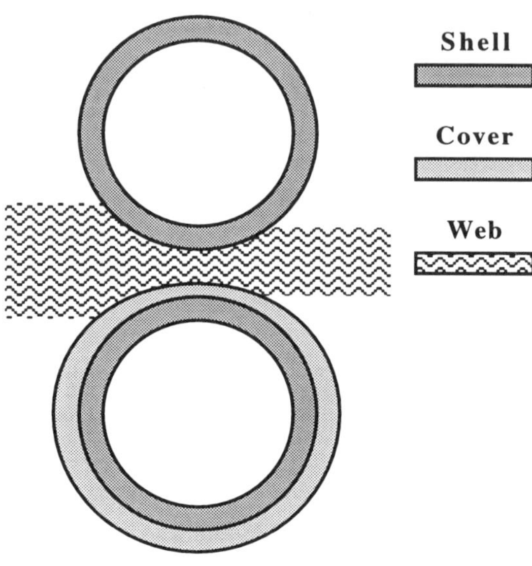

Figure 5.7
Hysteretic But Elastic Stress-Strain

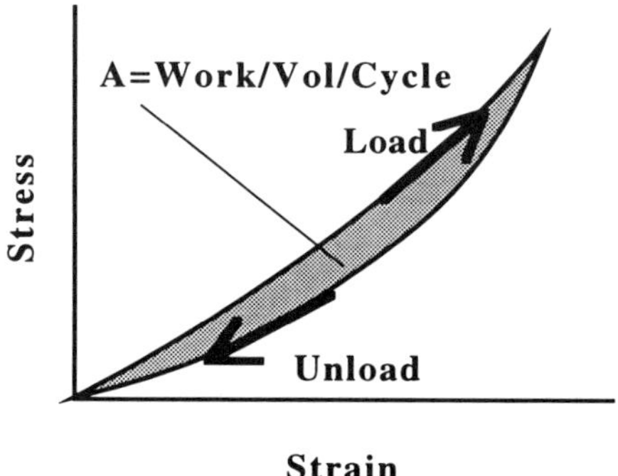

Figure 5.8
Elasto-Plastic Stress-Strain

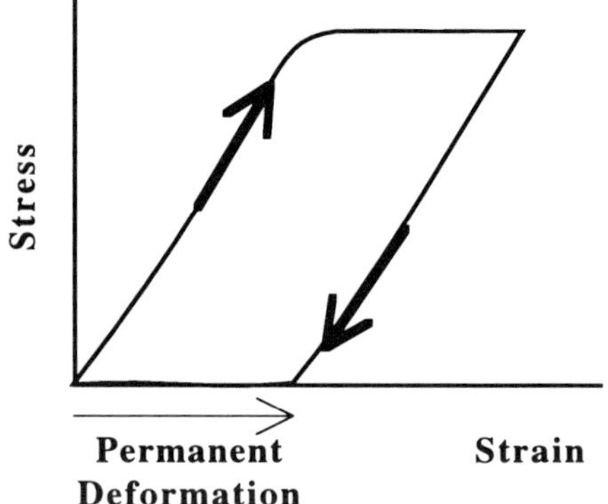

Nip Loading Mechanisms

Nip load is best expressed as the nip force between the rollers per unit width of the web, or in some cases per unit width of the roller. The units for nip in the metric system are N/m for light loads or kN/m for moderate or heavy loads. The units for nip in the English system are pounds per lineal inch, which is commonly abbreviated as PLI. Nip load may vary from less than 5 PLI for the winding of delicate materials, to over 2000 PLI for the supercalendering of paper, to considerably more for metal rolling.

$$N = F / L \quad (5.5)$$

where N = nip load (N/m or lb/in)
F = total force between rollers (N or lb)
L = length of web or roller (m or in)

The total nip force is determined by the vector sums of the distributed roller weight(s) and the external load typical provided at each end by a hydraulic or pneumatic cylinder. An example nip load control system is shown in Figure 5.9. Here, the pivoting bottom roller is loaded against the top roller by a pneumatic or preferably hydraulic cylinder. While a fixed bottom roller is easier to design, a loaded bottom roller provides a safer system that is quicker and surer to unload in the event of an accident.

In this system, a portion of the cylinder pressure is used to lift the weight of the bottom roll, with the remainder used to load the nip. This pressure is commonly displayed as PSI (pounds per square inch) on a pressure gauge on a control panel. However, the web cares about PLI, not PSI. Thus, a calibration curve between loading pressure and resulting nip load is required for any design or process engineering. This nominal (average) nip load is easily calculated from engineering statics from cylinder areas, geometries as well as roller and arm weight. In the case where roller weight is unknown, it can be weighed by the pressure gauge readout. This is calculated from the average of the lifting and lowering pressures. It is expected that the machine builder provide the calibration curve and that the process engineer verify it.

Figure 5.9
A Nip Loading System

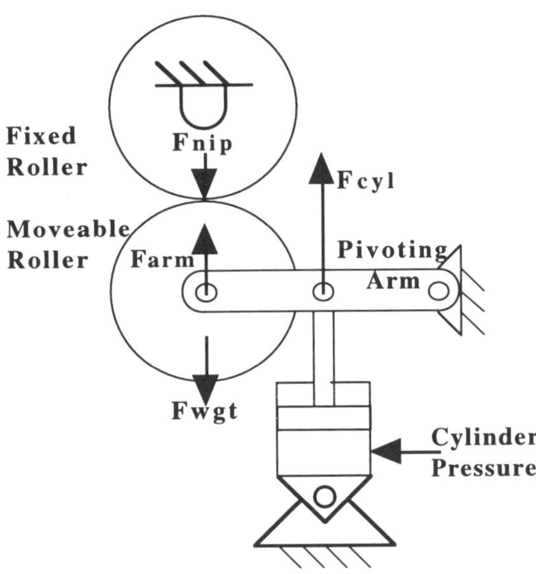

A common alternative to the pivoting roller is a roller guided at both ends in a linear slide. Slides are more compact than pivoting arrangements and can easily be extended to multiple nip rollers in a stack. However, it is difficult to achieve low friction without excessive looseness using slides. Also, it is more difficult to time the ends unless a rack and pinion or other device is provided. On a pivoting system, the parallelism of the moveable roller is ensured with a stiff torque tube at the arm pivot.

Improved load control performance can be achieved by counterbalancing the moveable roller weight with a fixed pressure on say the cap side of the cylinder. Then the rod side is pressurized to directly control the nip. Hydraulically pressurizing both sides of a cylinder also stiffens the system so that it better resists vibration and bounce. Also, it is preferable to put cylinders into tension rather than the compression as shown in the schematic. This eliminates any chance of buckling and improves the rod's bearing and seal life.

Nip Load System Friction

One problem that is common on many loading systems is excessive friction. In one particularly bad situation, a converter wanted to run a 2 PLI load, but the system had nearly 30 PLI of friction. The result was an out-of-control process where adjustments in the setpoint pressure did not result in predictable and significant changes in the actual nip loading.

An estimate of the system friction is easily calculated by observing the difference in pressure required to move the roller in the load versus unload directions. In the system of the previous figure, the cap side pressure would be slowly and manually increased until the pivoting roller is first observed to break loose and move upward, and the raise pressure then is recorded. Next, with the nip engaged, the pressure is slowly and manually decreased until the roller is first observed to break loose and move down, and the lower pressure is then recorded. The nip load friction is calculated as the pressure difference times the acting areas of the cylinder multiplied by the effective arm. Nip load friction can be similarly measured using a winch for loading and a force gauge to get differential loads directly.

Systems with excessive friction can't hold nip loads closely, thus cause a variation in the product along the length or with time. Friction also affects how low the nip load can be controlled. An example of the effect of nip load hysteresis on actual nip load as measured by load cells is shown in Figure 5.10. Here, the high hysterisis plot shows the actual nip load varies within the 3 kN/m friction band as measured by the static test. In contrast, the loading system was redesigned to lower the friction by two orders of magnitude for an immensely reduced nip variation.

The other effect of friction is that it raises the minimum nip load, which is half of the friction deadband plus a small margin for safety to make sure that the rolls remain in contact during run. The high hysteresis system was not able to run a nip less than about 1.5 kN/m without the danger of the nip opening up. In contrast, the precision nip system is able to hold loads easily down to less than 0.1 kN/m.

Figure 5.10
Effect of Hysteresis on Nip Load Control

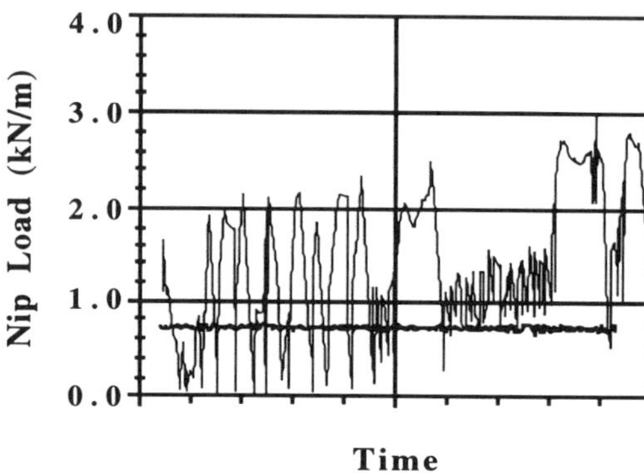

Thus, mechanical friction in a very real sense increases the variations of the nip load and consequently increases the variations in the end product. Mechanical friction also precludes running light or delicate nips.

The sources of this friction or hysteresis are as varied as the types of nip loading systems. However, in most cases the nip load is provided by regulated pressure on a cylinder. While there is some uncertainty in the regulators depending on their quality and precision, a larger concern is in the cylinder's rod and piston seals. This is one reason to avoid pneumatic systems as their typical 10% cylinder seal friction is much larger than their hydraulic counterparts.

Another large source of nip loading friction is the mechanical pivots or slides. Any binding here will have direct consequences on the quality of nip control. Finally, roller chains (for timing) or any other devices hanging off the loading system are also serious offenders.

Finally, a few general observations on nip load hysteresis. First, there is no truly effective control fix (such as closed-loop load cell feedback) to mechanical friction. Second, typically the friction is more the result of design compromises than of poor maintenance. Third, the kinetic friction (of vibration) is not significantly lower than the static friction as measured by our simple test. Fourth, the magnitude of friction will vary between nominally identical systems and even along the stroke of the same system, as well as with a host of other factors.

Load Versus Gap Control

Most often, the nip forced between rollers is controlled by a loading system. However, there are situations where the nip rollers are held to a specific value of gap (clearance) or penetration (interference). Figure 5.11 shows a schematic of a common way of setting a gap. Here, one wedge is fixed while the other is precisely positioned with a micrometer. The combination of a low wedge angle and the micrometer makes it easy to obtain precise settings. The micrometer may be zeroed after each roll grind by loading it against the mating roller and the operating setpoint checked using feeler gauges.

While the gap setting does influence the value of nip load, it does so only indirectly. For example, if the incoming product thickness or modulus increases, so does the resulting nip. The nip load will also increase with increasing cover hardness, and on the high spots on the roller due to radial runout.

Just as metal-metal nips are very intolerant, so is gap control. From Hertzian mechanics, the nip spring rate is very nonlinear and strain hardening as seen in Figure 5.12. Thus, small changes in gap settings can make disproportionately large changes in product nip load.

Finally, gap control systems do not have inherent overload protection. Thus, a wrap or wad of product going through the nip can cause the loads to soar because it will not move out of the way as does load control. Similarly, an extra thick material can also load the rollers up so high as to break mechanical drive ties between the rollers. To add nip force overload protection to gap control systems is simple: load a cylinder with slight overpressure against the gap wedges. To add torque overload protection to gap control systems is also simple: provide separate drive motors for each roller, or put a slip clutch between them if they are geared together (gears, chains, etc.).

A load control system can be, in most cases, easily retrofitted to do gap control. However, it may be somewhat more difficult to retrofit gap control to load control if a suitable cylinder loading system was not provided.

Figure 5.11
Gap Control Schematic

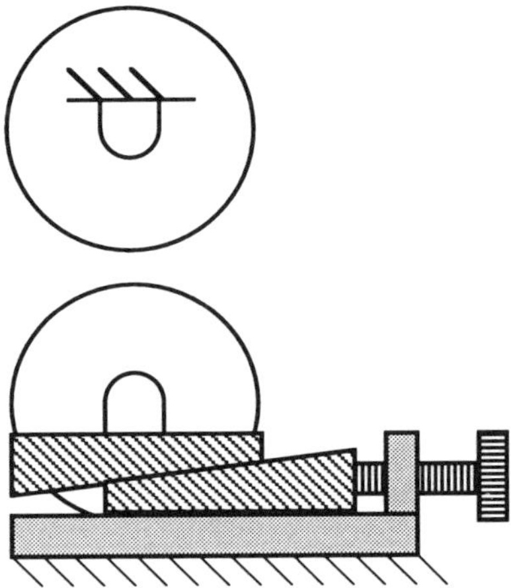

Figure 5.12
Nip Load versus Nip Deflection

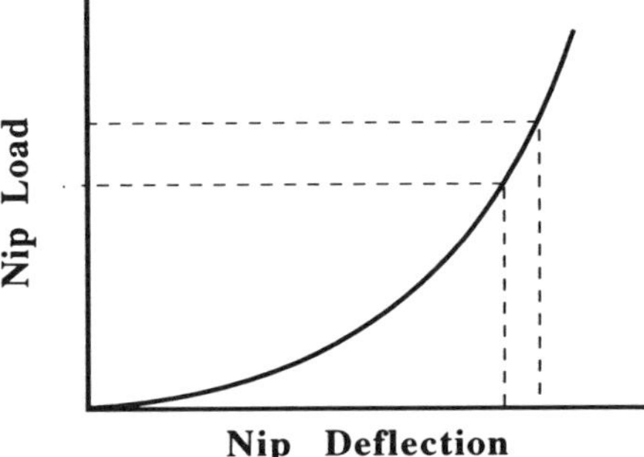

On a new machine, it would be wise to equip the nip to do either load or gap control. Pilot machinery or lines that run numerous products should always be equipped to do both. However, only one mode can be in effect at any particular time.

Because gap control is so intolerant, it should be reserved only for instances where inter-roll distance is clearly more important than load. One example is in a rolling application where the final thickness is to be held closely (at the cost of increased density, length and/or width) despite incoming caliper fluctuations. Another application is nested embossing where the two rollers can't be allowed to touch else the product will be cut.

Roller Deflection in a Nip

As discussed previously, nip pressure will vary with time (MD position) primarily due to nip control system friction and hysteresis. However, nip loads can also vary across the width (CD position) because of several types of geometrical errors. Figure 5.13 shows two nip rollers that are not parallel. Here, the nip is high on the left side such that it might not even contact on the right side. The cause can be either an unbalanced cylinder force or friction on one side, or it can be due to misalignment of the timing of the two sides.

Roller deflection, however, is usually the largest contributor to nip variations across the width. As seen in Figure 5.14, all end-loaded rollers will deflect and bow away from the nip, which causes a pinching at the ends. Indeed, the nonuniformity is so much on metal-metal nips than only about the outer 10-20% of the ends are in contact, truly leaving a gap in the center. Soft roller coverings can help even out the nip load variation to a degree; however, the tendency is still there. Nip roller deflection is a particular problem with slender rollers (length more than 10-15 times the diameter), and/or with hard nips.

Calculating the nip roller deflection and the resulting nip profile variation is very difficult. There are no closed form solution, such as the beam bending equation, which can describe this system. Rather, nonlinear numerical solutions such as FEM (Finite Element Modeling) must be employed. The primary nonlinearity is the variable spring rate of the nip as given in the last section. A further complication arises if the nip pressure is zero in the center area, which requires the use of gap elements.

A similar effect, but lesser known, is the shell wall deflection more commonly called oil-canning. As seen in Figure 5.15, the heads on the ends of the rollers preserve the shell's circular shape. However, the center is free to flatten much like a tire's contact with the road. The net effect is to cause a pinching on the ends much like the results of beam bending.

The deflection of the flat spot can modeled with full 3D FEM or can be estimated from[37]

Figure 5.13
Nip Roller Misalignment

Figure 5.14
Nip Roller Deflection

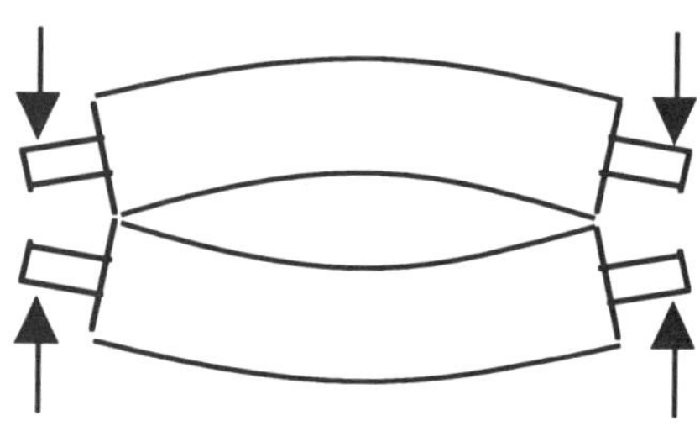

Figure 5.15
Shell Oil-Canning

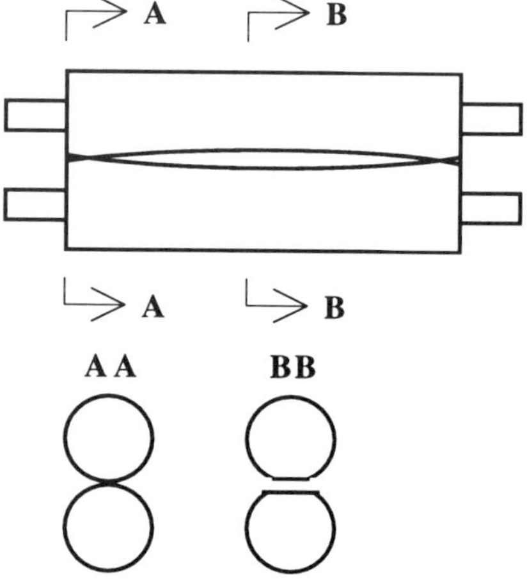

$$\Delta R = -0.2798 \frac{w R^4}{E I} \qquad (5.6)$$

where w is, unconventionally, the distributed nip force per unit <u>circumferential length</u>, R is the average radius of the thin shell, E is the modulus, and I is the area moment of inertia of the shell.

Crowning

One common method of reducing the nip variation due to roller deflection, and to a lesser extent oil-canning, is to crown one or both of the rollers. As seen in Figure 5.16, a crowned roller is barrel shaped. The amount of crown (diametral difference center versus ends) is calculated from the beam deflection formulas given in Chapter 3. The amount of crown is very small because, as we've seen, the deflection should not exceed 0.00015 x face for any nipped roller. Thus, the barrel shape is an imperceptibly small magnitude of a few mils (0.001") for all but the widest machines.

If only one of the two rollers is crowned, it must be at a value equal to the twice the total difference in deflection of the two rollers. The advantage of crowning a single roller is that it is cheaper (special cut on only one roller versus two) and more precise. If both rollers are crowned, they would both have a crown magnitude equal to the difference in deflection of the two rollers (due to nip load and gravity). The advantage of crowning both rollers is that it may, in some cases, reduce the number and types of spare rollers that must be kept.

The shape of a crown is nominally the shape of a beam deflected under a uniformly distributed load. Most modern grinding machines can take the deflection formula, or at least interpolate between tabulated values to produce the correct shape. However, many older grinders and articles refer to a segment of a cosine curve as given in Figure 5.17. It can be shown that 70 degrees on each side of a cosine curve is a close approximation to the deflection shape of a uniformly loaded simple beam, and is the most commonly used value. However, other portions of the cosine curve and even multiple segment shapes are sometimes used to compensate for higher order effects such as nonuniform temperature distribution or oil-canning. Covered rollers in nip are sometimes **dubbed** on the ends outside of the web width. This dubbing is a further relief in diameter which gradually relieves the contact stresses to improve cover life.

Figure 5.16
A Crowned Roller is Barrel Shaped

Figure 5.17
Deflected Shape is Similar to a Cosine Curve

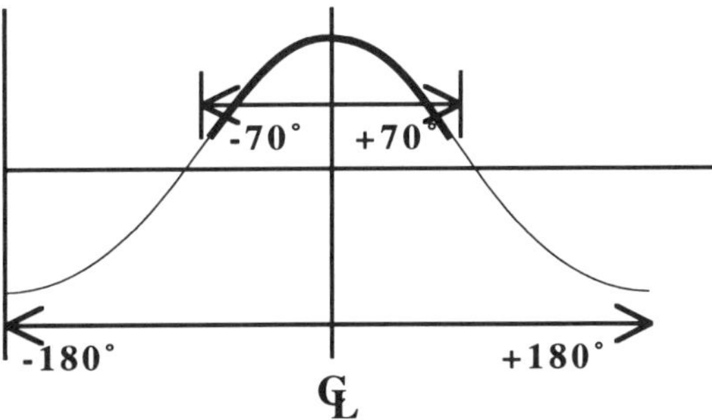

While a theoretical crown will go a long ways to evening out a nip, the results should be checked. First, the actual diameter as measured by a caliper or other means should be checked in at least 20 evenly spaced **stations** (cross machine positions) across the width against the target crown shape and magnitude. Second, the actual nip profile can be checked at the working load using nip impression papers and other techniques covered later.

There is a wealth of further information on crowning from technical associations[38,39] and machine builders.[40] However, a few generalizations may be appropriate. First, the rollers must be appropriately sized to reduce the magnitude of deflection and hence the magnitude and even the need for crowning. Second, cutting an accurate crown is a very difficult process, such that a compromised shape may cause more problems than not crowning at all. Third, most hard nips with significant end loads will require crowns, while most soft covered rollers may not. The decision to crown is merely one of your nip variation tolerances. Fourth, crowns are good for only one load (next section), unless there is provision for deflection compensation (later section).

Load-Crown Match

The magnitude of crown is calculated from the beam bending equations. Some of the deflection is due to the weight of the rollers, which will cancel out if the rollers are of the same material and geometry because the rollers deflect equally and thus will match across the face. However, the primary motivation is the externally applied forces at the ends of one of the rollers which actually provides the nip load. (Note that it doesn't matter whether the external load is in the load or unload direction as the nip will always be compressive).

The amount of deflection and thus the optimum crown magnitude is determined by the nominal nip load. There is a single value for load which determines the required crown, and thus a given crown matches only a single value of load. Any other load than the match load will not provide uniform nip loading across the width. This leads to following important conclusions which are illustrated by the deflected shapes of Figure 5.18 and the resulting pressure profiles of Figure 5.19:

If the nip load is ***lower*** *than the load-crown match, the nip load profile will be higher at the center than at the edges.*

If the nip load is ***equal*** *to the load-crown match, the nip load profile will be uniform across the width.*

If the nip load is ***higher*** *than the load-crown match, the nip load profile will be higher at the ends than at the center.*

A corollary to the last conclusion is that the match load for a zero crown (cylindrical roller) is a zero nip. Thus, any uncrowned nipped roller set will tend to pinch at the ends.

The problem, therefore, with simple nipped systems is that one can't change the nip load without adversely affecting the uniformity of the nip across the width. This may be a compromising situation when there is a number of different grades or products that must be run. Fortunately, there are deflection compensation methods given in the next section.

Figure 5.18
Deflection versus Load-Crown Match

Light Load and/or Heavy Crown

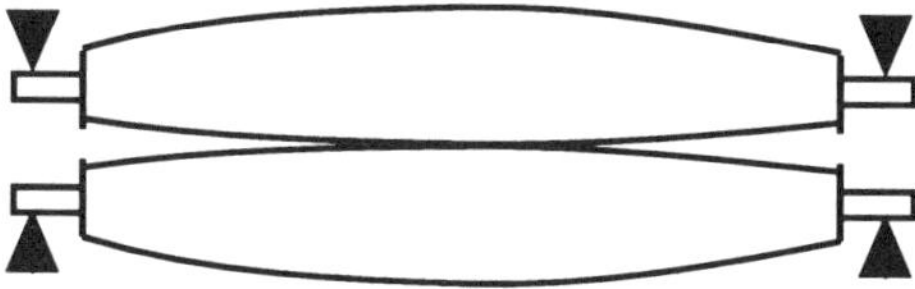

Matched Load and Crown

Heavy Load and/or Light Crown

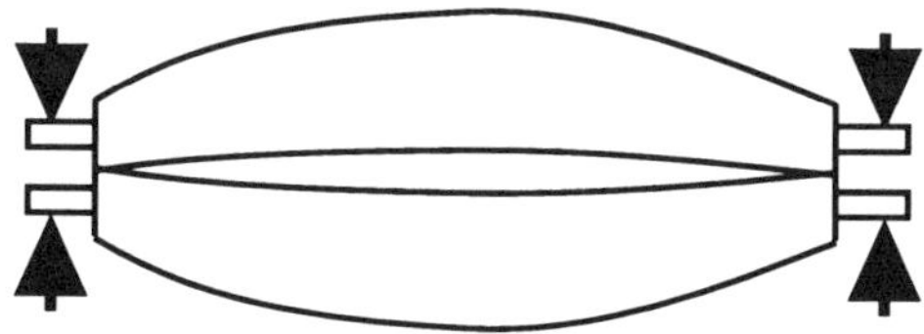

Figure 5.19
Nip Profiles versus Load-Crown Match

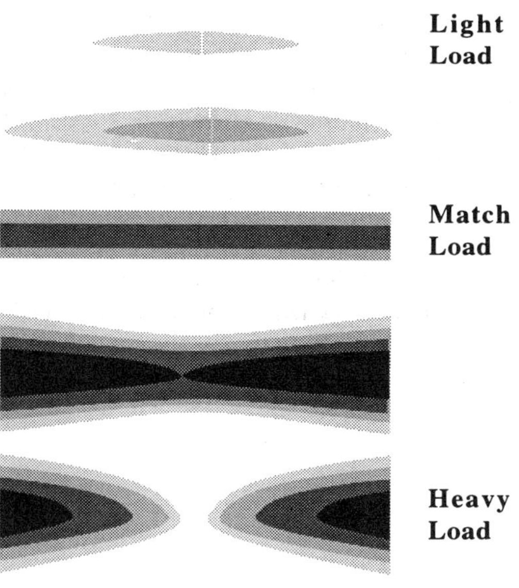

Deflection and Bearing Position

In most cases, roller beam deflection is the largest source of nip nonuniformity. As was covered in Chapter 3, increasing roller diameter is by far the most effective way to reduce deflection. Increasing wall thickness or changing to stiffer shell materials will only have a small proportional effect.

Bearing placement is another factor to which can substantially reduce deflection, if the design will allow. As seen in Figure 5.20, minimum deflection is obtained when the bearings are placed inboard of the shell ends approximately 22% of the roller face width. (Slightly different values will minimize critical speed and shell bending stresses.) Here, the magnitude of shell deflection with respect to the bearing location is the same on the ends as it is in the center.

Placing the bearings near the end of the shell, such as on most dead shaft rollers, will increase deflection 40 fold over the optimum inboard position!! The worst deflection is obtained with live shaft rollers where bearings are placed outboard of the shell.

Figure 5.21 shows the relative deflection (end mounted = 1.0) of rollers as a function of bearing relative CD location (end mounted = 1.0). This graph is based on strictly distributed loads (weight, nip, tension) acting across the entire face, and supported only on the bearings. Thus, the results may vary if the load distribution changes or other load types are present.

Nonetheless, one can see that the minima has a sharp gradient (near cusp) located at a bearing position of 22% of the face inboard. The large (44X) reduction is achieved both because the central span length is reduced and because the overhung load counteracts the central deflection.

There is probably no justification for placing bearings inboard of the minima as deflection increase and bearing loads will rise with any asymmetric loading. Also note: dead shafts and separate internal shells do not affect the deflection that the web sees. However, flimsy dead shafts and internal components will reduce critical speed.

Figure 5.20
Bearing Position Optimization
Best to Worst

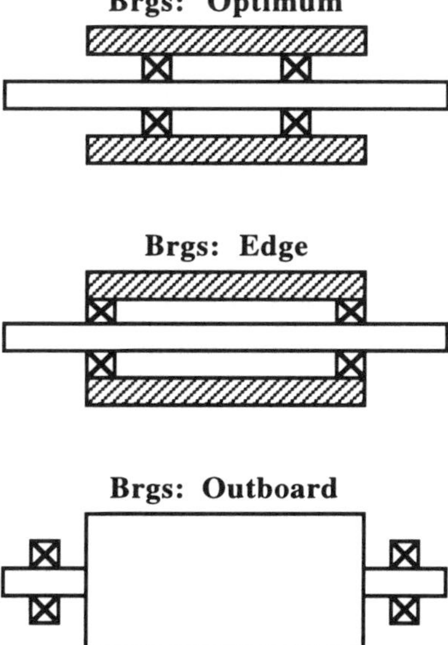

Figure 5.21
Relative Deflection versus Bearing Location

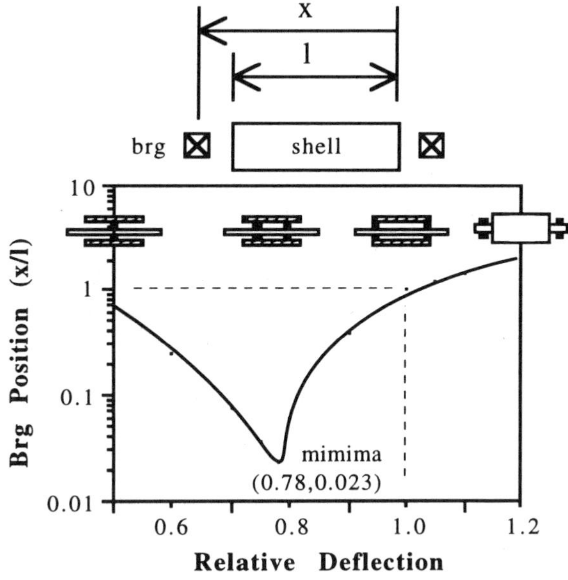

Deflection Compensating Techniques

Quite possibly, we may desire to change the nominal nip load without adversely affecting the uniformity of the nip profile. There are several techniques to do this as shown in Figure 5.22.

The first and simplest method is to use the weight of a top roller to largely provide for the nip load. This roller is made quite slender and flexible to conform to the minor and inevitable deflection of the enlarged fixed bottom roller.

Roller skewing has been called a poor man's CC (controlled crown) roller (see 2nd from bottom). Here, the ends of the one of the rollers is moved out of parallel with the other roller. The net effect of this is to shift the nip load profile from the ends to the center. The primary problem with this simple tool is that improper adjustments can be made by well-meaning but less than knowledgeable operators which can cause more profile problems than it corrects.

A bending moment applied to very husky journals can counteract some of the deflection due to nip load. The bending moment can be applied is with simple weights or a cylinder to load well outboard of the main roller bearings. This option must withstand tremendous loads.

A controlled "crown" roller, abbreviated as a CC roll, bends the shell against a very sturdy through shaft. In one version, the load is provided by hydraulic cylinders and the bearing between the load and shell is usually a hydrodynamic shoe. These are very expensive specialty rollers primarily used in paper calenders and printing.[41]

Nip roller deflection is minimized if some of the required nip is applied on the shell near the middle. This is commonly done by hydraulic cylinder loading of support wheels. Unfortunately, the contact loads, compromised bearing material and lubrication are very tough on the main roller shells and support wheels.

These deflection compensating techniques are usually rather weak and provide only a limited range of adjustment. Thus, one will usually crown one of the rollers to the lightest anticipated load and use the deflection techniques to achieve higher operating loads.

Figure 5.22
Deflection Compensating Techniques

Small Roller on Large Roller

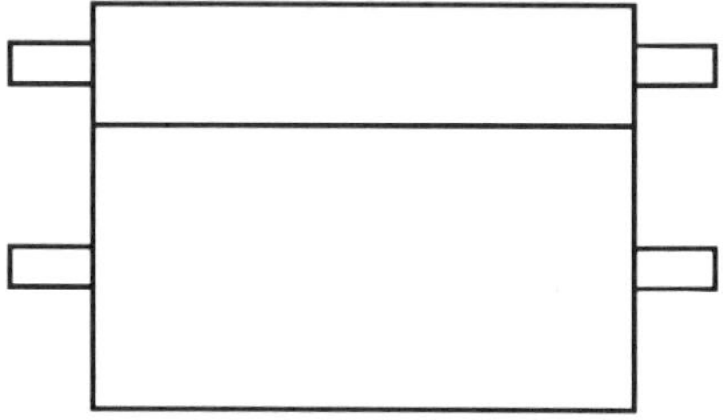

Roller Skewing

Bending Moment

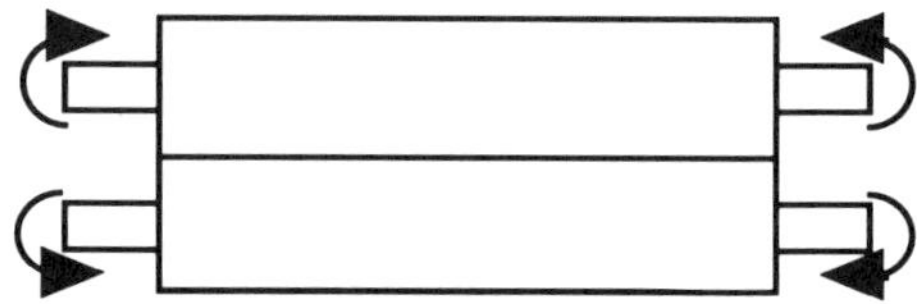

Controlled 'Crown' Roller

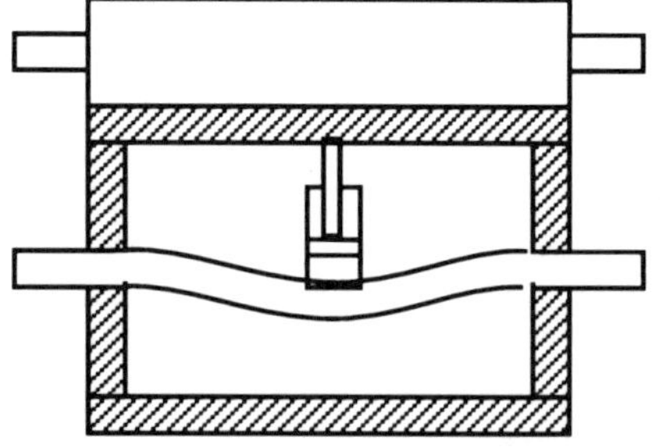

Intermediate Loaded Rollers

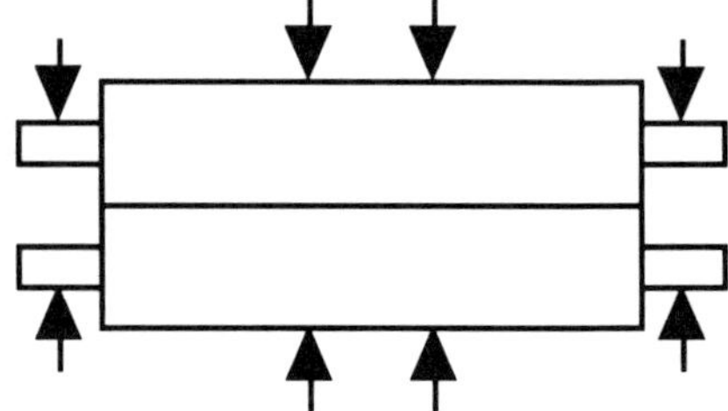

Roller Geometrical Errors

No roller is a perfect cylinder as seen in Figure 5.23. First, a roller will vary in radius as a function of rotational position but at a fixed CD (cross direction) position or station. This radial variation is measured with a dial indicator and is called TIR (Total Indicator Reading or Total Indicated Runout). If the runout is excessive, periodic marking (**barring**) and roller vibration can result. Barring caused by runout can be confirmed by a CD repeat distance corresponding to the circumference of one of the nip rollers. A more quantitative confirmation can be obtained as a spike in a FFT (Fast Fourier Transform) readout corresponding to an integral number times the circumference of a CD profile measurement such as caliper.

Vibration caused by nipped roller runout has similar dynamics as the motion of a cam. In the extreme, excessive runout can cause the mating roller to unload and leave the nip momentarily. From textbook cam analysis, we know that the oscillating forces are proportional to speed squared and the second derivative of the runout profile $r(\theta)$. Unfortunately, textbook cam analysis does not describe real world nips well because rollers are not rigid bodies (beam bending, cover/shell deformation, etc.), and because the radius is also a function of cross machine position.

Second, the diameter can vary across the width as measured with a caliper. These variations can cause bands or streaks in the product. A quantitative confirmation of this problem is the correlation between roller and web CD profiles.

The cylindrical accuracy required of nipped rollers is considerably higher than other process or transport rollers. Also, there are economical and practical constraints on how close can roller cylindricity be machined and maintained. The largest factor here, as detailed in a later chapter, is the surface material.

In general, metal rollers can be cut to within a few ten thousandths of an inch accuracy with a bit of care. However, covered rollers can only be cut to within a few thousandths because the soft surface will deform away from the tooling.

Figure 5.23
Roller Geometrical Errors

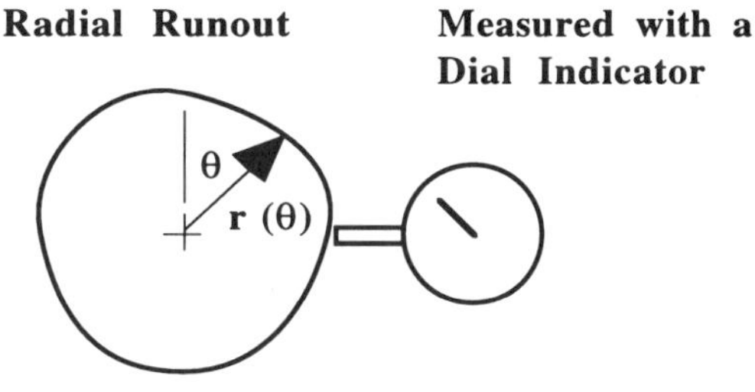

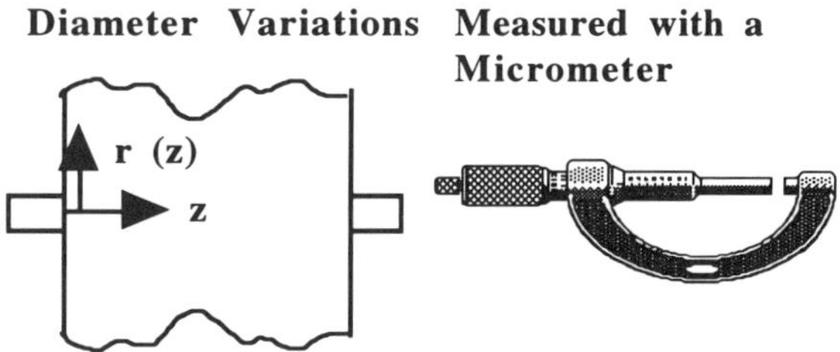

It is important note that the softer the cover, the less geometrical errors will affect nip and product uniformity. Indeed, metal-metal nips are so touchy that they should be avoided except for very thick products. Unfortunately, soft covered rollers may not provide the nip intensity needed for some applications. Also, soft covers are often less durable than harder covers. Cover hardness, measured by handheld indentors, is a very important roller and process specification.

In any case, roller geometrical tolerances must be specified for both ingoing (as received) and outgoing (to be reworked) cases to keep a nipped process under control. The most important of these tolerances are: nominal diameter range, maximum diametral variation, maximum station-to-station diametral variation and maximum TIR. Covered rollers should have additional specifications for nominal hardness range and maximum hardness variation.

Ultimately, product consistency will determine how close nipped rollers need to be. Unfortunately, the effect of nip variations on a product are too complex to be calculated. Also, the product itself is a noisy sensor or indicator. Thus, we will most often use nip impression methods to maintain nip consistency.

Nip Impressions

As we've seen, there are a number of factors that can contribute to nonuniform nips such as roller deflection and non-cylindricity. This makes a prediction or calculation of the pressure distribution in a nip very difficult. Fortunately, variations of nip pressure across the width, and to some extent MD position, can be easily measured with nip impression papers. These are, in simple terms, something like carbon paper and are supplied by a number of sources such as given in Table 5.3.

The pressure in a nip varies across the width and through the footprint (MD distance). The pressure must be sufficient to activate the color somewhere in the nip footprint else nothing can be learned. This minimum **activation pressure** varies considerably between the products. For example, Sto-Foil is an embossed aluminum foil whose embossment pattern is easily flattened at very tiny pressures. In contrast, the nip impression papers, such as supplied for paper supercalenders by machine builders, require hundreds of PSI to activate. Fuji Film is a unique product in that the depth of color change is calibrated to engineering pressure units. Also, it is supplied in several intermediate pressure ranges.

The **saturation pressure** is the pressure where the nip impression paper turns color to its maximum darkness or density. If most of the nip is above the saturation pressure at the middle of the footprint, the dynamic nip impression will be completely dark. The problem here is that the only thing learned is that there is a minimum (saturation) nip pressure everywhere. No information on nip pressure variation can be gleaned.

Thus, the optimally sized pressure range is such that the paper predominantly turns to an intermediate color and avoids the extremes of no color or saturation. Unfortunately, there are very few choices of pressure ranges. Also, beware that nip pressure versus color curves may change with humidity, temperature, and shear. Finally, the nip impressions can be quantified with a standard brightness meter as may be found in many web test labs.

Table 5.3
Examples of Nip Impression Papers

Name	Company	Location
Nip Width Kit	Beloit Corp.	Beloit, WI
Prescale Film	Sensor Products	East Hanover, NJ
Sto-Foil	Stowe Woodward	Southborough, MA
Nip Impression	Valmet	Appleton, WI

Nip impressions are commonly performed on ingoing and outgoing rollers as well as for in process troubleshooting. Unfortunately, good documentation is often overlooked. Unless the strip is well labeled, the information will be a one-shot disposable effort. Table 5.4 gives some of the most important items to be recorded on the impression strip. This strip can then be safely stored in the maintenance files in a folder for the roller or machine element.

It is very important that all safety precautions are observed when doing nip impressions because nips are among the most dangerous elements in web manufacturing environments. One rule that should never be violated is to make sure that all personnel are completely clear before loading a nip and/or putting the machine into thread speed. Both static and dynamic impressions can be taken without manually holding the paper. Rather, double-sticky tape or other means are to be used to hold the paper temporarily against one of the rollers.

There is a lot of additional useful information that can be found in the instructions for the kit, or in published articles.[42,43,44] For careful work, nip impression paper can be calibrated using a deadweight or compression testing machine and a brightness meter.

Table 5.4
Nip Impression Documentation

Date
Machine or section name
ID's of both rollers
Ingoing, in process, or outgoing
Nip control pressure settings both sides
Crown(s) or diametral variation for both rollers
Label front and back side of machine

Static Nip Impressions

A nip impression can be taken in the dynamic mode or in the less common static mode. The static mode is used to get information on nip widths and to diagnose crowning errors. In any case, the pressure range of the selected paper should be appropriate for the application.

Figure 5.24 shows the setup for a static nip impression. The shape of the ideal static nip impression is a uniform width rectangle. However, an actual impression may take on the characteristics of one of the four errors given in Figure 5.25. Instructions for some of the impression papers will give procedures and equations to estimate a better crown magnitude based on edge versus center impression width. If a skewed impression is found, alignment of the nip should be estimated as a consistent gap on the two ends of the roller, and confirmed with optical alignment. If, as is the usual case, the problem is with unbalanced loading, the pressure control and loading friction must be checked on both sides. Finally, banding can be from diametral profile problems. However, banding is better checked using the dynamic nip impression given in the next section.

Nip width can also be estimated by using the static nip impression. However, as seen in Figure 5.26, the actual nip width will be greater than the width of the impression. This is because the lower loadings near the edges of the contact are not enough to activate the impression there.

Figure 5.24
Static Nip Impression Procedure

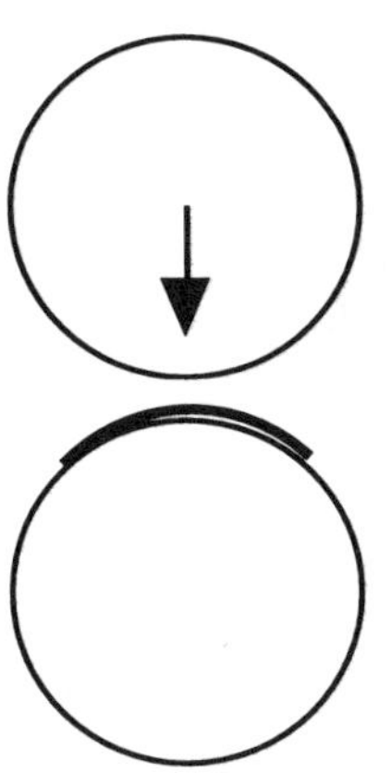

Cut paper the width of the machine and about 4-12" wide.
Center the paper under the nip and flatten against the smooth roller.
Tape edges if necessary.
Load nip to operating values and hold for a minute.
Unload nip, remove paper, examine impression.

Figure 5.25
Reading a Static Nip Impression

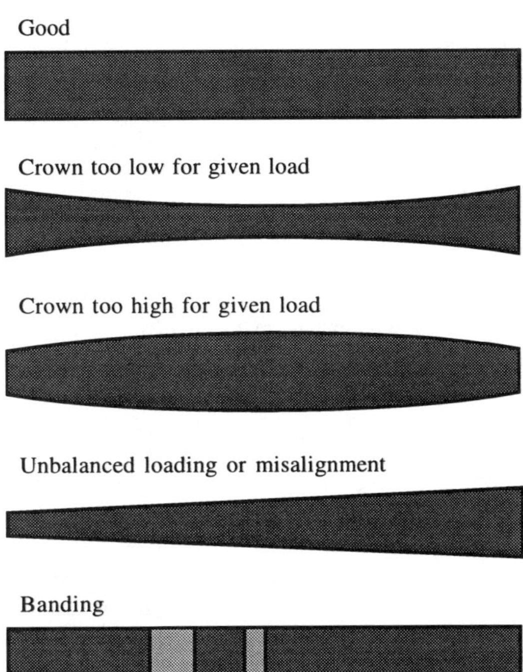

Also, the width of many hard roll or lightly loaded nips are so narrow as to make an accurate width measurement quite difficult. Finally, the actual width may be wider for materials that are much thicker than the nip impression paper itself. For these reasons, the dynamic nip impression may be the more useful.

Figure 5.26
Nip Width Actual (b) Versus Measured (b')

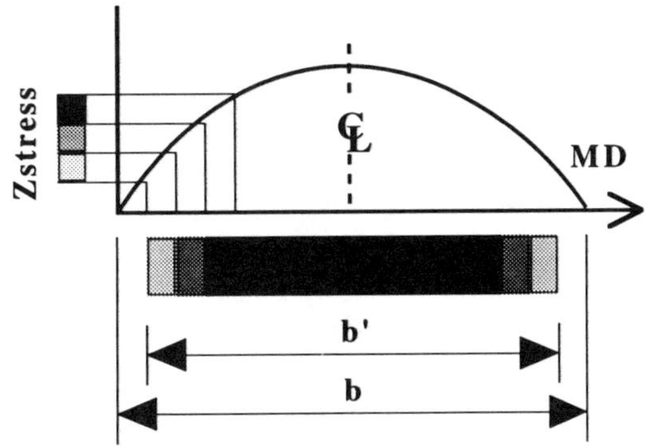

Dynamic Nip Impressions

The more common nip consistency measurement procedure is the dynamic nip impression mode. Unfortunately, this is a bit of a misnomer as there are no real dynamics involved as seen in Figure 5.27.

The first step, as always, is to select a paper with a pressure range appropriate to the application. This requires a knowledge of the lineal nip load (kN/m or lb/in) as well as an estimate of the nip width. Then, the procedure in the beginning of the chapter will estimate the peak load, which should be near the saturation load for the paper.

The nip impression paper is then applied similarly to the static mode, except that the paper is applied just upstream of the nip. For convenience, it is sometimes easier to have the leading tail under the nip. The nip is then loaded to the operating or test conditions. Now, while loaded, the machine is briefly jogged (at thread speed) until the paper is completely pulled through the nip. Finally, the paper is removed, annotated, and interpreted.

Interpretation, as illustrated in Figure 5.28, is similar to static impressions except that we are concerned with banding color instead of nip footprint shape. This is where a paper with a large range between activation and saturation becomes most useful.

The highly loaded areas are dark and normally correspond to excessively large local diameters on one of the rollers. The lightly loaded areas or light or uncolored and normally correspond to smaller than nominal local diameters on one of the rollers. The net result of these nip pressure profile variations may be wrinkles for transport nips and CD profile variations for process nips.

The dynamic method is much preferred for nip consistency checks in most cases simply because there is so much more area to read. It also can show MD variations (perhaps due to runout, bumps or dents) where the static impression can not. The only disadvantage is that the dynamic method can give no nip width information. However, if we are clever we can make both a static and dynamic impression on the same strip!

Figure 5.27
Dynamic Nip Impression Procedure

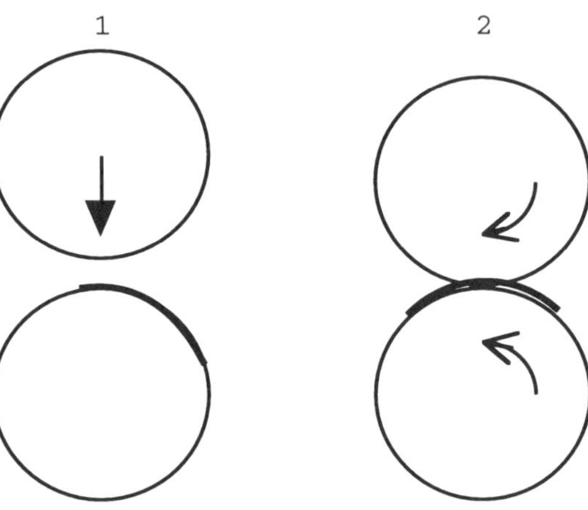

Figure 5.28
Reading a Dynamic Nip Impression

Good

Crown too low for given load

Crown too high for given load

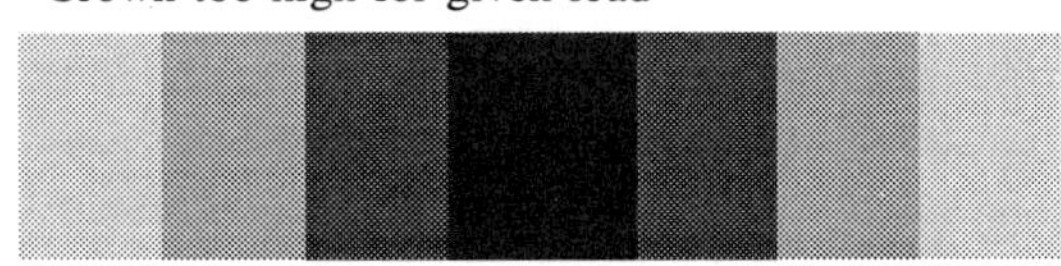

Unbalanced Loading or Misalignment

Banding

Bibliography

1. Pfeiffer, J. David. *The Mechanics of a Rolling Nip on Paper Webs.* Tappi J., vol 51, no 8, pp 77A-85A, August 1968.

2. Pfeiffer, J. David. *Nip Forces and Their Effect on Wound-in Tension.* Tappi J., vol 60, no 2, pp 115-117, February 1977.

3. Pfeiffer, J. David. *Surface Winding to Overcome the Strain Deficiency.* Tappi J., vol 73, no 10, October 1990.

4. Pfieffer, J. David. *Understanding Rolling Nip Action through Video Technology.* TAPPI Finishing and Converting Conf. Proc., New Orleans, pp 109-123, October 24-27, 1993.

5. Good, J. Keith and Fikes, M. *Predicting Internal Stresses in Center-Wound Rolls with an Undriven Nip Roller.* Tappi J., vol 74, no 6, pp 101-109, June 1991.

6. Good, J. Keith and Fikes, M. *Predicting Internal Stresses in Center-Wound Rolls with an Undriven Nip Roller.* Tappi J., vol 74, no 6, pp 101-109, June 1991.

7. Good, J. Keith and Wu, Zan and Fikes, M. W. R. *Stresses Within Rolls Wound in the Presence of a Nip Roller.* 1st Int'l Conf. on Web Handling, Oklahoma State Univ., pp 123-144, May 19-22, 1991.

8. Weiss, Herbert L. *Lay-on Rollers Improve Quality.* Package Printing, vol 31, no 6, pp 19; no 8, pp 32; no 9, pp 54.

9. Tajuddin, By. *Mathematical Modelling of Air Entrainment in Web Handling Applications.* MS Thesis, Web Handling Research Center at Oklahoma State Univ., December 1987.

10. Jones, D. R. *Air Entrainment as a Mechanism for Low Traction on Rollers and Poor Stacking of Polyester Film Reels and its Reduction.* ASME Winter Annual Mtg., Anaheim, November 8-13, 1992.

11. Good, J. Keith and Holmberg, Michael W. *The Effect of Air Entrainment in Centerwound Rolls.* 2nd Int'l Conf. on Web Handling, Oklahoma State Univ., June 6-9, 1993.

12. Bouquerel, F and Bourgin, P. *Irreversible Reduction of Foil Tension due to Aerodynamical Effects.* 2nd Int'l Conf. on Web Handling, Oklahoma State Univ., June 6-9, 1993.

13. Roisum, David R. *The Mechanics of Winding.* TAPPI PRESS, 1994.

14. Roisum, David R. *Nip Induced Defects of Wound Rolls.* TAPPI Finishing and Converting Conf. Proc., Chicago, pp 89-98, October 2-5, 1994.

15. Smith, R. Duane et al. *Roll and Web Defect Terminology.* TAPPI PRESS, 1995.

16. Satas, Donatas. *Web Processing and Converting Technology and Equipment.* Van Nostrand Reinhold Co. Inc., 1984.

17. Weiss, Herbert L. *Coating and Laminating Machines.* Converting Technology Co, Milwaukee WI, 1977.

18. *Gravure Process and Technology.* Gravure Assoc. of America, 1991.

19. Boresi, Arthur P. and Sidebottom, O. M. and Seely, F. B. and Smith, J. O. *Advanced Mechanics of Materials.* 2nd Edition, John Wiley and Sons, 1978.

20. K. L. Johnson. *Contact Mechanics.* Cambridge Univ. Press, 1985.

21. Batra, R.C. *Rubber Covered Rolls - The Nonlinear Elastic Problem.* J. of Applied Mechanics, vol 47, pp 82-86, March 1980.

22. Deshpande, Narayan V. *Calculation of Nip Width Between Elastomeric Cylinders.* Tappi J., vol 61, no 10, 115-118, October 1978.

23. Parish, G. J. *Calculation of the Behavior of Rubber-Covered Pressure Rollers.* British J. of Applied Physics, vol 12, pp 333-336, July 1961.

24. Moore, Robert H. and Moschel, Charles C. *A Tool for the Engineered Nip.* Tappi J., vol 77, no 3, pp 117-122, March 1994.

25. Smith, J. O. and Liu, C. K. *Stresses due to Tangential and Normal Loads on an Elastic Solid with Applications to Some Contact Stress Problems.* J. of Applied Mechanics, vol 20, no 2, pp 157, June 1953.

26. Liu, C. and Paul, B. *Rolling Contact with Friction and Non-Hertzian Pressure Distributions.* J. of Applied Mechanics, December 1989.

27. Eschmann, Hasbargen and Weigand. *Ball and Roller Bearings.* John Wiley & Sons, New York, 1985.

28. *Calendering and Supercalendering.* Lockwood Trade Journal Co, New York, 1964.

29. Browne, T.C. and Crotogino, R.H. and Douglas, W.J.M. *Measurement of Paper Strain in the Nip of an Experimental Calender.* J. of Pulp and Paper Science, vol 20, no 9, pp J266-J271, September 1994.

30. Rodal, Jose J. A. *Soft Nip Calendering of Paper and Paperboard.* Tappi J., vol 72, no 5, pp 177-186, May 1989.

31. Rodal, Jose J. A. *Modelling the State of Stress and Strain in Soft-Nip Calendering.* Fundamentals of Papermaking, Transactions of the Ninth Fundamental Research Symposium at Cambridge, September 1989, Mechanical Engineering Publications LTD, London, vol 2, pp 1055-1075, vol 3, pp 333-339, 345-346, 1990.

32. Rodal, Jose J. A. *Paper Deformation in a Calendering Nip.* Tappi J., vol 76, no 12, pp 63-74, December 1993.

33. Parish, G. J. *Measurements of Pressure Distribution Between Metal and Rubber Covered Rollers.* British J. of Applied Physics, vol 9, pp 158-161, April 1958.

34. Brink, R. *Instrumentation for Roll Nip Studies.* TAGA Proc., pp 103-117, 1963.

35. Keller, Sam F. *Measurement of the Pressure-Time Profile in a Rolling Calender Nip.* Canadian Pulp and Paper Assoc, 77th Annual Mtg, January 1991.

36. Lyons, Anthony V. and Thuren, Alan R. *Scale Up Procedure for Relating Pilot Calendering to Commercial Reality.* Tappi J., vol 75, no 10, pp 173-183, October 1992.

37. Roark, Raymond J. and Young, Warren C. *Formulas for Stress and Strain.* McGraw-Hill Book Co., New York, 5th edition, pp 231, 1975.

38. Anon. *Roll Crown Specifications and Definitions.* TAPPI Technical Information Sheets, TIS 405-10, September 1975.

39. Twitchen, D. *Paper Machine Roll Grinding Tolerances.* Canadian Pulp & Paper Association, Technical Section, Data Sheet G-20, May 1988.

40. Anon. *Crowning and Dubbing Information for Covered Rolls.* Beloit Corp., Manual J099-10015, January 1979.

41. GAA. *Gravure: Process and Technology.* Gravure Association of America, pp 292-301, Rochester, NY, 1991.

42. Kennedy, William B. *Nip Impressions.* Tappi J., vol 74, no 5, pp 277-280, May 1991.

43. Kennedy, William B. *Using Nip Impressions to get Uniform Water Removal.* Paper Technolgy, July 1992.

44. Genisot, Tony. *Nip Impressions - a tool for troubleshooting and maintaining quality.* Tappi J., vol 77, no 9, pp 246-250, September 1994.

Chapter 6

Shells

In this chapter we show how the geometry of the roller shell is primarily determined by the width of the machine, while geometrical tolerances are suggested by the particular application. Additionally, we discuss sizing wall thickness, selecting materials and the manufacturing of roller shells.

Introduction

The shell serves as a vital part of every roller. In most cases, it is the component which is in contact with the web and must have tight cylindricity tolerances. Indeed, a web's universe consists primarily of the portion of the roller surface it touches and the spans between roller surfaces.

In all cases the shell is the structural element that must be substantial or thick enough to avoid **oil canning** due to web tension and especially nip loads.[1] Also, the shell is the structural member which resists **beam deflection** within the web's width in both dead and live shaft simply supported rollers. The shell serves a lesser structural role in cantilevered rollers whose deflection is primarily resisted by the central through shaft and rigid frame mountings.

The roller surface, which is often the shell, must transmit the inevitable tension differences that exist across any roller. As described in Chapter 4, these upstream versus downstream tension differences should be sustained without web/roller slippage in almost all applications. The transition between traction and slippage will cause tension upsets as well as edge position upsets.

Tension differences may be intentional, as in the case of driven rollers, or unintentional as in the case of undriven rollers to overcome bearing drag always and inertia during speed changes. To get sufficient traction to avoid slippage, rollers may be roughened, covered and/or grooved.

On uncovered and uncoated rollers, the shell is one of two components that can **wear** (the other is the bearing). Thus, a high abrasion resistance, often associated with a high hardness, may be desirable for many applications. It is at best a nuisance to regrind rollers. Thus, rollers may be coated with an anodize (e.g., aluminum) or tungsten carbide (e.g., ferrous metals) to improve shell life.

Roller shells are a large contributor to roller **imbalance**.[2] True, the journals could be mounted eccentric in the shell. Regardless of the root cause, however, the shell has the largest mass and radii of any of the roller's components. The shell can also contribute imbalance due to ID (inner diameter) imperfections because shells may not be **machined** on the ID. Also, the shell wall may not be completely homogenous in density. While balance weights may be placed on the heads of narrow rollers, wider rollers may be balanced on the shell to reduce dynamic TIR imbalance.

In heated or cooled rollers, as discussed in Chapter 15, the shell serves as a conductor of heat from the roller to the web. In this role, a high **thermal conductivity** is desirable. Thermal mass can improve temperature uniformity or stability by providing a resistance to temperature changes due to transients. However, a high thermal mass can also be a liability because it prolongs the time it takes to come to temperature upon startup and cool down after shutdown.

A low coefficient of **thermal expansion** is desirable for two reasons. First, the front side of the roller should be mounted on floating bearings so that inevitable temperature changes will not overload the bearings axial rating. Second, a low thermal expansion helps maintain uniform diameters across the face despite nonuniform temperatures.

Thus, the shell may have many other functions beyond its primaries as the backbone of a roller and interface to the web.

Axial Shell Geometries

There are several widths that determine the axial geometry of a shell as given in Figure 6.1.

Minimum Deckle is the minimum width of the product to be run.

Maximum Deckle is the maximum width of product to be run, which is a very important parameter in sizing the diameter of a roller to meet a deflection criteria.

Face is the effective or useful width of a roller and may be about 2-4" wider than the maximum deckle to allow for web width and edge position variations.

Width is the width of the shell structure. The face may be slightly smaller than the width to allow for a dubbing, which is a relief of the outside diameter to protect the edges of a roller, especially when in nip.

Note how the ends of the shell may be counterbored to accept a precision fit of a bearing or head.

Figure 6.1
Axial Shell Geometry

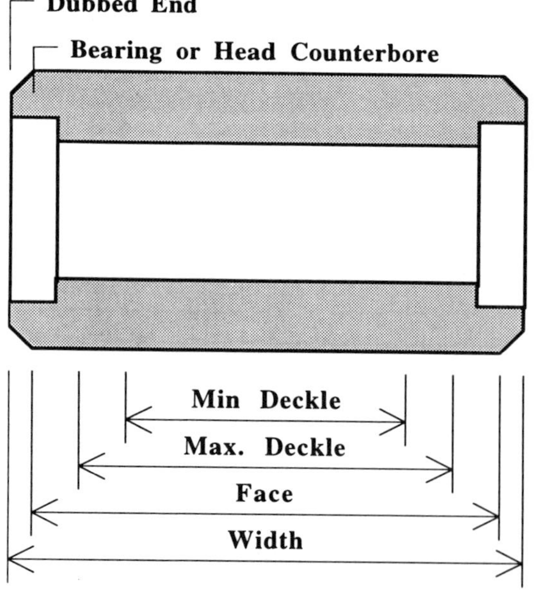

Radial Shell Geometries

There are several diameters that determine the radial geometry of a shell as given in Figure 6.2.

OD (outside diameter Do) is the factor that most strongly resists bending **deflection**.

Half of the difference in the OD and ID (inside diameter Di) is the wall thickness that must be sufficient to avoid shell wall deformation.

These diameters are used in the calculation of area moment of inertia of the shell which is used in the beam deflection calculations of Chapter 3 and the shell stress calculations given later in this chapter.

$$I = \frac{\pi}{64}\left(Do^4 - Di^4\right) \quad (6.1)$$

Finite element modeling may be required for exact stresses and deflections where the shells are grooved or drilled. A simple approximation for **grooving**, however, is to use the average OD for calculating I. An approximation for **drilled** shells is to use an effective modulus downrated by the area fraction of the holes. However, stress concentration factors, as found in many design books, must be applied when calculating stresses at the root of a groove or on the sides of holes.

Figure 6.2
Radial Shell Geometry

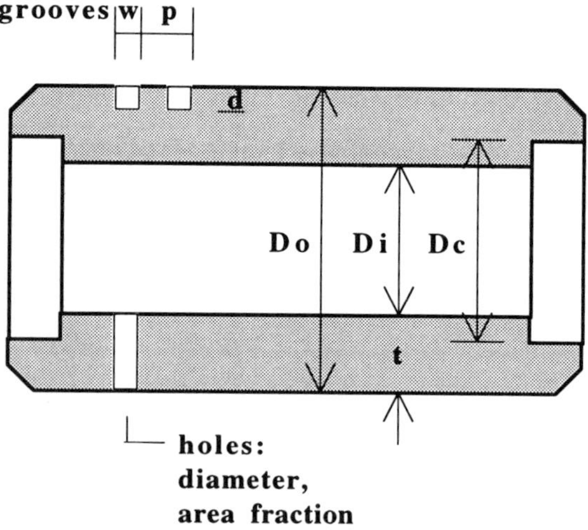

Shell Materials

Roller shells are made from a variety of materials, the more common of which are given in Table 6.1. The primary selection criterion for the shell stock is one of economics. For example, small diameter shells may be made from economical tubing while large diameter shells are typically cast because tubing is not available in those sizes.

The base stock for iron shells are castings. Some applications, such as high speed drums, may require the greater uniformity of centrifugal castings. Iron and cast steel is very common for large diameter rollers (>0.5m).

The base stock for steel shells are often a seamless heavy wall mechanical pipe or DOM (drawn on mandrel) tubing. A common application for steel is small idler rollers which need a greater wear and dent resistance than provided by aluminum.

The base stock for aluminum shells are again typically pipe or tubing. The strength/weight and modulus/weight ratios of aluminum and steel are quite similar. Thus, aluminum may not provide a lower **inertia** advantage than steel (for a given deflection criteria) for small idler rollers until the steel reaches a minimum useful wall thickness. Aluminum scratches and dents easily, and has no **fatigue** endurance limit that many other materials enjoy. Aluminum is more expensive than steels or irons, but is easier to machine, and is more resistant to oxidation and chemical attack.

Other metals used for shells includes brass, copper and stainless steels. Non-metals used for shells includes carbon-fiber-epoxy, fiberglass, granite,[3] nylon, phenolic, PVC and even wood. However, these materials would be used only in niche applications as they are not as structurally efficient as iron, steel, aluminum or carbon-fiber. Thus, they would suffer from high deflection and/or inertia.

Table 6.1
Shell Material Properties

The values are typical of listings from several sources such as ASTM. However, this chart is for illustration purposes only. Always check with the supplier for material property test values for use in any design work.

Material	Type	Density lb/in^3	Mod Elast MSI	Mod Rigid MSI	Coef Expans E-6in/in/F	Hardness Brinell	Tensile Max/Yield KSI	Elong %	Shear Max/Yield KSI	Fatigue Limit KSI
Alumin.	6061T6	0.098	10	3.75	12.9	95@500	38 / 35	10	24 / 20	14 @E5
	6063T5	0.098	10	3.8	13	50@500	22 / 16	8	13 / 9	10 @E5
Carbon Fiber		0.058	16	-	low	low	150 / -	-	-	-
Cast Iron	CI35	0.27	16	6.4	6.5	212	35 / -	-	35 / -	12.2
	CI40	0.27	18	7.2	6.4	235	38 / -	-	38 / -	13.3
	CI50	0.27	19	7.6	6.3	262	45 / -	-	45 / -	15.0
	CI60	0.27	20	7.9	6.15	215-225	52 / -	-	52 / -	16.2
Chilled	Iron	0.27	19.5	-	5.6	500-535	20 / -	-	20 / -	7.0
Nod. Iron	MT4A	0.27	24	9.5	6.5	163-192	65 / 45	10	52 / 27	19.5
	MT4D	0.27	24	9.5	6.5	170-207	70 / 45	10	56 / 27	21
	MT5	0.27	24	9.5	6.5	146-170	60 / 45	20	48 / 27	18
	MT6A	0.27	24	9.5	6.5	207-269	80 / 60	3	64 / 36	24
	MT7B	0.27	24	9.5	6.5	255-285	100 / 65	2	80 / 39	50
	MT7E	0.27	24	9.5	6.5	250-275	80 / 60	1	64 / 36	24
CastSteel	Annealed	0.283	30	11.2	6.5	140	65 / 35	24	49 / 21	23
CastSteel	A27	0.283	29.5	11.2	6.4	152	73 / 45	12	55 / 27	24.5
WroughtSteelPipe		0.283	30	11.4	6.5	95	60 / 35	22	45 / 21	19.2

Heads

Figure 6.3 shows a few of the more common methods of mounting a head to a shell and journal/shaft.

The preferred style for any application depends on a number of factors and especially the economics of roller construction. However, as we will be seen in the next section, stresses in the head area must be considered for safe operation.

Figure 6.3
Head Mounting Styles for Rollers

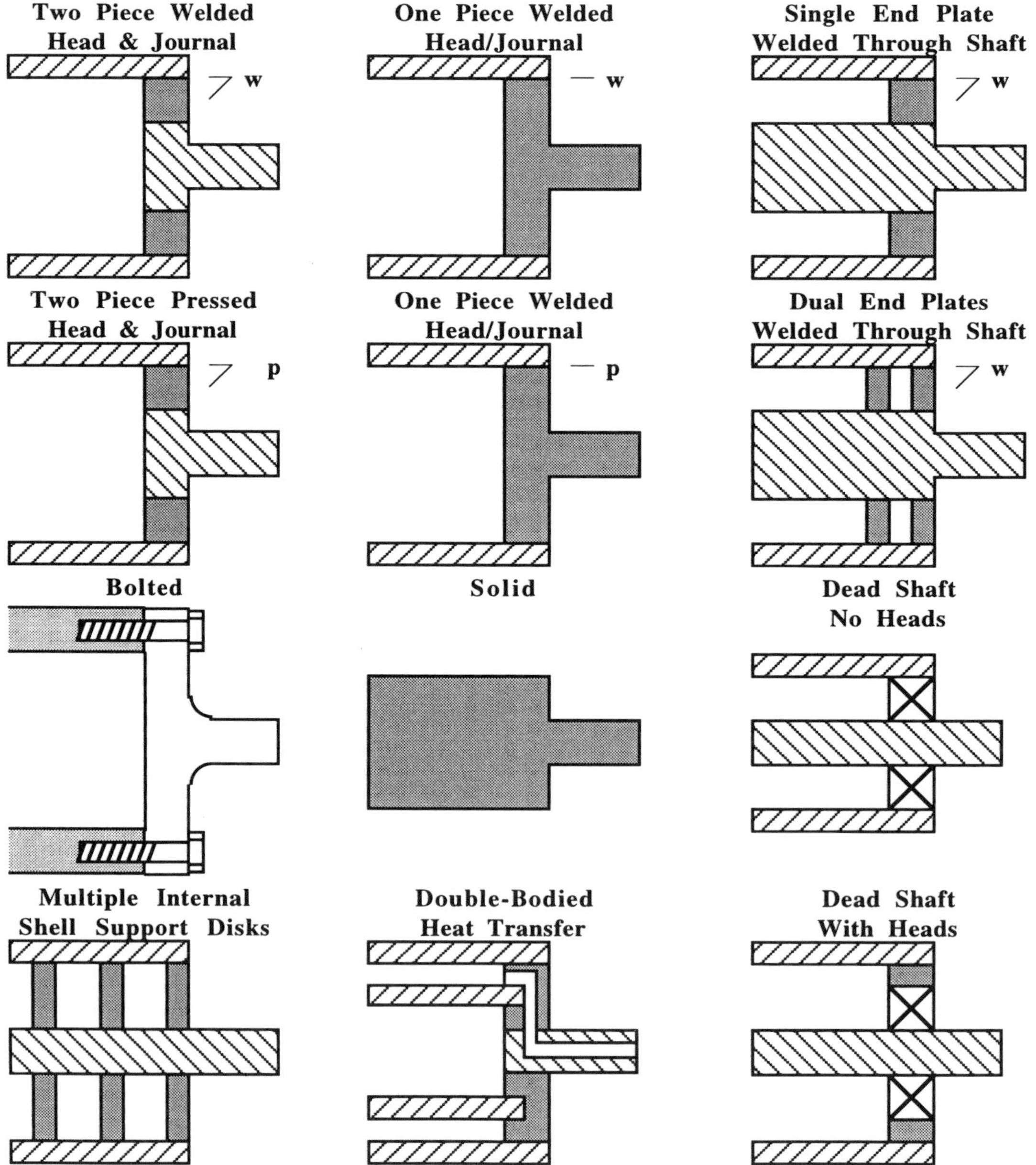

Shell Stresses

There are at least six critical regions in a roller where stresses should be calculated or modeled and compared against some failure criteria. These stress regions are shown in Figure 6.4.

Region 1 is the position of maximum bending stress at or near the center of a roller due to applied loads of roller weight, web tension and a nip if present. While a roller properly sized for deflection will rarely have high stresses at the center, it should nonetheless be checked.

To calculate the **bending stress**, we first need to calculate the bending moment at the center of the roller. The procedure for calculating this moment is by the shear and bending moment diagram techniques that are taught in engineering statics and thus will not be covered here.

However, many simple systems can be approximated as the resultant bearing loads uniformly distributed across the face of the roller.

Center Bending Stress Example English Version

Calculate the bending stress at the center of a 7" diameter aluminum roller finished to a 6.88" OD which has a 6" ID. The face of the roller is 80" and the bearing to bearing centerline distance is 88". This unnipped roller carries its own weight and a maximum 2 lb/in tension with a 180 degree wrap pulling down. These are the same values used in a deflection calculation in Chapter 3.

The shell weight is calculated as:

(6.2)

$$s = \rho \frac{\pi}{4}\left(D^2 - d^2\right)$$

$$s = 0.098 \frac{\text{lb}}{\text{in}^3} \frac{\pi}{4}\left(6.88^2 - 6^2\right)\text{in}^2 = 0.8723 \text{ lb/in}$$

The distributed load is calculated as:

(6.3)

$$w = 2t + n + s = (2)\,2 + 0 + 0.8723 = 4.872 \text{ lb/in}$$

Figure 6.4
Critical Stress Regions in a Roller

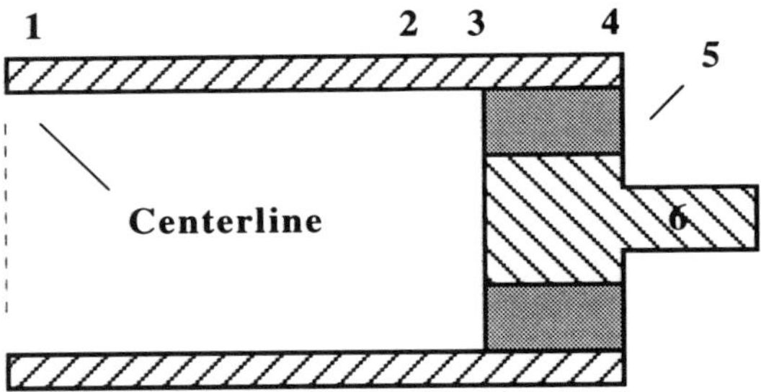

The area moment of inertia of the shell is calculated as:

(6.4)

$$I = \frac{\pi}{64}\left(D^4 - d^4\right) = \frac{\pi}{64}\left(6.88^4 - 6^4\right) = 46.37 \text{ in}^4$$

The bending moment at the center is calculated as:

(6.5)

$$M = \frac{w}{2}\left(\frac{BF}{2} - \left(\frac{F}{2}\right)^2\right)$$

$$M = \frac{4.872}{2}\left(\frac{88x80}{2} - \left(\frac{80}{2}\right)^2\right) = 4677 \text{ in-lb}$$

The bending stress is calculated as:

(6.5)

$$\sigma = \frac{Mc}{I} = \frac{(4677 \text{ in-lb})(3.44 \text{ in})}{46.37 \text{ in}^4} = 346 \text{ lb/in}^2$$

As typical of most any roller conforming to deflection standards, the shell bending stresses are quite small. However, a **safety factor** of both 5 on fatigue or 10 on yield should be used unless there is considerable prior testing and experience which may permit slightly smaller values.

Stresses in regions 2-5 are quite complex because they include stress concentrations and assembly stresses and will often require FEM modeling. However, Tonglin *et. al.* has closed form solutions for some head styles which are press fit.[4] Journal stresses in region 6 will be covered in the next chapter.

Composites

Composite rollers are the most common material alternative to the traditional irons, steels and aluminums. Even so, however, they number at best only a couple of percent of all applications. This is in part due to their **cost** which may be 1.5-2X as much as their metal counterparts. Also, their recent introduction less than a decade ago means that much of the existing machinery was built before composites were widely available. However, the biggest reason may be that justifications for this more expensive alternative have not been clearly presented.

Composite rollers are perhaps a bit of a misnomer because only the shells are composite while the heads, journals and bearings are usually steel. While the word composite can have a wide variety of meanings, in rollers it is usually taken to mean a carbon fiber and epoxy shell as opposed to fiberglass, for example, which is also a composite. The carbon fiber may be braided or may be filament wound around an inner mandrel and embedded in an epoxy or other matrix which is later cured.

Composites were not developed by web machine or component companies. Rather, the technology was borrowed from aerospace and defense and applied to rollers and shafts. Indeed, even today most of the composite shells start as tubing purchased from a supplier by a roller manufacturer.

The largest physical difference between composites and metal rollers is their **weight**.[5] However, one can never make a useful comparison using the same geometries. Composite wall thickness may be a little thicker than metal alternatives to avoid oil canning of the shell due to machining or application loads. As a very rough guideline, metals may be 1.5-2X as heavy as composites for the same deflection standards. Thus, the composite should always be considered for winding shafts which are handled manually.[6,7,8,9,10,11] However, most rollers are only rarely changed out so that light weight does not confer any significant advantage except when mounted on load cells or other specialized applications.

Also, while composites have perhaps a 4:1 strength to weight ratio over steel and aluminum, this is also not a significant advantage. This, as we saw in the previous example, is because almost all rollers are **deflection** or **critical speed** limited rather than strength limited.

What may be important, but only for some applications, is the stiffness to weight ratio. Here, composites are about twice as structurally efficient as both aluminum and steel. The applications where this is important is where the weight of the roller shell is large compared with external loads such as tension and nip, such as on many idler rollers running a light web. If tensions or nips are large, the lighter composite roller will not be a significant advantage.

Composites have advantages over metals in some niche applications. First, composites are more resistant to attack by chemicals in the environment, such as acids. Second, composites have a higher damping factor and may reduce the amplitudes of some types of roller **vibration**, but not at resonance where there is no flexure of the roller. Composites have a very low coefficient of **thermal expansion**. However, I'm not sure how this would offer a practical advantage since a floating bearing is always required anyway because of machining tolerances and potential frame movement.

Composites have a **fatigue** endurance limit, as does steel, that is not enjoyed by aluminum. Additionally, composites lend themselves to many methods of fatigue damage detection such as dye penetrants and ultrasonics.[12] However, there are two easier methods. First, fatigue damage will often show as a light hazy spot on a roller as fibers are broken or are separated from the matrix. Second, fatigue damage can show as a reduced stiffness (deflection under load).

Composites also have disadvantages beyond their higher cost. First, they have little dent or abrasion resistance when compared with metal. Second, they've had a reputation of occasionally failing at the bonds to the heads at the end of the rollers. Third, they can be more difficult to balance because it is difficult to manufacture them as homogeneously and as geometrically precise as metal.

Case for Composites

The previous section did little to clarify when composites should be used except for in a few niche applications such as manually handled core shafts, some load cell rollers, and in the presence of some types of corrosion.[13,14] The more common argument is that composites, by virtue of their lower **inertia**, will reduce **tension fluctuations**. While the tendency is certainly true for some types of tension excursions, the argument is subjective at best and incomplete at worst. Here, we will show the application range for the most common use of composite rollers, that is the idler roller.

To understand how composites might reduce tension deviations, we can look at a schematic of a drive section as given in Figure 6.5. The only point where tension is truly controlled is at the load cell. Elsewhere, the web tension stairsteps upward over every undriven roller as it progresses down through the machine section. The size of the stairstep is composed of a drag and inertial part as discussed in Chapter 4. During run, the composite roller does not provide a significant advantage except as a possible small reduction in **bearing drag** due to a reduced shell weight. However, during an acceleration a composite will have a reduced downweb tension rise due to a lower inertia. As a general guideline, I suggested that the total tension rise in drive section be less than 10% (of setpoint tension) during run and that the maximum tension difference between the accel and decel curves be less than 25%.

Now we can map out the application range for composite rollers as in Figure 6.6. On the x axis is roller length, which by a deflection standard largely determines diameter and thus inertia. On the y axis is web strength. We must allow only a certain fraction of the web's strength as a tension excursion. Also, the map is constructed on the basis of the # of rollers in a section and the acceleration rate of the machine. In general, composites are not indicated unless there is frequent speed changes at high accelerations rates (such as on a roll-to-roll rewinder). Continuous machines can afford to run slow accel/decel rates to reduce the demands on drive and rollers.

Figure 6.5
Tension Variations in a Drive Section

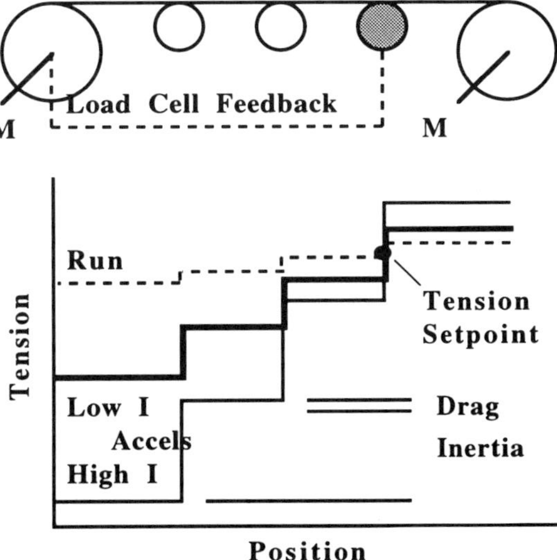

Figure 6.6
Application Map for Composite Rollers

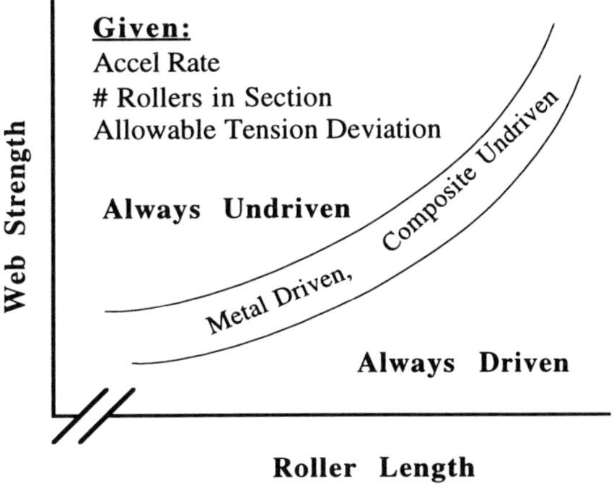

Transport rollers on the lower right are always driven because the high inertia would slack or break the web on speed changes. Idler rollers on the upper left never need to be driven because the web is strong and the inertia tiny. Composites lie on the boundary of never and always driven. That is, where their structural efficiency is just enough to bring tension control within a standard, but where a metal roll would require a drive. Thus, a composite roller is not a low inertia alternative to metal. Rather, it is a (always) cheaper alternative to drives when the application is calculated to lie in that region.

Bibliography

1. Luthi, Oscar. *Ring Stresses and Deflections of Thin-Walled Rolls and Application.* Tappi J., vol 56, no 8, pp 121-124, August 1973.

2. Rothenbacher P. and Vomhoff E. *Mass Centering of Cast-Iron Rolls.* TAPPI Finishing and Converting Conf. Proc., pp 31-34, October 1984.

3. Fernside, Richard L. *Granite Roll Analysis.* Canadian Pulp and Paper Assoc. Annual Mtg. Proc., January 1990.

4. Tonglin, Shu et al. *Assembly Stresses and the Strength of Hollow Iron Rolls.* Tappi J., vol 78, no 3, pp 185-190, March 1995.

5. D'Amour, Michael R. *Composite Materials Lighten the Load.* Converting Magazine, pp 65, 68-69, June 1993.

6. Anon. *Carbon Fiber Core Shafts.* Converting, November 1990.

7. Anon. Carbon Fiber Solves Critical Shaft Handling Problems. Nonwovens Industry, September 1991.

8. Cox, Jacqueline. *Composite Shafts Give a Break to the Papermaker's Back.* American Papermaker, October 1991.

9. Cox, Jacqueline. *Carbon Fiber Shafts.* American Papermaker, pp 25, October 1994.

10. Fortin, M.B. *Carbon Fiber Composite Tubes Provide Solutions to Heavy Shaft Handling Problems.* TAPPI Finishing and Converting Conf. Proc., Houston, pp 245-257, October 6-10, 1991.

11. Zuck, Robert A. *Using Composites in Shafts Turning into Weighty Issue.* Paper, Film & Foil Converter, vol 67, no 1, pp 60-61, January 1993.

12. Allevato, C. and Williams, J.D. *Acoustic Emission Evaluation of Yankee Dryer Shell Materials.* Tappi J., vol 75, no 7, July 1992.

13. Sonada, J. *The Merit of Introducing Composite Roll.* Japan Paper Tech., October 1989.

14. Zuck, Robert A. *Composites Gaining Ground in Converting Applications.* Paper, Film & Foil Converter, vol 68, no 1, pp 41-42, January 1994.

Chapter 7

Journals and Shafts

In this chapter we primarily focus on the loads imposed on journals, the resulting stresses and the required strengths. Strengths are described both statically for overloads and dynamically for fatigue.

Journals and Shafts Must Be Short

Live shaft journals and dead through shafts must pass loads and stresses from the framework support to the head or shell. Additionally, a driven live shaft journal also has to transmit braking or motoring torques between the drive and the head or shell.

The shear diagram for a cantilevered roller is given in Figure 7.1. Here, the overhang, *a*, should be reduced as much as possible to reduce stresses and deflection. Conversely increasing b will be useful to decrease framework bending, vibration and improve the ease of realignment. Indeed, if *a* is not at least 18" (0.5 m), the roller can not be precision aligned.

Figure 7.1
Shear Diagram for a Cantilevered Roller

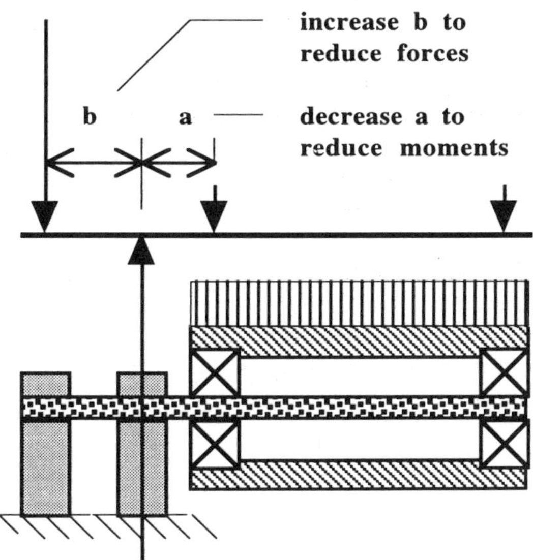

The shear diagram for the dead shaft and live shaft rollers are given in Figures 7.2 and 7.3 respectively. It is a very important design practice to shorten the distance, *a*, as much as possible between support and bearing or bearing and face. Shortening journals and shafts will reduce bending moments and vibration on both roller styles as well as deflection across the deckle for the live shaft roller. A designer who does not make all efforts to shorten journals can be considered quite unskilled.

Figure 7.2
Shear Diagram for a Dead Shaft Roller

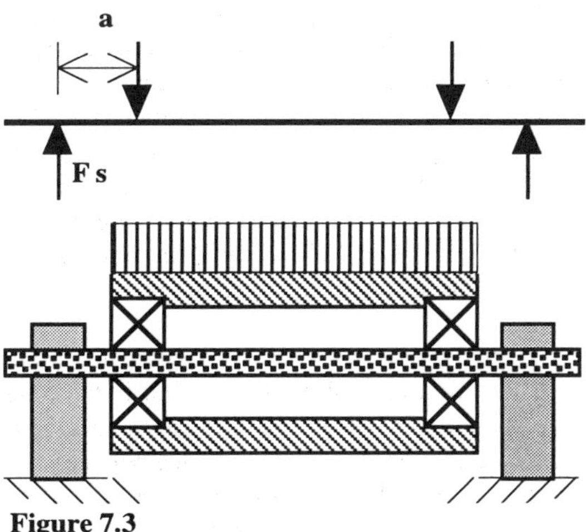

Figure 7.3
Shear Diagram for a Live Shaft Roller

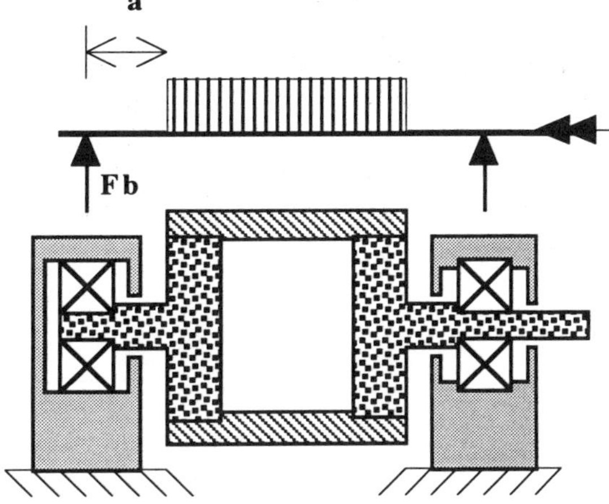

Bending

Bending stresses at a journal or shaft are the largest cause of structural failure of rollers. Obviously, this situation is to be avoided by proper design and operation. A structural failure will cause an unexpected and often long downtime for a web machine. More importantly, however, a roller structural failure is a safety hazard. Even if the roller does not fail catastrophically, a bent journal will put a live shaft roller out of balance and out of alignment as well as increase runout. A bent through shaft will cause dead shaft roller misalignment.

In order to calculate bending stresses, we must first determine the bending moment at all critical positions along the journal or shaft. The support load that creates the bending moment is the resultant (vector sum) of roller weight, web tension and nip as described in Chapter 3. The bending moment is this support load times the distance from the support to a critical position. The critical positions on a shaft or journal are at any change in geometry (often diameter). The bending moment diagram for a simple journal is given in Figure 7.4. However, while through shafts will have a simpler geometry, some journals can have an order of magnitude greater complexity. Nonetheless, the principles given here are the same for any application.

There are several important parameters at critical positions in a journal or shaft. First is the load reaction, which we are seldom able to adjust significantly. Second is the distance between the load reaction and a key position which should be kept as short as possible. Third is the diameter which needs to be increased sufficiently to reduce stresses, and will affect the minimum ID of the bearing that can be used. Finally, there is the diameter ratio and the radius of the corner which will determine the stress concentration factor as shown on the next page. While radii should be as generous as possible, there is a maximum radius that will clear the corner of a bearing.

Figure 7.4
Bending Moment Diagram for a Journal

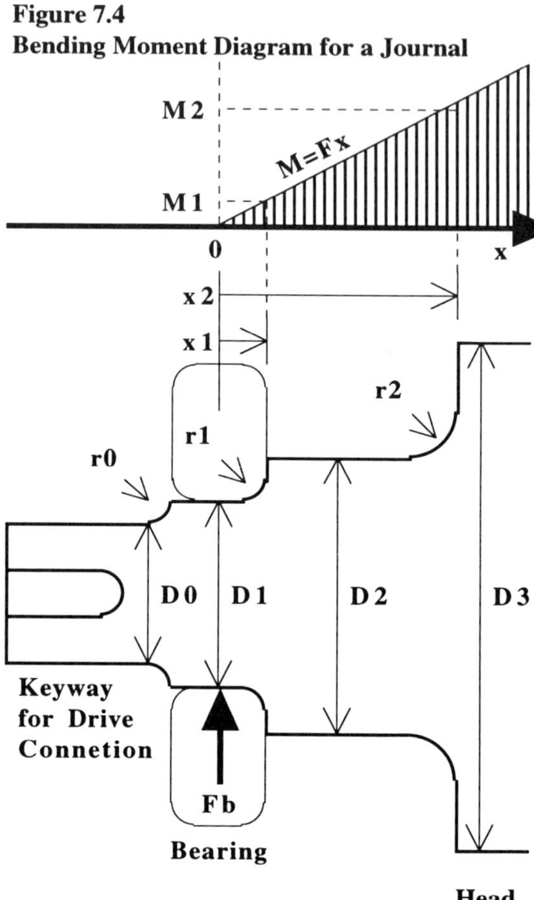

There are several factors that are not usually considered in journal stress analysis. First, we anticipate no axial load for unnipped applications as the roller should be equipped with a floating bearing to prevent constrained thermal expansion. (Thrust load can reach nip x width x friction for nipped rollers.) Second, while a bearing may be pressfit onto a journal, these stresses are usually small and compressive and thus often ignored. However, torque transmissibility and tension of female halves of a pressfit should be calculated.

We will have to consider torques imposed by driven live shaft designs, but will be covered in a later section. Locations outboard of the support need only consider torques and shear stress while those inboard are a combined stress problem.

Bending Stress Example Problem

Find the maximum bending stress at just one of the critical locations, at r2 of the previous figure. The 100 lb per side roller resultant reaction is carried by a journal that steps from a D2 of 1" to a D3 of 3", with a 0.125" radius. The distance between the bearing centerline and that critical location is x2 = 4". We assume no bending loads imposed by the coupling to the drive.

First, we calculate the bending moment at that location.

(7.1)
$$M = F x$$
$$M_2 = F_b x_2 = 100 \text{ lb x } 4 \text{ in} = 400 \text{ in-lb}$$

Next we calculate the area moment of inertia at the smaller of the diameters at the step.

(7.2)
$$I = \frac{\pi}{64}\left(D_o^4 - D_i^4\right)$$
$$I_2 = \frac{\pi}{64}\, 1^4 = 0.0491 \text{ in}^4$$

Since there is a geometry change, we need to calculate the stress concentration factor for bending. The required parameters are

$$\frac{r_2}{D_2} = \frac{0.125 \text{ in}}{1 \text{ in}} = 0.125$$

$$\frac{D_3}{D_2} = \frac{3 \text{ in}}{1 \text{ in}} = 3$$

Now from any good strength of material handbook[1] or machine design text[2] we can enter a graph for bending stress concentration factors to pick off Kb. For our problem, Kb turns out to be 1.75, meaning that the changes in geometry caused an additional 75% increase in stress. Stress concentration factors increase dramatically when the radius ratio, r/D, gets less than 0.1, and thus should be avoided. Stress concentration factors increase mildly with the diameter ratio Dn+1/Dn

Now, we can calculate the bending stress at the critical position #2 where *c* is D/2 as:

(7.3)

$$\sigma_b = K_b \frac{Mc}{I}$$

$$\sigma_{b2} = 1.75 \frac{(400 \text{ in-lb})(0.5 \text{ in})}{0.0491 \text{ in}^4} = 7{,}128 \text{ lb/in}^2$$

At this point, the reader may be tempted to compare this stress with tabulations for the material's strength. Since most journals are made of high grade steels, the calculated bending stress is but a fraction of the yield or tensile strength. However, we must resist that temptation for the moment.

First, we must make sure that we calculate the stresses at all locations, not just one. Second, we have not considered the combined stress problem if any torsion was present. Third, we have not considered fatigue, which will be covered in a later section. Fourth, we have not considered safety factors.

The safety factors for journals must be quite conservative. I suggest that novices or new applications use the more conservative of a safety factor of 10 on yield stress or a safety factor of 5 on fatigue strength or endurance limit. With more field experience, the safety factors may be reduced a little. However, there is no real strong impetus to do so in most cases as the journal costs are a tiny fraction of the cost to build a roller and seldom does the journal size cause us a significant bearing selection problem.

Many larger engineering departments will automate the calculation of journal or shaft stresses via spreadsheet or other computer program. Then, the designer merely needs to enter the force and torque reaction along with as many triplets of diameter, radii and distance as need be checked.

Torsion

While torsion induced shear stresses are not often high on most driven roller journals, they do add insult to injury and must be considered. The procedure is much the same as for bending. An example considering the same section as the previous will be used. However, determining torque loads is a little more difficult. It often means finding the performance charts of the electric motor, pneumatic brake or other drive element. The torque that should be used for design purposes is the maximum torque the device will generate, even if a lesser value is anticipated in usage, unless there is a slip-clutch or other weaker link in the system. For our example problem, we will use a maximum torque of 500 in-lb.

First we must determine the torsional stress concentration factor in much the same manner as we did with bending in the previous section. Then, the calculation for shear stress of a stepped round solid shaft under torque loading is:

(7.4)

$$\tau = K_T \frac{16\,\text{Torq}}{\pi D^3}$$

$$\tau_2 = 1.5\,\frac{(16)(500\ \text{in-lb})}{\pi\,(1\ \text{in})^3} = 3819\ \text{lb/in}^2$$

Especially here, we must again avoid the temptation to rush and compare shear stresses with the yield strength for the material. First, we need to check all critical positions in the journal. Stresses at keyway or couplings will not be covered in this book so that readers should turn to their favorite machine design texts. Second, the most relevant strength for a pure torsion problem is the shear yield strength, which is not as widely tabulated as other values. Third, shear stresses from torsion and tensile stresses from bending can not be simply superposed or added together. It is a combined stress problem.

Combined Stresses

The tensile stresses from bending and the shear stresses from torsion must be combined via Mohr's circle. Figure 7.5 shows the stresses for our sample problem. The circle for pure bending has its left edge at the origin because there are no shear or hoop stresses. The circle for pure torsion is centered at the origin. The maximum shear stress for the combined loading is given by:

(7.5)

$$\tau_{max} = \sqrt{\left(\frac{\sigma_{bending} - \sigma_{hoop}}{2}\right)^2 + \tau_{torsion}^2}$$

A failure criteria for <u>ductile</u> materials is the Maximum Shear Stress Criterion. This states that the stress as calculated by (7.5) should be less than the material's shear strength (or tensile strength divided by 2) divided by a safety factor. Alternatively, the Maximum Distortion Energy Criterion is also used for ductile materials. <u>Brittle</u> material failure is compared against the Maximum Normal Stress Criterion or Mohr's Criterion. <u>Fatigue</u> failure is usually compared against the Soderberg Criterion.

If the reader is not very comfortable with these strength of materials topics, he/she should probably not attempt journal design or stress analysis.

Figure 7.5
Mohr's Circle for Combined Stresses

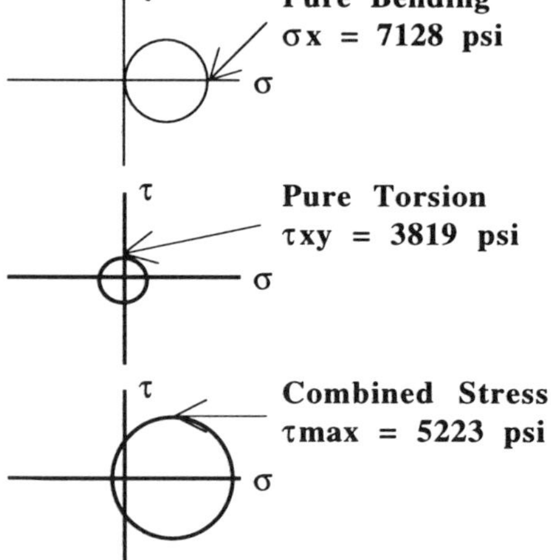

Fatigue

Bending stresses, due to roller weight for example, are compressive at the top of the journal or shell and tensile at the bottom. Thus, the member sees a reversing axial stress once per revolution. Alternating stresses such as this can weaken materials after numerous cycles. This can cause the member to fail at stresses well below the yield strength.

The number of cycles accumulated by a rotating element can be truly astronomical. For example, a slow speed and low duty converting roller may see at least a 150 million cycles in its lifetime.

$$\text{Cycles} = \left(100\frac{\text{ft}}{\text{min}}\right)\left(\frac{12\text{in}}{\text{ft}}\right)\left(\frac{\text{rev}}{\pi 6\text{in}}\right)\left(\frac{60\text{min}}{\text{hr}}\right)$$

$$\text{x}\left(\frac{8\text{hr}}{\text{day}}\right)\left(\frac{200\text{day}}{\text{year}}\right)25\text{year} = 1.5\text{x}10^8 \text{cycles}$$

However, a wind/unwind shaft for a typical mill roll may accumulate 10 billion cycles!

$$\text{Cycles} = \left(\frac{44-4}{2}\text{in}\right)\left(\frac{\text{rev}}{0.003 \text{ in caliper}}\right)\left(\frac{4\text{sets}}{\text{hour}}\right)$$

$$\text{x}\left(\frac{24\text{hr}}{\text{day}}\right)\left(\frac{360\text{day}}{\text{year}}\right)40\text{year} = 9.2\text{x}10^9 \text{cycles}$$

We must calculate the number of cycles because the *fatigue strength* of a material depends on this number. An example of a fatigue strength plot for a structural steel and aluminum is shown in Figure 7.6 which illustrates a few key points.[2]

First notice the large reduction in strength as a function of load cycles. Second, notice that the curve for steel has a knee, beyond which there is little or no strength reduction. This is called the fatigue endurance limit. Steels and composites generally have an endurance limit and are thus safer to use, while aluminums do not and are thus risky. Third, precious little high cycle fatigue strength information is available (>10^6 cycles), even for commonest materials. This makes it even more imperative to use materials which have a known endurance limit in any converting application that sees reversing (tensile) stresses that are but a tiny fraction of the material's strength.

Figure 7.6
Fatigue Strength Example

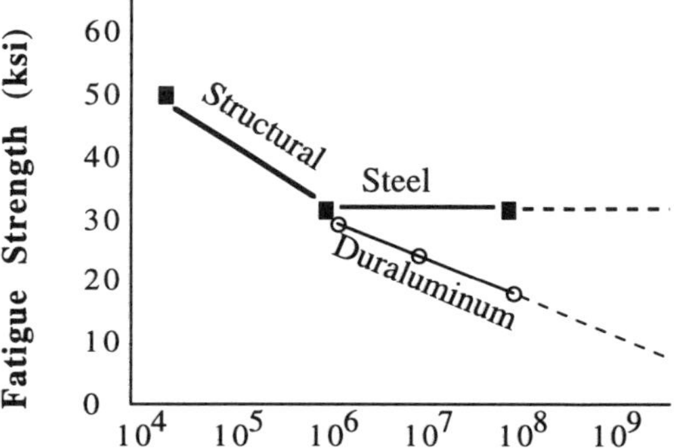

This is by no means the end of the complexity of designing for fatigue. If there is a mean stress, the problem may be treated using Gerber's Parabola, the modified Goodman Line or the Soderberg Line. Fatigue damage from the occasional overload can be severe and must be accounted for by the Palmgren-Miner law. Finally, a reliability approach to material variations can make use of the Deviation Multiplication Factor. Further information on these subjects can be found in strength of materials[3] or fatigue[4,5] books.

There are several factors that can exacerbate the risk of fatigue failure. These include the already mentioned tensile mean stress and the occasional overload. Also, even the most minor corrosion will greatly reduce fatigue strength. Finally, the fatigue strength of brittle materials can be devastated if there are sharp scratches, machining marks or other flaws.

A fatigue failure is not always catastrophic, but tends to completion over the course of minutes to months. Thus, inspection is one method to reduce the risk of failure. First, any increase in imbalance or TIR runout should be investigated because it may have resulted from a bent journal. Second, dyne penetrants can be used on most materials and ferrous metals can be Magnafluxed. Third, composites can be visually checked for whitish spots or reduced stiffness (deflection under load).

Fits

The tightest machining tolerances of a roller will likely be at the through shaft and especially the journal. The portion of the journal or shaft which carries the ID of the bearing is especially critical. At the very least the bearing manufacturer will specify a minimum and maximum shaft OD tolerance, minimum width and maximum radius at a shoulder.

If the OD is too small, the bearing may be loose or eccentric. Some bearing mounting arrangements have a mild pressfit on the inner race to keep it from spinning on the shaft due to bearing drag torques and other loads. If the OD is too large, the bearing will carry an increased radial load and thus life will be shortened. The diametral tolerances of small bearings are typically a few ten thousandths of an inch.

The quality of a pressfit of the inner race on a journal or shaft can be checked by blueing the shaft. After removal, at least 50% of the blueing should have been scraped away. If the shoulder radius is too large, the inner race will bear on a point instead of distributing the load on its side and also will not be positioned appropriately.

Sometimes the journal or shaft must carry drive torques or other significant loads through a pressfit. The dimensional tolerances of force or shrink fits are based on the class (FN1 light fit to FN5 force fit) and nominal diameter. There are four design requirements that need to be checked against appropriate safety factors. First, the compressive radial stresses in the shaft can't be excessive. Second, the tensile hoop stresses in the hub can't be excessive. Third, the fit must be able to carry the applied torque. Fourth, the shrink temperature and/or pressfit force for assembly must be calculated. An interference of 0.001" per inch of diameter is quite heavy. Details of these design requirements can be found in machine design books or handbooks.[6,7]

Bibliography

1. Roark, Raymond J. and Young, Warren C. *Formulas For Stress and Strain.* McGraw-Hill Book Co., New York, 5th Edition, 1975.

2. Deutschman, Aaron D. and Michels, Walter J. and Wilson, Charles E. *Machine Design, Theory and Practice..* Macmillan Publishing Co., New York, 1975.

3. Boresi, Sidebottom, Seely & Smith. *Advanced Mechanics of Materials.* John Wiley & Sons, New York, 1978.

4. Sandor, Bela. *Cyclic Stress and Strain.* University of Wisconsin Press, Madison, WI, 1972.

5. Society of Automotive Engineers. *Fatigue Design Handbook.* AE-10, 1988.

6. *Mark's Standard Handbook for Mechanical Engineers.* McGraw-Hill Book Co., NY, 8th Edition, 1978.

7. *Machinery's Handbook.* Industrial Press Inc., New York, 21st Edition, 1979.

Chapter 8

Bearings

In this chapter we discuss the common bearings styles used for rollers, their mountings and lubrication. Also covered are sizing considerations for load capacity and life as well as maintenance issues.

Introduction

There is a wealth of design and application information on bearings. The catalogues and other literature provided by bearing manufacturers are an excellent resource because they are quite detailed and complete. Also, bearings enjoy a standardization of sizes, specifications and terminology that is quite unique compared to other converting components. This is largely due to the long and consistent efforts of the AFBMA (Anti-Friction Bearing Manufacturers Association). Thus with this wealth of information, we need only capture a few of the highlights in this chapter.

There are several important design considerations for the bearings of rollers. Perhaps the most important is that the bearings must support radial loads due to roller weight, tension, nips and roller imbalance. The radial load is specified for both static (stationary) and dynamic (rotating) conditions. The static load rating minimizes the brinelling (denting) of bearing parts when a roller is stopped or stored for long periods of time. The dynamic load rating is an important factor used to calculate bearing life and reliability.

However, bearings must also support some axial thrust. Thrust is usually quite small for unnipped rollers, but can be as high as μN for nipped applications. The radial and axial loads are combined into a single equivalent load using factors. Finally, the bearings must not rotate faster than their RPM limits. This is primarily to avoid causing a breakdown in lubrication due to the heat generated. However, it may also be to keep the balls/rollers tracking properly despite centrifugal effects.

These and other factors determine the life or reliability of a bearing. Life is a very important design consideration for bearings because most web equipment is operated continuously for decades. Though worn bearings can certainly be replaced, the life on any particular bearing must be quite high because web machines may have hundreds or even thousands of bearings. Moreover, bearings may fail without notice or warning and cause a machine to shut down unexpectedly. A seized bearing can also damaged connected elements such as journals, couplings and drive trains.

While there are ways to estimate bearing condition without removal and inspection, they are not practiced as often as they perhaps could be. These methods include FFT vibration analysis, bearing temperature, bearing noise, bearing play and bearing or contaminant particulate in the lubrication. If bearings in a certain section of a machine are causing undue problems, the design or maintenance causes must be tracked down.

From a web handling point of view, bearings must perform two vital functions. First, the bearings must hold the axis of the shell in precise geometry without excessive play or runout. Radial clearance or play compromises alignment and can allow excessive vibration. Axial clearance can cause web edge position shifting. The concentricity or runout of the bearings must be controlled so that a bearing can be replaced without requiring the shell to be remachined (in the bearings) and/or be rebalanced.

Second, the bearing assembly friction can't be excessive for the application. As discussed in Chapter 4, bearing drag is an element in the uncontrolled tension rise as a web progresses downstream through a machine. Low bearing drag is especially important for light webs on narrower machinery. Bearing drag is a function of bearing style, shields and seals, and lubrication.

Bearing Types

A commonly used bearing style for small transport rollers is the **ball bearing**. As shown in Figure 8.1, the bearing consists of an inner race, an outer race and a number of balls. The radius of the balls are slightly smaller than the races so that loads are supported on a few point contacts. Not shown are a separator to keep the balls evenly spaced, as well as shields and seals which contain the lubricant and exclude contaminants. While some cantilever roller bearings are duplexed (adjacent and paired) for rigidity, most rollers have two bearings that are widely spaced.

The **spherical roller bearing** is commonly used for process rollers and most large rollers.[1,2] The spherical roller bearing has a much higher load capacity than ball bearings because of the line as opposed to point contact. Another advantage is that it can accommodate some angular misalignment between the shaft and bearing mount.

Another high load capacity bearing style is the hydrodynamic journal bearing. A journal bearing is a shaft which floats on a cushion of oil above pads fixed to the housing, which is developed merely by the relative motion of the shaft. Hydrodynamic bearings are principally used for very large rollers in older machinery. These are also sometimes referred to as babbitt bearings after the material which is often used for the pads. Indeed, some of the oldest machinery still uses wooden pads soaked with oil, grease or wax.

Other bearings that have been used occasionally for rollers are given in Table 8.1. Air or gas bearings are used for pilot plant equipment and instrumentation where frictionless rotation is needed. Bushings (such as oil impregnated bronze) could be used on very low speed and low duty applications.

As the reader may have gathered, there are so many types and variations of bearings that a detailed discussed could not be justified here. Thus, they should refer to their favorite bearing manufacturer catalogues, books,[3,4] handbooks,[5] or machine design textbooks[6] for further information.

Figure 8.1
SS Roller Mounted on Ball Bearings

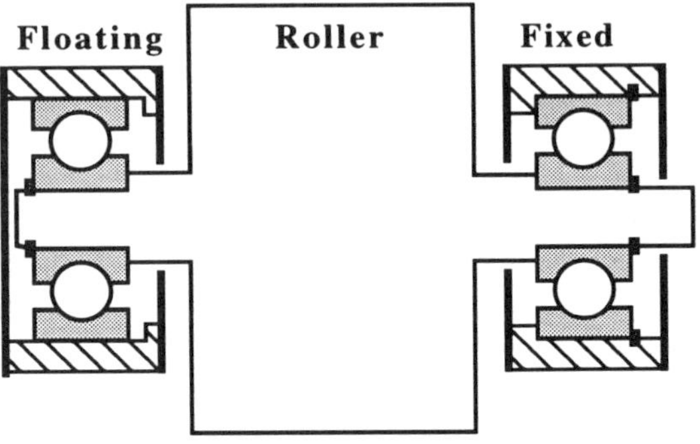

Also note in the figure how the drive side (right) has both the inner and outer races fixed to hold the roller's axial position. However, the tending side (left) allows the outer races to float in the housing. The purpose of the **floating bearing** is to keep the bearing set from being axially loaded due to thermal expansion mismatch between the roller and the frame and accommodate machining tolerances. The required axial float is:

$$y = \alpha \Delta T L + \text{Mach'g Tol.} \qquad (8.1)$$

where α is the coefficient of thermal expansion of the shell, ΔT is the maximum temperature range and L is the bearing to bearing distance. Also note that the floating bearing must be properly positioned within its housing during assembly so that the expansion and contraction is accommodated.

Table 8.1
Bearing Types Used in Rollers

Air Bearings
Ball Bearings - Angular Contact
Ball Bearings - Radial Contact
Ball Bearings - Self Aligning
Bushings
Journal Bearings - Hydrodynamic
Journal Bearings - Hydrostatic
Roller Bearings - Cylindrical
Roller Bearings - Needle
Roller Bearings - Spherical
Roller Bearings - Tapered
Thrust Bearings

Sizes and Options

Most ball and roller bearing sizes are standardized so that bearings from different manufacturers are interchangeable.[7] The size is referred to as a two digit **dimension series** where the first digits represents the width series and the second digit representing the diameter series. While sizing a bearing may seem intimidating at first because of the bewildering variety of bearings, there are a few guiding principles that can help narrow the options.

The first step in **sizing** a bearing is to stress out the journal or shaft as given in the last chapter which then determines the minimum ID of the bearing. Next, the radial and axial loads application loads are calculated. From these loads, the minimum bearing series (and/or increased shaft diameter) can be determined from life calculations given later. The relative axial to radial load capacity is primarily determined by the bearing style. While undersizing a bearing can be disastrous, oversizing is also undesirable. Large bearings are more costly, have more friction, may be difficult to package and may not achieve their anticipated life if they are very lightly loaded.

There are many ways of **mounting** bearings on a shaft or in a housing. The most common means include press or shrink fits, snap rings and fine threaded lock nuts. It is imperative that neither race rotate or slip axially on their mounts. An exception is the required axial travel of the floating bearing on one of its races. If angular misalignment is higher than the bearing's capacity, such as on some winder shafts, a self aligning bearing or mount may be required.

The bearing **housing** must then be secured to a framework. While there are a number of methods, the pillow block and flange mountings shown in Figure 8.2 are quite common. The reason that they are popular mountings is that they are commercially available and are thus less expensive that custom designed bearing housings. Regardless of the mounting method, however, there are two requisites. First, the mounting must be quite secure and rigid (machined mating surfaces). Second, the housing must be moveable in the vertical and MD directions for realignment.

Figure 8.2
Bearing Housing Mounting Examples

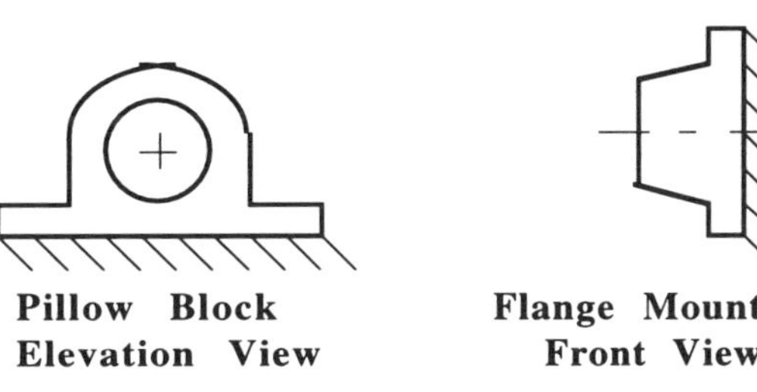

Pillow Block Elevation View **Flange Mount Front View**

The **grade** of a ball bearing is designated from a ABEC 1 (widest tolerances) to a ABEC 9 (super precision). Grades for other bearing styles will have similar nomenclature. Bearings used for rollers are more commonly a mid grade rather than a low grade. Bearing components are available in different steels or materials for unusual applications such as high temperatures (above 350 °F) or corrosive environments.

Seals are required in most applications to keep lubricant in and to keep contaminants out of a bearing. These seals may be either located in the bearing or in the bearing housing. The choice of location and type of the seal depends on a number of factors including the type of lubricant, nature of the contaminants, speeds, temperatures and so on.

There are basically three categories of seals: contact, labyrinth or clearance, and combinations of contact and labyrinth. Contact seals vary from simple felt stripping to spring loaded lip assemblies. Contact seals may be appropriate where submerged, where the housing is pressurized or in difficult environments. However, contact seals require extremely smooth shaft finishes and are speed limited due to heating of the seal by friction. Indeed, contact seals are responsible for more friction than the rolling bearing elements and sometimes even the pumping of viscous lubricants such as grease. Thus, contact seals are not appropriate for many light web grades. Labyrinth seals do not contact the shaft and thus are almost friction free. However, labyrinth seals should not be used with pressurized housing, submerged seals or in some environments.

Contact your bearing representative for more information on options and applications.

Lubrication

Nearly all bearings require some sort of lubrication to extend the life of the elements by coating their surfaces with a thin protective boundary film. Exceptions to this include air bearings, some specialized low friction instrument rolling bearings and some very low speed/duty journals in bushings. While bushings and bearings can be obtained which are pre-impregnated with a lubrication, they are not suitable for most web roller applications. This is because the lubricant will be exhausted or degraded far quicker than the useful life of the machine, which would entail replacing the bearings instead of merely relubricating them.

The choice of lubricant depends on many factors including economics, bearing life, friction, temperature and access. The common lubrication methods are given in Table 8.2. As seen here, oil is a higher performance lubricant than grease, but comes at a higher installed cost.

At the low end of performance, a regreasable bearing or housing provides an inexpensive means of lubricating applications at low speeds and temperatures. However, **grease** may not be suitable for idler rollers for light web grades because of the enormous drag. The tendency of most maintenance people is to overgrease (more than 1/3 to 1/2 of housing capacity) the bearings that are easily accessible. The results are an even higher roller drag, potential seal damage and a housekeeping mess.

The regrease intervals are specified by the bearing or machine manufacturer based primarily on bearing speed and to a lesser extent bearing bore. However, grease also degrades merely with time or can become contaminated and need be completely flushed out and replaced. The type of grease (base, hardness and additives) to be used will be specified by the bearing manufacturer or machine builder.

At higher speeds, **oil** is required because it can extend the bearing's life, has a lower drag, and can carry heat and contaminants away from the bearing. However, the initial cost of oil lubrication is higher than grease if for no other reason than the complexity of the bearing housing increases.

Table 8.2
Lubrication Methods

Listed in order of increasing cost, bearing life/reliability, speed capability and decreased friction.

Grease - Prepacked
Grease - Regreasable
Oil - Bath or Sump
Oil - Continuous or Drop Feed
Oil - Jet or Air/Oil Mist

The simplest oil lubrication system is a bath or sump. Here, the bearing sits in oil nominally up to the midpoint of the lowest ball or roller. Raising the oil level above this point can cause a churning of oil (high drag, frothing) as well as causing the oil to pour out the shaft seals. The level of oil is determined by a level plug or better yet a sight glass. The seals for oil systems are on the housing which holds the bath of oil rather than the bearing which does not have enough oil storage capacity. The increased size of the bearing housing of oil versus grease systems allows a greater surface area to get rid of heat generated by the bearing and seals.

As speeds or duties increase, a continuous or drop feed oiling system may be desirable. Here, oil is metered into the top of the bearing and is drained away at the bottom of the housing. The continuous lube system will often have a means of calibrating flow or a sightglass to make sure that an appropriate amount of oil is flowing. At very high speeds or duties, oil may be injected in a small jet or blown in an oil/air mist onto the bearing and again drained away at the bottom of the housing. Continuous or jet lube systems have the greatest capacity to remove heat from a bearing by virtue of conduction through the mounting, convection off the housing and thermal transport with the outgoing oil.

The type of oil (base, viscosity, etc.) will be based in part on lubrication method, temperature, duty and speed. Consult the bearing manufacturer or machine builder for recommended lubrication and lubrication intervals.

Load & Life

The design life of many web machine components in general and bearings in particular should be between 10 and 40 years. The actual value used is a function of the complexity and desired reliability rather than the anticipated useful life of the machine. Complex machines which must run continuously and reliably (such as in the paper industry) need long component design lives. However, simple converting machines which run only a single shift may, for economic reasons, get by with a slightly shorter lives.

The term life in this context has a very specific meaning. That is, the length of service in which some fraction of identical components will fail. The nomenclature commonly used with bearing is L10 life, where the number indicates the failure rate in percent. If for example, 2 of 20 identical bearings (10%) might be expected to fail after 200,000 hours of service, the L10 of those bearings in that application will be about 22 years or its equivalent in revolutions (function of speed and roller diameter).

However, before we can calculate bearing life we must first calculate the applied radial load (from Chapters 2 & 3) and the potential axial load (0 for unnipped, μN for a nipped roller fixed bearing). We check the maximum radial load in service, in storage or in handling (impact factor) against the **static basic load rating** tabulated for the bearing. If this value is exceeded, the load carrying ball will permanently indent the race to more than 0.0001 x ball diameter, which will greatly reduce the bearing's life. The most common situation that can generate this damage is when dropping or bumping a bearing housing into a stiff surface (frame or floor) during handling.

The radial load should add in imbalance effects for all but slow speed rollers as given in Chapter 11. Similarly, the dynamic effects of bouncing nips may need to be included. Finally, variable loads (such as with widely varying tensions and grades) can use the highest values, which is very conservative, or use a time weighted load which will not be described here.

Next, we must calculate the equivalent load based on radial and thrust loads for many ball and roller bearings as:

(8.2a) $$P = X\,V\,F_r + Y\,F_a$$

and

(8.2b) $$P = V\,F_r$$

where the higher of the two equivalent loads, P, is used and

F_r is the radial load
F_a is the axial load
X is a radial load factor from bearing tables
Y is the axial load factor from bearing tables
V is the rotation factor (1.0 for inner ring rotation, 1.2 for outer ring rotation)

Finally, we can calculate the bearing life as

(8.3) $$L'_n = a_1 a_2 a_3 \left(\frac{C}{P}\right)^b$$

where L'_n is the adjusted life in revolutions based on several factors found in tables or charts provided by the bearing manufacturer. First, the a_1 factor is used for bearing failure rates different than 10% (thus L_n instead of L_{10}). For example, a L_1 life is less than 1/4 of the L_{10} life. Second, the a_2 factor is used adjust life based on better quality bearing materials that could be used to extend life upwards as much as a factor of three. Finally, the a_3 factor will derate life based on difficult conditions such as shock, low lubricant viscosity or high bearing temperature. The ratio C/P is the **dynamic basic load rating** found in tables divided by the equivalent load calculated earlier. The exponent, b, is 3.0 for ball bearings and 10/3 for roller bearings. Thus, a doubling of the applied load will decrease bearing life almost an order of magnitude.

Once the life in revolutions is calculated, life in years can be calculated from roller radius, speed and duty. This example illustrates the most common bearing size procedure. However, always use the bearing manufacturers suggested methods as they may differ slightly.

Bearing Drag

As described in Chapter 4, bearing drag will cause web tension to rise through a drive section. Drag can become troublesome on light web machines with many idler rollers. Drag will be proportionally more significant on machines which run continuously because inertial tensions should have been reduced with slow acceleration rates. If the tension rise is too high through a section, such as more than 10% of the lightest tension, an intermediate driven roller may need to be added to bring the tension back down. This greatly adds to the expense and complexity of a machine. Obviously, the first step would be to reduce roller count to an absolute minimum. Beyond that, however, one may need to reduce drag by redesigning the bearings.

The coefficient of friction of bearings is seldom listed for bearings because it depends on so many factors such as the type of seal, lubrication, load, speed, temperature and so on. As a very rough approximation, however, the running coefficient of friction as calculated at the bearing bore diameter will be around 0.001-0.002 with starting frictions perhaps 50% higher. However, contact seals have much higher frictions than the bearings themselves and thus cause friction to increase up to 0.01.

The total drag (bearing friction plus windage) can be calculated by a coast down test as given in an example in Chapter 4. This requires only a stopwatch, speed readout and a known roller geometry. From a maintenance perspective, high drag bearings (which can indicate distress) are quite visible during a web break as those rollers are the first to come to a stop. They may also run at higher bearing temperatures. Idler roller bearing friction can be checked crudely with a sharp hand spin. Idlers in the weight range of 5-50 lb should typically coast for at least 10 seconds after the sharp hand spin. Minimum coast times for other rollers can be established by calculation or experience.

If total bearing friction (rolling elements, lubrication and seals) is of concern, then it would be wise to work with the bearing manufacturer to investigate alternatives.

Bearing Drag Example Problem

How many idler rollers can be tolerated in a drive section for a 10 PLI strong, 60" wide web? This is a precision application so that the web tension should not vary more than 10% due to drag (or 25% due to inertia and drag). The 100# idler rollers are 7" in diameter, 80" long, and have sealed grease bearings with a friction coefficient as measured at the 1" bore of 0.01.

We must first calculate the bearing load which is roller weight plus tension. Since no information was given, we will need to make an estimate. A 180 degree wrap at 1 PLI tension (10% of its strength) on a 60" wide web will load the bearings by 120#. However, while wraps on the sides or tops of the roll will increase the resultant load, tension will relieve weight on rolls wrapped from underneath. Thus, we will estimate a conservative 150# bearing load (100# weight, 50# tension).

The drag friction at the roller surface is then:

(8.4)

$$F_D = \mu N \left(\frac{D_{\text{brg ID}}}{D_{\text{roll OD}}} \right) = (0.01)(150\#)\left(\frac{1\text{in}}{7\text{in}} \right) = 0.214\#$$

The maximum drag force for 10% of a minimal 1PLI tension variation across a 60" wide web is:

$$(8.5) \qquad F_{max} = (0.10)\left(1\frac{\text{lb}}{\text{in}}\right)(60\text{in}) = 6\text{ lb}$$

Thus, the maximum number of idler rollers is:

$$(8.6) \qquad \#\text{Rollers}_{max} = \frac{6\text{ lb}}{0.214\text{ lb}} = 28$$

In practice, however, we will never want so many idlers in a section. First, it violates the first rule of web handling machine design which demands minimum roller count. Second, the textbook friction value is quite a bit optimistic when compared with the real world. Third, the inertial tensions may be so high as to preclude starting and stopping the machine safely.

Maintenance of Bearings

There are really only three tasks that a roller bearing must perform. First, it must hold a roller in a fixed x,y position, which means the roller must have no measurable radial play. This can be checked with a dial indicator and hand lifting small rollers or using a wooden prybar on intermediate sized rollers. Second, the bearing must hold a roller in a fixed z position, which means the roller must have no measurable axial play. Again, axial play in small rollers can be checked with a dial indicator and prybar. Third, the roller must rotate with minimal friction. The simplest check of friction is to time how long it takes a roller to coast to a stop. Small and intermediate sized rollers should spin at least 10 seconds with a sharp hand spin.

However, a roller also must have a cost effective life. It is good practice to use a (L10) design life of 40 years, even if the machine's useful life is expected to be less, to ensure a high overall process reliability. Thus, bearing failures should be relatively rare.

The most expensive bearing of all is one that seizes unexpectedly during service and shuts down a process line. Thus, the emphasis here will be to estimate bearing condition so that they can be replaced near the end of their life, but well before catastrophic failure.

The most reliable way of determining bearing condition is to pull it off, clean it up and inspect the rollers, races, cages and clearance. Even the tiniest pits or scratches or other visible flaws means that the bearing is very near the end of its life. The usual progression of failure is inner race; ball or roller; outer race and finally the cage. Unfortunately, it is not always practical to pull bearings off for inspection.

One way to get a reasonably reliable bearing condition evaluation is to use a FFT vibration analyzer which looks at key frequencies corresponding to the bearing geometry. This can be done with portable equipment,[8] or with permanently mounted continuous monitors[9] if process reliability is at a premium. These are just sophisticated versions of listening to a bearing through a wooden handled screwdriver.

Also, one can have the lubrication analyzed to determine if bearing metal particulate is present. This is also useful to diagnose certain causes of premature bearing failure due to water or dirt contaminants, due to thermal degradation of the lubrication and so on. However, experienced maintenance personal can sometimes also diagnose these types of problems by sight, feel or smell of the lubricant.

Finally, many bearings tend to run hotter during their initial break-in and toward the end of their lives. The point here is not whether the bearing temperature is in spec. Rather, it is to determine if the bearing temperature is warmer than in the past or when compared with its neighbors. This is perhaps the easiest check because all one has to do is to instruct the oiler to touch and monitor bearing housing temperatures during his normal duties.

If a bearing fails in less than 10 years of service, its reason for premature failure should be determined. This is not to satisfy an academic curiosity. Rather, it is to determine if anything can be done to prevent it or its cousins from failing again and possibly shutting down a line. Since bearing failure analysis is beyond the scope of this short section, the interested reader should look through the many guides and articles written by bearing manufacturers for more information.

In general, premature bearing failure can be the result of inappropriate design or maintenance. Design problems include: radial or axial overload, overspeed, no provision for thermal expansion of the free bearing, poor fits, and inappropriate seals or lubrication. Maintenance problems include: inappropriate lubrication or lubrication schedule, misalignment, unbalance, poor fits, impact loading of the bearings during handling and so on.

In summary, it is the role of the maintenance department to prevent unexpected bearing failure by appropriate practices and continuous checking. If bearing life is short, the maintenance department should also be the vanguard of problem solving.

Bibliography

1. McKenzie, M.R. and Borbas, J.S. *The Effect of Shaft and Housing Support on the Life of Spherical Roller Bearings.* Canadian Pulp and Paper Association, 77th Annual Meeting, Montreal, January 29-30, 1991.

2. Wallin, K.E. *Roller Bearing Load Ratings and Reliability*. SPCI, World Pulp and Paper Week Proc., Stockholm, April 10-13, 1984.

3. Harris, Tedric A. *Rolling Bearing Analysis: Theory and Analysis.* John Wiley & Sons, New York, 1966.

4. Shigley, Joseph E. and Mischke, Charles R. *Bearings and Lubrication Workbook.* McGraw-Hill Inc, New York, 1990.

5. Baumeister, Theodore et. al. *Marks' Standard Handbook for Mechanical Engineers.* McGraw-Hill Book Co., New York, 6th Edition, 1978.

6. Deutschman, Aaron D. et. al. *Machine Design, Theory and Practice.* Macmillan Publishing Co., New York, 1975.

7. *Standards of the Anti-Friction Bearing Manufacturers Assoc.*, New York, 1972.

8. McLain, D. and Hartman, D. *New Instrumentation Techniques Accurately Predict Bearing Life.* Pulp & Paper, February 1981.

9. Liddle, Ian and Reilly, Steve. *Expert Systems Offer Precise Analysis, Diagnosis of Mill Rotating Machinery.* Pulp & Paper, vol 67, no 2, pp 53-55, February 1993.

Chapter 9

Coatings and Covers

In this chapter we look at a variety of roller shell coatings and covers, their manufacturing methods, and the motivations for such treatments. Coatings are often used to modify abrasion resistance, traction, adhesion or other surface characteristics Covers are typically used to give tolerance to nip imperfections, increase nip dwell time, or increase traction.

Why Coat

Steel, iron, aluminum and composites are the most common shell materials because of their good balance of strength, ductility, stiffness, machinability and economy. Unfortunately, these materials are not very hard or abrasion resistant compared to many webs. Abrasive materials include paper (clays, fillers and contaminants), fiberglass, metals, some polymers and many web coatings. Also, rollers may have contact with other hard materials such as a mating nipping rollers, metal cores and shafts of surface winders, score slitters and operator's knives.

In these and other situations, it may be more economical in the long term to coat a roller than to periodically regrind and/or replace it. The most common means are to tungsten carbide coat steel or iron and anodize aluminum. These coatings are much more abrasion resistant than their base shells.

Another application of roller coatings is to control (often increase) traction with the web. In this role, tungsten carbide is most common. It is a very versatile treatment because a wide range of mechanical roughnesses can be obtained varying from mirror smooth to very coarse depending on the application parameters and a possible post grinding operation.

Finally, coatings can improve adhesive, coating or contaminant release from rollers. This reduces cleanup time during accidents, grade changes and machine shutdowns.

Why Cover

The most common application of covers are on nipped rollers. Example converting applications that may use covered nips include calendering, coating, embossing, laminating, printing, pull rollers and winding. While a metal-metal nip might also be used, it is quite stiff and intolerant to imperfections in roller cylindricity and web basis weight or caliper profiles. If one or more of the nipping rollers is covered, the cover will absorb the imperfections without causing large pressure variations as was discussed in Chapter 5. This is important because many of these processes are quite sensitive to peak pressure in the nip.

However, a secondary parameter in many of these processes is time because the mechanics are viscoelastic. The dwell time in a nip is proportional to nip width (footprint) and inversely with machine speed. If machine speed is increased for productivity, the nip width may need to be increased as well to maintain dwell by either increasing nip load or decreasing cover modulus (related to hardness). Indeed, for the same parameters a covered nip may be 10-100 times as wide as a metal-metal nip.

There are tradeoffs when adding a cover, however. First, covers can take much less nip load and pressure than metal-metal nips. The load rating of covers (which also depends on speed) can vary from only a few PLI (lb/in) to more than a couple thousand PLI (0.005-10 kN) Secondly, covers are expensive and often have more limited lives. Finally, covers are susceptible to damage from solvents or heat.

A final common application of covers is to increase web/roller or web/roll traction. Covers can often double the coefficient of friction of plain metal shells. Thus, covers might be used on driven rollers, spreaders, drums and the like.

Anodizing

Anodizing is commonly used to increase the hardness and abrasion resistance of aluminum shells. Anodizing is the addition of a thick oxide (Al_2O_3) layer to an object (often aluminum). Commonly the roller shell is the anode of an electrolytic cell, though electroless coating is also possible.

While natural oxidation of aluminum is only a millionth of an inch thick, anodized coatings might be a few tenths of a mil to a few mils thick (0.005-0.050 mm). While the natural oxidation of aluminum shells is typically a light gray, anodizing is available in a variety of colors such as clear, green, blue, and charcoal. This variety of colors certainly lends distinction between many competing coatings. However, the performance of different colors should be nearly identical. The reason is that the color is imparted merely by trace elements. For example, a trace amount of Ti3+ ion will impart a color of a blue sapphire gemstone.

Aluminum oxide is quite hard and abrasion resistant. Indeed, it is the ingredient of industrial abrasives and sandpaper. Aluminum oxide is, however, a nonconductor so that static electricity may not be transferred from web to roller and thus to ground. Also, surprisingly, aluminum oxide gives no substantial improvement in the resistance to attack by strong acids and bases.

While anodizing is quite hard, it is usually too thin to impart significant dent resistance to soft aluminum shells. Fortunately, anodizing is strongly attached or adhered to the base metal so that debonding should not be a problem (tungsten carbide can spall, covers can debond).

Aluminum oxide is just one of several ceramic compounds that are used to coat roller surfaces. Other ceramics include oxides of chromium, titanium or zirconium.

Plating

The printing industry is one of the largest users of plated rollers, particularly for the rotogravure image carrier.[1] The image carrier is most commonly a steel sleeve or mandrel upon which a copper layer is plated in two stages. This copper layer is then engraved or etched with the image. Since copper is very soft, however, it will not hold up well against doctor blades or some abrasive printing inks. The solution is to plate the copper after etching with a few microns of chrome. After the run, the image carrier can be dechromed and refinished almost indefinitely.

Chrome plating is also used in specialized applications which require a hard but mirror smooth surface. Ironically, a high smoothness can either decrease web-roller friction (less tooth) or increase friction (greater surface contact and adhesion). Very smooth surfaces are also easier to clean and thus may be desirable in coating or printing applications.

Most commonly, chrome is plated electrolytically in a bath of de-ionized water with chromic and sulfuric acid. Chrome plating differs from copper plating because the plating comes from the solution instead of the anode. Of course, chrome is both harder and more resistant to oxidation than copper.

Plating often involves two process materials and/or steps where the first bonds to the base and serves as a primer for the finish plating. Care must be taken to mask off parts, such as bearings, which are not to receive plating. Plating may also involve a finish polish step to achieve high smoothness.

In any case, plating requires tight process control, is time consuming, often involves hazardous materials, and thus is expensive.

Tungsten Carbide

The paper industry is one of the larger users of tungsten carbide coated rollers. The most common examples are steel or iron winder or reel drums that are subject to wear and/or require good web traction.[2] Most new winders are supplied with such coatings and many of the older ones have been coated in place.

Tungsten carbide is even harder than aluminum oxide and is used on many cutting tools. In most cases, the roller coatings will outlast the machines. However, tungsten carbide may wear smooth in a few years on particularly difficult applications (high speed paper winder drums, or doctored rollers).

The tungsten carbide coating is also useful because the surface roughness can be adjusted from less than 1 Ra (with post polishing) to several hundred microinches. While this can also be obtained with conventional machining, tungsten carbide is more versatile and reproducible.

Smooth surfaces are used when the primary purpose of the coating is for wear resistance. The higher roughnesses are used primarily to increase web-roller traction. However, the rough tooth may pick at fibrous materials, mark soft materials and can get filled in with dust or contaminants. Ironically, rough coatings may actually reduce traction with smooth webs.

Finally, tungsten carbide is much more resistant to corrosion than most steels and irons. However, this is a side benefit because stainless steels, electroless nickel and superalloys are most often selected for corrosive duties.

While the coating itself is strong, the bond can be relatively weak. There have been several instances where the coating has flaked or spalled off of a shell due to poor surface preparation or application. Another occasional complaint is from operators who quickly wear out the knees of their bluejeans when working regularly around a rough coated drum. Perhaps the worst problem is that the coating has been touted to solve everything from roll structure problems to machine vibration.

Tungsten carbide is produced by reacting hydrocarbon vapor with tungsten at high temperature. The primary constituent is WC, but may contain W_2C, W_3C, and W_3C_4 and traces of non-tungsten carbides. Also present are oxides and numerous small voids.

Surface preparation is vital to a strong bond. First, the metal shell is degreased with a strong solvent or heated to more than 700°F. Then, the shell may be ground to obtain cylindricity and surface finish. Finally, the shell may be acid etched to remove oxides.

Prior to the application of tungsten carbide, the surface is heated to a uniform but modest temperature for two reasons. First, it will increase the bond strength of the impacting tungsten carbide to the shell. Second, it will minimize the tensile stresses developed in the coating as it cools.

Tungsten carbide can be applied by a couple of mechanisms. The most common method is by flame spray. Here, a tungsten powder is metered into an acetylene or other hydrocarbon flame. The flame then heats the powder and carries it into the shell. An alternative proprietary method uses a detonation gun which operates cyclically in pops rather than continuously.[3,4] Reportedly, the detonation gun achieves supersonic particle velocities, reduced void volumes, higher bond strengths and other benefits.[5]

Another recent innovation is the impregnation of Telfon or other release agents into tungsten carbide coatings. This may be the best of both worlds for some applications. First, one can have mechanical tooth for traction. Second, one can have release in processes involving sticky coatings or adhesives. The release is so good that most adhesive tapes will simply fall off the roller.

The process engineer has several choices when selecting the coating. First, the surface roughness must be selected based on required traction. Second, the coating thickness must be specified. While thick coatings may be more durable and last longer, cylindricity will suffer unless there is a subsequent machining step. Finally, existing equipment can either be coated in place or taken out and sent to a vendor.

Thermal Coating Processes

There are three common methods for thermal spray deposition of coatings onto roller shells. In each case, the coatings are melted or heat softened and propelled toward the target roller.

Wire arc and wire combustion uses sophisticated welding technology to apply wire metal materials to the roller. This technology is used to restore, fill in, or add large amounts of material (up to 1/4" thick). Wire arc uses electricity to atomize the wire while wire combustion uses an electric ignition source and gases to melt the wire. Wire combustion provides a smother as-sprayed surface but does not deposit material as thick as the wire arc method.

Plasma spraying is used to apply ceramic and metal powder materials such as the tungsten carbide described earlier. High energy plasma can apply a stronger and denser coating, but is much more expensive than the more common low energy plasma. Coatings are typically applied in thicknesses between 3 and 12 mils, but can be sprayed up to about 0.150" before microcracking occurs due to cooling contraction.

High velocity oxygen fuel (HVOF) is used primarily to spray metal materials. HVOF simply heat softens (rather than melts) the metals and propels them at speeds greater than Mach 2. Coatings applied with this technology include tungsten carbide for wear resistance and stainless steel for corrosion resistance.

Because the application for thermal coating of roller are too many to be listed in entirety here, we will merely discuss a few illustrative cases. Coatings (as opposed to sleeves) can be used to provide durability to low inertia composite rollers and low weight composite coreshafts. Ceramics are used for specific electrical properties, such as dielectric coverings with improved life for bare roll, flame treating and corona treaters. Ceramics can be used to improve life of rollers that are in contact with abrasives or doctor blades such as found in many applications. Stainless steels are used to impart corrosion resistance and can be formulated to provide high release for easy cleanup of coatings and adhesives. Table 9.1 gives a general comparison of common roller coatings.

Table 9.1
Coating Comparison Chart
Courtesy of American Roller

Property/ Material	Aluminum Oxide	Chromium Oxide	Titanium Oxide	Zirconium Oxide	Tungsten Carbide	Stainless Steel	Super Alloy
Class	Ceramic	Ceramic	Ceramic	Ceramic	Metal Matrix	Metal Matrix	Metal Matrix
Hardness (Rockwell C)	55-63	65-72	45-55	45-55	65-70	30-45	30-45
Finish μinch Ra	8-500	3-500	3-500	10-500	<1-800	4-1000	4-300
Abrasion Resistance	Good	Excellent	Fair	Fair	Excellent	Fair	Fair-Good
Chemical Resistance	Good	Excellent	Good	Excellent	Fair	Excellent	Excellent
Electrical Conductivity	Poor	Fair	Good	Poor	Good	Good	Good

Cover Compliance and Nip Pressure

The primary effects of covering a steel shell are shown in Figure 9.1. All else being equal, the covering will decrease the peak ZD nip stress and increase the nip width or footprint. Recall from Chapter 5 that the area under the pressure versus MD distance is determined by the lineal nip load (PLI or kN/m). Thus, if the nip loads are equal, the areas under the curves must also be equal.

If our objective were solely to reduce the peak, we could also achieve this merely by reducing the nip load. The only issue is whether the mechanicals and controls for our nip are sensitive enough to achieve low but uniform loadings. Thus, the motivation for covering usually lies elsewhere.

The other effect of soft coverings is to increase the nip width or footprint. This may be desirable if dwell time is desired without decreasing line speed. Dwell time is a factor where webs, coatings and adhesives creep. Nip dwell time could also be a factor in heating/cooling if the web is not wrapped significantly across the temperature controlled roller(s).

However, the greatest motivation for covering is often because it evens out nip pressures despite variations in roller cylindricities, web caliper or basis weight profiles, end load differences or misalignment. Figure 9.2 is a schematic of the nip pressure profile across the width of a web for covered and uncovered rollers.

Without a cover, the pressure profiles at the ends of the rollers are considerably higher, perhaps due to nip roller deflection. Thus, a soft cover may eliminate the need for crowning one or both of the rollers.

Also, there are pressure nonuniformities due to other factors such as roller runout, roller diametral variations and web profile problems. With a cover, however, the pressure profiles become more uniform because geometrical errors are absorbed better by the cushioning effect of the cover instead of resulting in pressure dips and spike.

Figure 9.1
Effect of Covers on ZD Nip Stress

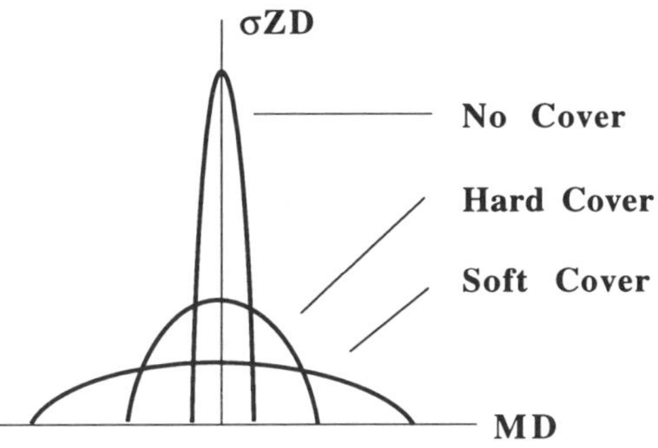

Figure 9.2
Covers Even Out Nip Pressure Variations

Calendering, coating, laminating and printing are examples of converting processes that are very pressure sensitive Thus, it is common to cover one or both of the nipped rollers. Though these applications may also employ metal-metal nips, the demands on the web and machinery are quite high. For example, a metal-metal calender roller may need to be cut to within a few tenths of a mil (0.001") to achieve acceptable nip uniformly. Conversely, a covered calender roller for the same application may need only to be cut within a few mils. This illustrates how the cover may be an order of magnitude less sensitive to profile variations. This forgiveness is requisite for most applications using gap/interference control, but is useful in load control as well.

About the only class of application where nip pressure is of only minor concern is with rolling. Here, pressure is an outcome of achieving a desired final web caliper in gap/interference control.

Covers and Speed Control

Covered nip rollers are usually driven. The primary reason is to prevent the nip rolling resistance from generating a large increase in web tension. Secondary reasons are to compensate for other sources of drag (bearings and windage) or to intentionally change web tension across the nip (change in web strength or for printing registration). If the covered nip is torque or tension controlled, there is no issue. However, we should be very reluctant to ever put the covered nip roller in any form of speed control (draw, geared or belted to the mating roller etc).

Speed control would be better termed RPM control because a drive seldom knows directly the roller surface speed or web speed. Rather, the drive calculates the surface/web speed based on a measured RPM and an assumed roller radius. If the roller is relatively rigid, the radius is well determined. If, however, the roller is covered, the radius and thus surface speed are not so predictable.

Figure 5.6 showed a covered roller nipped against a hard roller. The questions becomes what radius does one use for calculating speed? Is it the radius at the nip or the free radius of the roller? The answer, in general, is neither. Rather, the effective radius is usually somewhat greater than the free radius. The reason is that most covers are relatively incompressible ($\mu \cong 0.5$). Thus, the cover has to speed up going through the nip so that the flow rate of material is the same at the nip as elsewhere.

Figure 9.3 shows the effective diameter ratio as a function of nip loading and for three different cover Poisson Ratios.[6] Many covers, such as rubbers and polyurethanes, are relatively incompressible ($\mu \cong 0.5$) and thus act bigger than their free radius and thus turn faster than expected. The higher the load, the greater the difference. At the other end of the spectrum are foams which have Poisson ratios near zero. Here, the roller turns slower than expected. Other materials, such as foamed rubbers, have Poisson Ratios near 0.3 and are relatively insensitive to nip load.

Figure 9.3
Effective Diameter versus Nip Load

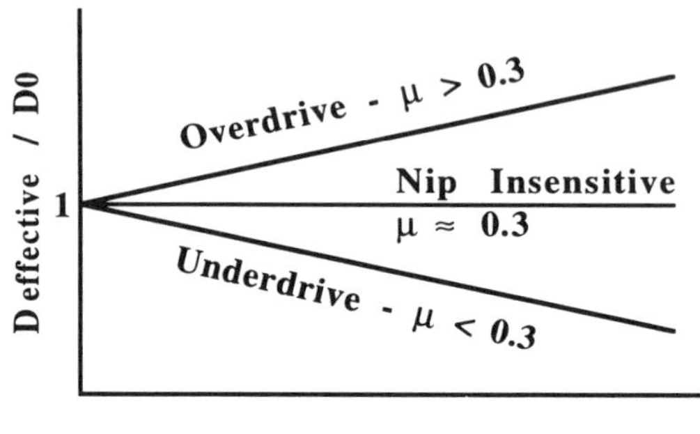

The magnitude of the ratios between actual and expected speeds can be as much as several percent. This is enough to break or slack most stiff webs and will shift tensions considerably on even the most extensible materials. Thus, we must compensate for the effective diameter on speed controlled soft nipped rollers, which might be done with very careful speed measurements.

However, what happens if the nip varies intentionally (process setting change), or unintentionally (poor control of actual nip load)? Obviously, the speed and thus web tension will then also vary. This coupling of nip to web tension for speed controlled soft nips is undesirable and very difficult to deal with. It would be far better to avoid the difficulty entirely by a more appropriate and robust drive control strategy.

The first choice would be to tension control the nip roller drive. If two drives are required, separate motors could share the load (equally or weighted to the grippier or more well wrapped roller). Mechanically coupling the rollers together is undesirable. The developed torque differential, regardless of how small speed ratio error is, would be the nip load x coefficient of web/roller friction x width x radius. This could well be high enough to break components. It is also not advisable to torque control both rollers as torque will also vary with nip load.

Question: What is the effect of nip load variations across the width of a covered roller? Answer: Speed variations, tension variations, and possible shear wrinkles.[7,8]

Cover Modulus

As seen in previous sections, softer covers reduce the peak nip stress, increase the nip width and tend to even out the nip. However, softer covers are typically less durable (lower load, speed and abrasion resistance). However, the material property that most affects these performance characteristics is modulus rather than hardness. Modulus is the slope of the stress-strain curve in a uniaxial tension or compression test. Hardness is the resistance to indentation, which is a convenient but primitive indicator of modulus.

Modulus is not a constant. Rather, the stress-strain curve of most cover materials are nonlinear and strain hardening (concave upward). Thus, more complete testing will generate load and unload curves for several values of peak strain. If a single value of modulus is reported, it should specify the value of strain at which the modulus was obtained and whether it was from the load or unload portions of the curve. A measurement of modulus is ASTM D412.

Modulus also depends on temperature. Thus, while testing is most commonly performed at room temperature, the material may be significantly softer at elevated operating temperatures. Thorough testing will investigate temperature dependence if the application is expected to exceed 100 or 150 °F.

Finally, cover modulii are strain rate dependent (has a real and imaginary component). This means that the apparent modulus will be higher at the fast loadings of real operating conditions than it will be at the slow strain rates of typical lab testing. For example, a nip load cycle (nip width divided by machine speed) on a specific part of a cover may be only a few milliseconds, while the load cycle in a test lab may be closer to a minute. Fortunately, there are several types of tensile test instruments that are able to generate load cycles as quick as 100-1,000 Hz.

From this same equipment, one can and should also obtain yield strength, ultimate strength, ultimate strain (elongation), hysteresis and possibly fatigue strength.

Cover Hardness

As we have seen, cover material properties such as modulus are complex and may require involved testing. Thus, a simpler and more convenient indicator of modulus, such as hardness, is needed. Hardness is not the relative resistance to scratching as used in mineralogy. Rather, it is the resistance to an indentation by a weighted probe. Most of the commercial test instruments for hardness, often call **durometers**, are based on this principle. The tests are quick and easy, and the instruments are small and inexpensive.

Durometers have seven scales, but the most common is the "A" scale which extends from zero (very soft) to 100 (very hard). For example, the sole of a shoe may vary from less than 50 for a soft foamed rubber to upward of 100 for a bone hard material such as compressed leather. Hardness may be measured by tests such as ASTM D2240, D531 and D1415.

The testing should be performed on a smooth section of specimen with a minimum thickness of 1/4" (6mm). The testing must be performed at an agreed upon temperature of say 20 °C or 75 °F because of the significant temperature dependence of most cover materials. The test data is quite noisy so that at least 5 points should be read and averaged. Finally, gauges should be calibrated frequently and checked against each other.

One of the common hardness test instruments is the Shore Durometer manufactured by the Shore Instrument and Manufacturing Company. Thus, a reported value of 85 **Shore A** indicates a reading of 85 on the A scale of the Shore instrument. A similar gauge is also marketed by the Rex Gauge Company.

Another common hardness instrument is the **P&J** (Pusey & Jones) Plastometer made by Noram Quality Control and Research Equipment Ltd. The P&J measures the indentation of a 3.1 mm diameter ball under a one kg load. The scale is reversed (low values are harder) and the Plastometer values (0.01 mm deflection) are not directly correlatable to the Shore A.

Hysteresis and Heat

At sufficiently elevated temperatures, the properties and life of cover materials are reduced. Suppliers should provide a temperature rating for coverings, which may depend in part on other factors such as nip load and machine speed. If this temperature is exceeded, the cover may chemically or physically break down.

The temperature of a cover can be elevated by one or more mechanisms. First, the roller shell may be intentionally heated by electrical induction, electrical resistance, steam, hot water, hot oil and so on (see Chapter 15). If the roller is heated by a fluid, the cover will be of similar but lower temperature as the exiting fluid temperature. This is because the air and often the web will help cool the cover surface below the core temperature of the roller.

Another mechanism for heating the roll is more unintentional. That is, a hot web will heat the roller cover to a temperature probably closer to the web temperature than the ambient. Hot webs are byproducts of many manufacturing processes such as drying and extruding.

However, one of the most common means of temperature rise on covers is due to hysteresis heating in a nip. On a macro level, this heating is the product of nip rolling resistance which is converted almost entirely into heat. The higher the nip load and/or speed, the greater the heating power, and thus the higher the cover temperature. (While higher speeds also carry away more heat via web and air, the cooling may not quite keep up with the higher heat generation.)

On the micro level, the heating is the result of cyclic stresses in the nip. As seen in Figure 9.4, the load and unload curves of the stress strain curve are not coincident. The area between the displaced load and unload curves represents the energy (heat) per unit volume per load cycle (revolution). Note, however, that this material is elastic in the sense that it is not permanently deformed. It merely does not return as much energy as was put into it during mechanical straining.

Figure 9.4
Hysteretic Stress-Strain Curve

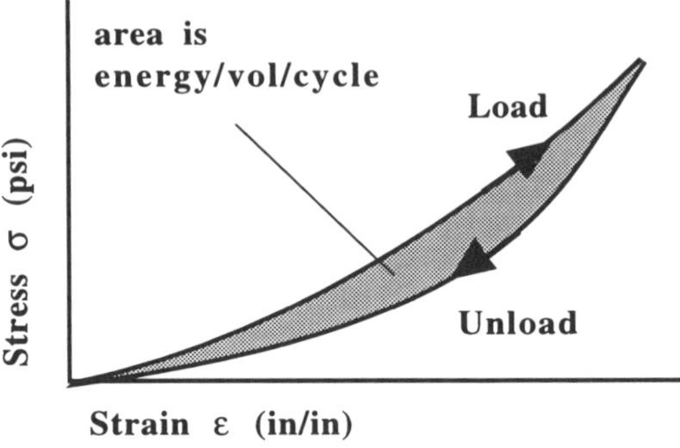

The hysteresis of materials varies enormously. On the one hand, pure natural rubber and special compounds such as those which make superballs, have very low hysterisis (i.e., high resilience) and are quite bouncy. On the other hand, most roller cover materials, such as modified rubbers and polyurethane, are quite hysteretic and are quite dead. Measures of resilience (lack of hysteresis) include ASTM D945 and D2632.

While a hysteretic cover may be useful in reducing nip vibrations, it is generally considered an undesirable property because it may cause cover heating on heavily nipped application. In these cases, the load rating must drop as a function of increased machine speed to keep cover temperatures moderate.

Cover temperatures are very difficult to predict.[9] Fortunately, temperatures are easily measured with contacting pyrometers, infrared guns and RTD's. However, the most practical method is to use an appropriate thermometer immediately after the line has stopped.

If the cover temperature is excessive at a required load and speed, it might be cooled with circulating water. However, chilled water is a messy and usually unwelcome addition to a converting line. A better approach would be to shop around for alternative cover materials with better heat ratings. Covers are something like cooking: no two desserts are the same even if using the same recipe (ingredients). Slight changes in the mix or process may reduce hysterisis, or increase strength at elevated temperatures.

Cover Failures

Covers are usually much shorter lived than other roller and converting components. Life may vary from a few days to a few years depending on the application. In this section, we briefly describe several common failure modes.

Abrasion and Wear

Most covers, particularly those made of polymers, are soft and thus subject to wear from harder web materials. Wear occurs because of inevitable microslippage at any nipped or unnipped roller. Wear is more pronounced with abrasive webs, such as filled paper, and inks or coatings that contain minerals. Unfortunately, abrasion tests such as ASTM D1630 and D2228 are not good predictors of actual service life.

Barring

Barring is a series of CD oriented defects equally spaced around a nipped roller. Barring is the result of nip vibration.[10] Barred covers must be ground down well below the lowest areas because they have been compacted.

Chemical/Solvent Attack

Chemicals and particularly hydrocarbon solvents may attack the surface or diffuse into the interior of a cover. Attack by solvents are accompanied by measurable changes in size (swelling), modulus and other properties. Occasionally the solvent will benignly penetrate the cover and cause the bond to the shell to fail. A test for attack is ASTM D471.

Cracks

Cracks can result from fatigue, overstress or oxidation.

Cuts

The most common cause of cover cuts are from operator's knives when they try to free a wrapped roller. Circumferential cuts are often caused by something intruding into the roller while it is spinning.

Delamination

Delamination refers collectively to four distinct failure modes that include a cover cohesion failure, a cover/adhesive adhesion failure, an adhesive cohesion failure and an adhesive/roller adhesion failure. Examination of the initiation site of the failure will determine the mode and guide an appropriate solution. Two common modes are due to residual stresses of manufacturing or due to the shear stresses in the interior caused by a rolling nip and described in Chapter 5. Adhesion can be tested by ASTM D413 and D429.

Dub

Failure to dub (relieve the diameter of the roller ends) of a covered nip roller can cause those areas to crack or spall.

Glazing

Glazing is the hardening and/or polishing of the outer surface of a cover. Glazing will reduce the coefficient of web/roller friction considerably, and will compromise the operation of rollers which demand traction, such as bowed roller spreaders.

Hardening and Oxidation

Many cover materials will harden or oxidize with time. This chemical aging can be accelerated by ozone (from electrical sources), other chemical compounds and high temperature.

Loose Cover

A loose cover may indicate a delaminated cover or the loss of fit on a sleeve due to stretch or centrifugal forces

Set

A stationary nipped cover may take a set or permanent dent. Thus, nips should be opened when a machine is stopped. More obviously, a roller should never be supported by contact with the cover during storage and handling. A test for set is ASTM D395.

Cover Materials

Most covers are polymers, and thus can be formed by almost any method also used to manufacture plastics. These methods include casting, winding, extruding, molding and so on. Many of the polymers are cured, which can induce large residual stresses.[11] Covers can be formed on the base roller shell or on a mandrel to make a sleeve. The most common rubber polymers are described below.

Acrylic (polyacrylic or polyacrylate) is based on the polymerization of ethyl or butyl acrylate. It possesses good heat resistance (to 350° F), and good oil resistance. Unfortunately, it has poor resilience, is difficult to work with and is expensive.

Butyl rubber is a copolymer of isobutylene and small amounts of isoprene. It is highly stable, highly ozone resistant and has good heat resistance ((to 350° F). Unfortunately, it has extremely low resilience and swells severely in petroleum based solvents or oils. Butyl has largely been displaced by EPDM in roller applications.

EPDM or EPT consists of a backbone of ethylene and propylene. It has excellent chemical resistance, heat resistance (to 350° F) and ozone resistance. While it has outstanding resistance to polar type solvents such as ketones, it has no resistance to solvents and oils of the aliphatic type. EPDM finds applications where chemical, heat or ozone resistance is required.

Fluorocarbon Elastomers are copolymers of vinylidine flouride and hexafluoropropylene. They have outstanding chemical, solvent and heat resistance (to 500° F). Unfortunately, they are very expensive.

Hypalon is chlorosulfonated polyethylene. It has excellent chemical and ozone resistance, good heat resistance (to 300° F), good physical properties, and fair oil resistance.

Natural Rubber is chemically polyisoprene. It is still used in applications requiring good strength, but lacks resistance to oil, ozone and heat.

Neoprene is polymerized chlorprene (2-Choro-1,3-butadiene). Neoprene is noted for good all around properties including toughness, good aging, chemical resistance, fair ozone resistance, and high resilience. A typical application for neoprene is a feed or pull roller.

Nitrile or **Buna N** is a copolymer of butadiene and acrylonitrile. Its main attribute is good oil resistance, and good resistance to chemicals and water. Typical applications for nitrile include inking rollers, and squeegee rollers in wet copy machines.

Polyurethane or urethane consists of a backbone of a diol which has been reacted to form urethane linkages. If the backbone is a polyester, the polymer has excellent oil and solvent resistance, but poor hydrolytic stability (reversion resistance). Conversely, polyether based urethanes have better hydrolytic stability but poor oil and solvent resistances. Urethanes can revert or soften under adverse conditions. Typically, urethanes are used in rough applications where good physical properties and abrasion resistance are required.

SBR, formerly known as GR-S is a copolymer of styrene and butadiene. It has been the general purpose rubber for many years, but has limitations of poor oil resistance, heat aging and ozone resistance.

Silicone rubber is usually polydimethylsiloxane. It is noted for two main attributes - its outstanding release characteristics and its heat resistance to 500° F. It also has good chemical resistance and excellent ozone resistance. However, it has drawbacks including high costs and poor physical properties. Silicone is most often used in very high temperature applications.

Thiokol is a polysulfide polymer and was the first synthetic rubber developed. It is known primarily for its outstanding resistance to a broad range of solvents. However, it has poor strength, poor resilience and is limited to about 212° F.

Cover Properties

Physical properties are summarized in Table 9.2 and chemical resistance properties are summarized in Table 9.3. For comparison's sake, test data is presented on compounds of 50 durometer. Note that properties will vary depending on the formulation and manufacturing.

Acknowledgment

I would like to express an extra thanks to American Roller for providing me information for several sections of this chapter.

Table 9.2
Physical Properties for Cover Compounds - Typical Examples
Courtesy of American Roller

Property/ Material	Neoprene	Buna N	EPDM	Silicone	Urethane
Durometer	50	50	50	50	50
Tensile Strength (psi)	2500	1350	1200	600	4500
300% Modulus (psi)	480	340	400	-	440
Elongation (%)	730	600	590	250	530
Tear, Die C (pli)	207	121	152	82	193
Permanent Set (%)	16	11	10	1	1
Compression Set (%)	26	38	52	6	32
Picco Abrasion (grams loss)	0.0829	0.1230	0.0797	0.3288	tiny
Taber Abrasion (grams loss)	0.1900	0.7547	0.5408	0.4917	0.0094
Resilience (%)	57	37	62	65	47

Table 9.3
Chemical Resistance for Cover Compounds - Typical Examples
% volume swell after immersion for 70 hours at room temperature
Courtesy of American Roller

Chemical	Neoprene	Buna N	EPDM	Silicone	Urethane
ASTM #1 Oil	-2	-1	16	1	0.6
ASTM #3 Oil	3	0.3	70	12	2
Reference Fuel B	54	13	170	210	14
MEK	113	209	-2	70	150
Toluene	235	143	141	175	70
Hexane	14	-6	134	170	2
Ethyl Acetate	41	118	7	62	200
Cellosolve	6	34	0.5	2	100
Methyl Chloride	500	720	25	120	370
Trichloroethylene	310	230	200	170	210
Diethylene Glycol	-1	-0.3	-0.3	-0.6	3
Distilled water	3	1.0	-0.7	0.4	2
10% KOH	34	8	5.0	-0.2	disintegrates
10% H2SO4	5	7	3	0.2	disintegrates

Bibliography

1. *Gravure Process and Technology*, Chapter 8. Gravure Association of America, Rochester, NY, 1991.

2. Thun, D.P. *Traction Coat on Rewind Roll Ends Slippage Problems.* TAPPI Finishing and Converting Conf. Proc., 1982.

3. Gill, J.J. *Detonation Gun and Plasma Coatings in Surface Engineering.* Symate '86 Conf. Proc., Bern, Switzerland, 1986.

4. Tucker, R.C. *Detonation Gun Coatings.* J. of Metals, vol 38, no 2, February 1986.

5. Phillips, Bruce A. and Knapp, James K. *The Application of Thermal Spray Coatings on Calender Rolls.* TAPPI Finishing and Converting Conf. Proc., pp 143-155, October 2-5, 1994.

6. Stack, K. D. and LaFleche, J.E. and Benson, R. C. *The Effects of Nip Parameters on Media Transport.* 3rd Int'l Conf. on Web Handling, Oklahoma State Univ., June 18-21, 1995.

7. Stack, K. D. and Benson, R. C. *The Effects of Axial Variation in Nip Mechanics.* Proc. of the 2nd Int'l Conf. on Advanced Mechatronics, Jeiji University, Tokyo, August 1993.

8. Benson, Richard C. and Chiu, Han Chieh and LaFleche, John and Stack, Kenneth D. *Simulation of Wrinkling Patterns in Webs Due to Non-uniform Transport Conditions.* 2nd Int'l Conf. on Web Handling, Oklahoma State Univ., June 6-9 1993.

9. Moore, Robert H. and Moschel, Charles C. *A Tool for the Engineered Nip.* Tappi J., vol 77, no 3, pp 117-122, March 1994.

10. Chen, Yian N. and Boos, Gunther. *Calender Barring on Paper Machines - Theoretical Model Development.* Tappi J., vol 58, no 7, pp 93-97, July 1975.

11. Rodal, Jose J.A. *Roll Covers in Soft-Nip Calendering and Supercalendering.* TAPPI Finishing and Converting Conf. Proc., San Diego, CA, pp 1-14, Oct 15-18, 1995.

Chapter 10

Surface Topology

In this chapter we look at the topology of the roller surface that is in contact with the web. Roughness, grooving, patterning and other features are discussed.

The Importance of the Surface

The topology of the roller surface is important because it is the interface between the web and the manufacturing or converting machine. This interface is what communicates loads and stresses between the web and the roller. Thus, a primary concern of surface topology is to maintain web/roller traction.

If traction is lost, the roller is free to and will certainly turn at a different speed than the web. This will accelerate roller wear because wear is approximately proportional to relative velocity and normal load. Also, scratch sensitive grades such as coated paper, photographic film and printed webs may be damaged by web/roller slippage. Furthermore, the roller's control over web edge position is severely compromised or entirely lost during slippage.

Slippage may be an acceptable mode of web/machine interaction. However, the applications where this is desirable are very few. Transport examples include web flattening and a few of the more unusual spreaders (see Chapter 17). Process roller examples include some types of coating such as the reverse roll coater.

Most rollers are intended to be in traction so that control will be compromised during slippage and particularly when transitioning between traction and slipping or vice versa as described in Chapter 4. The coefficient of web/roller friction along with tension, wrap angle and speed primarily determine the tension differential that can be sustained across a driven or undriven roller before the web breaks loose and slips.

The coefficient of traction is determined in large part by the topology of the roller and web surfaces. In general, the traction will increase as the roughness of the roller and web surfaces increase. However, traction may actually increase if both surfaces are extremely smooth because of the additional microcontact area in conjunction with other phenomenon such as adhesion or static. If roughness is increased too much, smooth webs may be marked and fibrous webs may be picked. Also, rough rollers tend to foul quicker and are more difficult to clean.

At medium or high speeds, roller topology can counteract a loss of traction due to air/fluid entrainment. The mechanism for this is quite simple. The topology allows the fluid pressure to bleed around the roller instead of trying to lift the web away from the roller. The usual course of treatment for air entrainment is to groove a roller. However, as seen in Chapter 4, the grooving patterns in common use are a poor fit with the needs - something like using a tractor tire on an automobile. Also, grooving is erroneously thought to provide a spreading function when in fact it tends to contract or MD trough wrinkle the web.

Some roller shells are radially drilled. A suction roller is a process example where vacuum is used to pull a liquid (such as water in the paper industry) out of the web, or air through a web. A vacuum roller is a transport example where the vacuum (if properly designed) can increase web roller traction on a driven roller.

Finally, roller topology may be patterned for a variety of particular process applications. On a fine scale, rotogravure printing makes use of tiny microetched cells to hold and deliver a fixed volume of ink in some types of printing. On a very coarse scale, embossers and diecutters have very deep topology to pattern or cut a web respectively.

Roughness and Waviness

The surface of even a mirror smooth roller is rough when magnified sufficiently. The roughness of a roller can be measured with a profilometer.[1,2] This device drags a lightly loaded diamond stylus, something like a phonograph needle, over a short distance (< 1 inch) of the surface of the roller. The movement of the needle is captured by a piezoelectric crystal which generates a signal for processing.

Figure 10.1 shows a schematic of a roller surface representative of a profilometer trace. The profilometer may calculate several statistics from this reading which are used to quantify roughness within the sample length.

Ra is the arithmetical average of the absolute value of the distance from the profile to the average.

RMS is the root mean squared value of the profile and is a commonly used measure.

Rz is the average maximum peak to valley height over a portion of the profile

Rt is the maximum peak to valley height.

There are several limitations of the roughness measurement. First, the sample size is extremely small so that readings may vary with position, particularly on a worn roller. Second, the roughness will tend to be much greater in the CD (roller axis) than in the MD (roller circumference) because machining and wear tend to make MD oriented features. This is referred to as lay. Third, the readings do not always give information on texture or shape, such as whether the peaks are rounded or pointy. Fourth, roughness readings can only assist in an empirical correlation with traction, but can with some difficulty predict air/fluid entrainment.[3,4,5,6]

The as manufactured surface roughness depends largely on the machining method, but also on tooling, speeds and feeds, and shell or cover material. Smoother finishes will cost more because of the additional time and steps.

Figure 10.1
Profilometer Trace of Surface Roughness

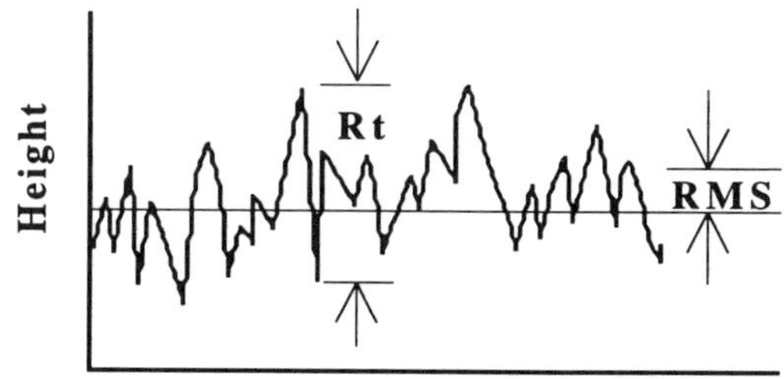

For example turning will produce Ra roughnesses from 2-500 μinch (0.05-12.5 μm), while the finish grinding step will produce a smoother finish of 1-250 μinch (0.025-6.3 μm). Finer finishes will require electropolishing, polishing or superfinishing.

A common as manufactured surface roughness of many rollers might be either 32 or 16 μinch (0.0.80 or 0.40 μm) for coarser or more tolerant webs. Smoother surfaces of perhaps 4-12 μinch may be required on certain processes such as calendering, coating, extrusion and printing. The surface roughness after usage may be either increased (crack, dent, foul, score or groove) or decreased (fill, polish or glaze).

While roughness describes a surface on a very fine cycle scale, **waviness** measures surface deviations on an intermediate cycle scale. Waviness will not usually affect traction or nip profiles significantly because most applications are sufficiently conformable. However, waviness can reduce the effectiveness of doctoring and the life of blades. Perhaps the most capable tool for surface profile measurement and characterization is by laser interferometry.[7]

On the most coarse scale, deviations from perfect cylindricity are measured by roller deflection as well as TIR (total indicator reading), station-to-station diametral differences and maximum diametral difference (taper).[8,9] These subjects are covered elsewhere in this book.

Grooving

Whether plain, coated or covered, rollers are often grooved with a bewildering variety of patterns. One machine builder alone has more than 150 grooving patterns. Sometimes these patterns are given near mystical attributes.

Grooving does not, as widely believed, provide any spreading or wrinkle removal function.[10] The apparent sideways movement of the spiral is merely a barber pole optical illusion because the roller surface moves strictly in the MD. Anyone who has flipped a spiral grooved roller around has found absolutely no change in the behavior of the web.

In fact, the tendency of grooving is always toward contraction or MD trough wrinkle generation. As seen in Figure 10.2, a web is pulled in slightly when it conforms to the grooves, especially on the web's edges. Thus, it is very important that grooving width be less than 10-20 times the caliper of the thinnest web to reduce the tendency of grooving wrinkles.

Rollers should be grooved solely for one purpose: to maintain web/roller traction at speed by allowing air or liquid to pass around a roller instead of lifting the web. As seen in Figure 10.3, fluid is carried or pumped by both the web and roller into the gap between the two. The fluid thickness, H, can be calculated[11,12] as given in Chapter 4. As an order of magnitude approximation, air film thickness is about 1 mil per 1,000 FPM for light nonporous webs. However, liquids have such a large viscosity that webs may float off rollers at just a few FPM.

With this calculation and the methods of Chapter 4, grooving can be designed to match the application needs. From Figure 10.4, we note that the total area of the grooves or texture must be greater than the film height times web width. Then we select a grooving width so that it is no more than 10-20 times the thinnest caliper web to avoid grooving wrinkles. The pitch should be as narrow as practical and the depth makes up the difference. The result of this designed approach is similar to the texture of a phonograph record for most applications. Thus, the grooving used in the web industries is much like using a tractor tire on an automobile.

Figure 10.2
Grooving Tends to Wrinkle Webs

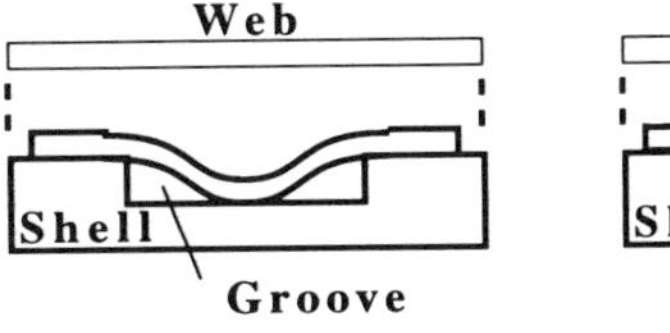

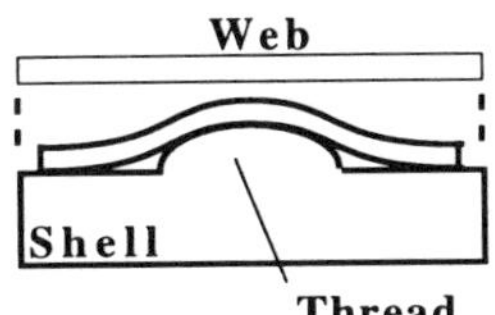

Figure 10.3
Air Entrainment on a Roller

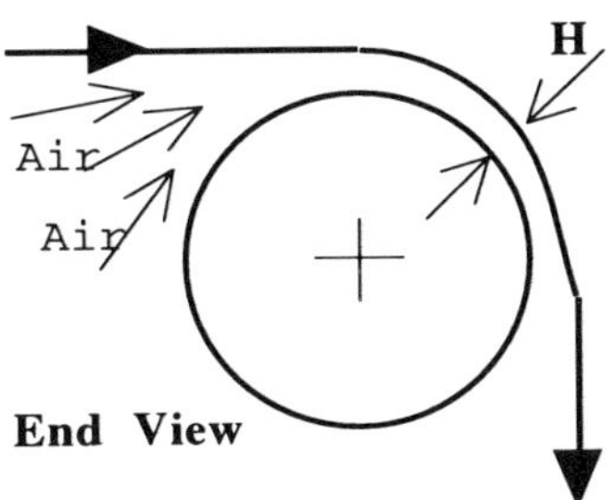

Figure 10.4
Grooving Geometry

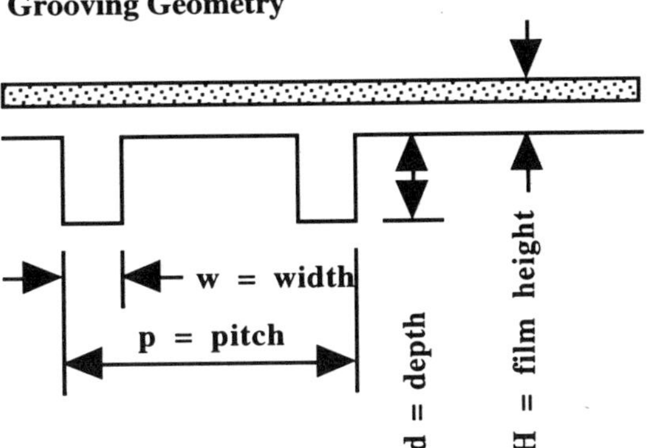

Some grooving has such a wide pitch that it is oriented more in the CD or with the roller axis than in the MD. This type of grooving is called either **chevron** (for winder drum traction) or **herringbone** (erroneously believed to spread). There seems to be no substantial motivation for these grooving styles. Also, these grooving styles can be noisy at high speeds.

Web Flattening

A web which is spread will have positive CD stresses and strains. As defined in Chapter 17, a web will be wider (discounting hygrothermal strains) at a spreading element than at its upstream roller.[13] Spreading is used most commonly to prevent or remove some types of wrinkles,[14,15,16] though it is rarely effective for baggy webs as widely believed.[17,18] Spreaders are also used to open up slit lanes (during winding) and even permanently widen the web (some nonwovens and films).

Web flattening is similar to spreading, but has a subtle distinguishing difference. While spreading will tend to make a web wider than it would naturally be, web flattening will make as web as wide as it would naturally be. Web flattening is still a very useful tool nonetheless. It will help prevent either MD trough wrinkles or diagonal shear wrinkles from developing on that element.

Web flattening works on a principle that webs abhor compound curvature. A web which wraps around a bar or roller already has one plane of curvature. Thus, it would resist developing a troughing curvature in the other plane. What would prevent the wrinkle from sliding sideways and flattening itself out is the friction developed between the web and roller. Thus, as seen in Figure 10.5, a wrinkle will be sustained on a roller when the web/roller frictional resistance equals or exceeds the CD compression at the wrinkle.[19,20]

Web **troughing** is defined as out-of-plane deformations of the web in an open span which are primarily oriented in the MD and roughly approximating a sine wave in cross section. This curtain like appearance is a symptom of a marginal performance but is seldom a problem in itself. An exception is drying where one wants to keep the web flat so that it doesn't lock in any out-of-plane features. Web **wrinkling** is when the trough crosses a roller, and may leave a permanent impression on the web. **Creases and foldovers** are when wrinkles collapse on the roller, and are usually a rejectable defects.

Figure 10.5
Forces of a Wrinkle Crossing a Roller

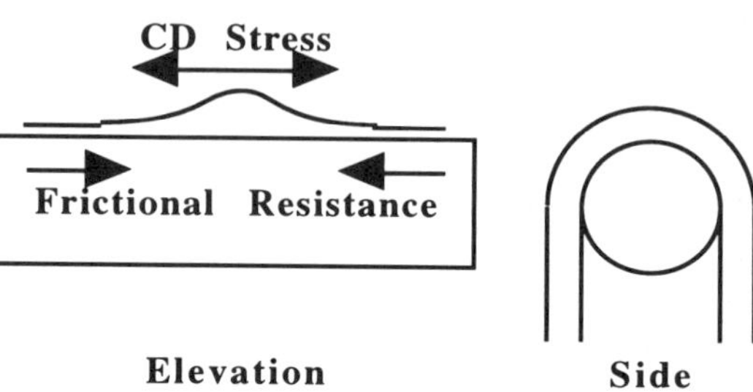

Thus, while most processes can tolerate web span troughing, wrinkles over a roller are not acceptable. Obviously, we could add a spreader in that location to effectively reduce both troughing and wrinkling. However, a less expensive way to reduce the tendency of a trough to cross a roller as a wrinkle is to use web flattening. To do this, we need to change one or more of the traction and other factors on the roller as summarized in Table 10.1.

While web flattening can be very effective, there are a few caveats. First, one should consider the sliding or traction modes of web/element interaction. If the element is in pure sliding, there will be some tension rise in the web there. Traction is a somewhat more difficult mode because it is a balancing act. If traction is too high, the web will not slide sideways and flatten. If the traction is too low, gross slippage will take place everywhere so that the web and roller will not travel at the same speeds. What is desired is that the web track along part of the width but slips as necessary in portions to relieve any developing wrinkles. However, a roller that is in partial slippage across the web will compromise edge position as conditions change.

Table 10.1
Factors for Web Flattening

Roller Diameter - Increase
Roller Speed - Faster or Slower than web
Tension - Decrease
Wrap Angle - Decrease
Web/Roller Friction - Decrease
smooth roller surface
metal instead of covers
web speed increase

Hardness and Durability

Desirably, the surface of a roller will be resistant to abrasion wear, cuts and scratches, and permanent indentation. Resistance to these factors means an increased durability and thus an increase in life between resurfacings. Durability can be an important issue for metal surfaces which are covered in this section, and covers which were treated in an earlier section.

Just as with covers, metal shell durability is weakly associated with hardness. Just as with covers, hardness is also measured with indentors. However, while harder metals are sometimes not as tough, there is no counterpart with covers. The primary differences between metal and cover hardness is that the indentor ball is much smaller so that the unit pressure is higher, and that the deflections are much smaller.

Common metal hardness testers include Brinell, Rockwell, Vickers and the Shore Sclerosope.[21] The **Brinell** test impresses a 10mm diameter ball under a 3000kg (or less for softer metals). The hardness number is the load divided by the area of the indentation as measured by a microscope. The **Rockwell** hardness is popular because readings don't require a microscope. Instead, the hardness is the difference in deflection between high and low indentor loadings. There are 5 scales, A-E, which are used for measuring very hard to very soft metals. The different scales use different indentors and loads varying between 60 and 100 kg. The **Vickers** hardness is similar to the Brinell test except that the indentor is a pyramid instead of a ball or cone. The hardness is the indentation area divided by the load. The **Shore Scleroscope** is an impact device where the rebound height of a diamond hammer is measured. The scales of all of these tests are such that a greater number implies a harder material. Conversions between the scales can and have been done empirically.

If **abrasion** is of primary concern, it should perhaps be tested rather than inferred by hardness. Abrasion tests involve a controlled speed difference between two objects loaded together for a period of time. Wear is measured by a depth of material removal or a decrease in weight.

Large Scale Topology

Some process rollers, such as rotary die-cutters and embossers, have very deep patterns cut into their surfaces. These patterns may be generated by a number of machining processes. The cost of adding a pattern can be quite large and depends on the volume of material to be removed, the difficulty of material removal, and the intricacy of the pattern.

Casting
A patterned shell can be cast using a mandrel supported in the center of a split mold. Casting is common for polymers with a wheel type geometry (a narrow width roller).

EDM
EDM (electro-discharge machining) is used to cut patterns into very hard metals. The discharge is between a carbon pattern and the workpiece.

Etching
Acid can etch patterns into a roller shell which is stenciled to mask off the land areas from attack. Etching is also called chemical milling. Etching which is sped up via electrolysis is called electrochemical machining.

Knurling
A patterned wheel highly nipped against a shell can texture soft materials such as aluminum.

Laser Engraving
Cutting lasers can engrave a pattern on materials which are sufficiently dark to absorb the laser's energy. It is a very quick and cost effective method for cutting soft nonmetals.

Machining
Patterns can be cut into a blank shell using conventional rotary bits and tooling. The tooling path is usually determined via a computer programmed NC (numerical controlled) machine.

Soldering
A raised pattern can be silver soldered onto a blank shell. This time consuming art is used to make watermark rollers for currency and other specialty papers.

Bibliography

1. *Surface Texture.* American National Standards Institute, ANSI B46.1, 1978.

2. International Standards, ISO R-468, 1880 and 3274, 1974-75.

3. Forrest, Albert W. Jr. *A Mathematical and Experimental Investigation of the Stack Compression of Films.* 2nd Int'l Conf. on Web Handling, Oklahoma State Univ., June 6-9, 1993.

4. Forrest, Albert W. Jr. *Wound Roll Stress Analysis Including Air Entrainment and the Formation of Roll Defects.* 3rd Int'l Conf. on Web Handling, Oklahoma State Univ., June 18-21, 1995.

5. Forrest, Albert W. Jr. *Air Entrainment During Film Winding with Layon Rolls.* 3rd Int'l Conf. on Web Handling, Oklahoma State Univ., June 18-21, 1995.

6. Good, J. Keith and Xu, Y. *Computing Wound Roll Stresses Based Upon Web Surface Characteristics.* 2nd Int'l Conf. on Web Handling, Oklahoma State Univ., June 6-9, 1993.

7. Knapp, J.K. and Lampman, R.D. *Characterization of Calender Roll Surface Features Using Interferometric Profiling.* TAPPI Finishing and Converting Conf. Proc., San Diego, CA, pp 15-24, Oct 15-18, 1995.

8. Maniatty, George S. *Roll Grinding Measurements and their Effect on Paper Making.* TAPPI Finishing and Converting Conf. Proc., New Orleans, pp 311-320, Oct 24-27, 1993.

9. Maniatty, George S. *Roll Grinding Measurements and Their Effect on PaperMaking.* Paper Age.

10. Roisum, David R. *The Mechanics of Web Spreading - Parts I & II.* Tappi J., vol 76, no 10, pp 63-70, October 1993 and vol 76, no 12, pp 75-86, December 1993.

11. Knox, K.L. and Sweeney, T.L. *Fluid Effects Associated with Web Handling.* Ind. Eng. Chem. Process Des. Develop., vol. 10, no 2, 1971, pp 201-205, 1971.

12. Tajuddin, By. *Mathematical Modelling of Air Entrainment in Web Handling Applications.* Oklahoma. State Univ., WHRC, M.S. Thesis, 1987.

13. Roisum, David R. *Web Spreading.* Converting Magazine, November 1995.

14. Roisum, David R. *Why is My Web Wrinkled?* Converting Magazine, Web Works column, pp 16, February 1994.

15. Roisum, David R. *What Causes Machine Direction Trough Wrinkles?* Converting Magazine, Web Works column, pp 22, March 1994.

16. Roisum, David R. *What Causes Diagonal Shear Wrinkles?* Converting Magazine, Web Works column, pp 22, June 1994.

17. Roisum, David R. *What Causes a Baggy Web?* Converting Magazine, Web Works column, pp 22, April 1994.

18. Roisum, David R. *How Can I Fix a Baggy Web?* Converting Magazine, Web Works column, pp 20, April 1995.

19. Friedrich, Craig Richard. *Stability Sensitivity of a Web Wrinkle on a Cylindrical Roller.* Oklahoma State Univ., WHRC, Ph.D. Thesis, 1987.

20. Friedrich, Craig Richard. *Stability Sensitivity of Web Wrinkles on Rollers.* Tappi J., vol 72, no 2, pp 161-165, February 1989.

21. Deutschman, Aaron and Michels, Walter J. and Wilson, Charles E. *Machine Design, Theory and Practice.* Macmillan Publishing Co., Inc., New York.

Chapter 11

Vibration and Balance

In this chapter we discuss the causes, cures and effects of web machine vibration. In most machines, roller unbalance is a large driving force for vibration so that the basics of balancing techniques are covered. In some machines, it is the runout of nipped roll(ers) that is the predominant source of vibration.

Adverse Effects of Machine Vibration

Web machine vibration can be a very troublesome issue that can cause a bewildering variety of deleterious effects as shown in Table 11.1. In one way or another, rollers are central to most vibration issues, and thus deserves a brief coverage in this book.

The production superintendent will most likely lament the loss of productivity as the machine has to be slowed in order to keep the process from pounding itself apart. This unfortunate situation is most easily resolved during the machine design stage of a project. The technology exists to design and manufacture web machinery to operate smoothly well in excess of 10,000 FPM. Unfortunately, vibration is not so easily resolved on existing machinery unless the problem source is an easily corrected maintenance item such as reducing clearance, balancing rollers and optical re-alignment.

Another common result of web machine vibration is the loss of product quality and perhaps even waste and web breaks. Only rarely is this connection immediately obvious such as barring due to nip vibration, or web breaks due to sheet flutter. More often, the connection is insidious and overlooked.

Maintenance people have long known that machine vibration will take its toll on the life of components such as bearings, couplings and so on. For example, if dynamic vibration loads are merely increased from a modest 10% of static to double the roller weight, the life of ball bearings will decrease 8 fold!!

Table 11.1
Adverse Effects of Vibration

Maintenance increase due to component wear
Noise
Operator discomfort and illness
Productivity loss due to the need to reduce speed
Safety issues: roller failure and thrown rolls
Waste due to product variations and web breaks

Thus, if one expects to achieve anything near the desirable 40 year useful life of machines, the vibration levels must be limited. Unfortunately, maintenance people do not always realize that they have a key role in this area by eliminating excessive clearances and periodically checking balance and alignment.

It is also uncomfortable for operators to work in the vicinity of vibrating machinery. High frequency vibration is the source of noise which can cause permanent hearing loss at regular exposures of as little as 85 dBa. While hearing protection does offer significant protection, noise will impair communication. Noise also increases operator stress and fatigue. Low frequency vibration, such as 10-20 Hz, is common on web machinery and does not pose a threat to hearing. However, these frequencies can cause a sympathetic vibration of the human abdomen which can result in mysterious illnesses if moderate, and nausea if extreme. The path of transmission of this type of vibration is from the machine, through the floor and foundation, and up the feet and legs of the operator.

Last but not least, vibration can be deadly. Though rare and avoidable, roller journals have failed by fatigue causing them to escape their containment in a machine. A more common risk is the threat of thrown rolls on some shaftless winders which might have several episodes every year. These and many other reasons suggest prudence when approaching an operating web machine.

Vibration Sources

Machine vibration is an extremely complex system issue that does not usually lend itself to a single cause or correction. Nonetheless, there are several major and recurring themes of web machine vibration as listed in Table 11.2.

A **bent shaft**, journal or roller is the result of an obvious accident and/or poor design. It will increase both runout and unbalance. The diagnosis is an increase in TIR, especially at one end. The cure is usually replacement as rework may be too costly and/or ineffective.

Brakes, clutches and other components may **chatter** due to the tendency for frictional stick-slip. Sometimes this chatter will go away after one or both surfaces are refaced.

Clearance of bearings, slides and similar elements does not in itself cause vibration. However, undesirable amplitudes may result from the extra freedom. Clearance must be reduced per supplier specifications to a minimum value, but not so much as to increase binding loads and wear. Compliance (flexibility) allows large amplitudes and is corrected by stout framework, foundations and component design.

Electric motors generate a small oscillation of torque at integral numbers of rotation speed due to discrete poles and windings, and due to air gap variations etc.

Misalignment can cause vibration at integral multiples of the rotational speed, mainly through its interaction at nips, couplings and so on.

Machinery with **nipped roll(ers)** are unusually suspect to machine vibration. First, any runout on nipped roll(ers) will cause them to oscillate much like a rotating cam follower. Second, any circumferential hardness variations in the cover of a roller or in the surface of a wound roll will have a similar effect as does runout. These two sources are a common cause of barring. Finally, nipped rollers may have a very low system resonant frequency because nips are typically much softer than other machine elements.

Table 11.2
Common Causes of Vibration

Bent shaft
Chatter of brakes, clutches, gears, etc.
Clearance, compliance or looseness
Electric motor torque variations
Misalignment
Nip roll(er) cover hardness variation
Nip roll(er) runout
Nonconstant speed: chains, gears, u-joints, etc.
Other machinery nearby on insufficient flooring
Resonance
Roller unbalance
Unwind/windup roll

Vibration from **nearby machinery** can be transmitted large distances through unsubstantial flooring and foundations.

Some mechanical elements are not **constant speed** devices. Examples include roller chains, gears, universal joints and other couplings. In these examples, the speed and thus torque varies with a small sinusoidal dither. This is often easy to detect by looking at the pitch of the product variations or the shape of the polar TIR plot.

Resonance, sometimes inappropriately called critical speed, can raise vibration amplitudes as much as 10X at certain machine speeds depending on the amount of damping present. We will show that the critical speed of rollers is a usually meaningless calculation as most modes of real machine vibration are system resonances that are far lower than predicted by simple beam equations. Indeed, most machinery resonates in operation and resonance is unavoidable on wide high speed machinery.

Roller unbalance is a common source of vibration that is easy to diagnose (frequency of vibration is equal to rotational frequency) and is usually easy to correct by balancing. This will be covered in more detail later in this chapter.

Unwind or windup rolls as well as diecutting and embossing rollers can cause the sheet to flutter due to surface speed variations. However, if the web doesn't break in that span, it will usually survive the transit as the tension variations will be greatly attenuated after going over just one or two rollers.

Vibration Terminology

Amplitude can be defined as the magnitude or extent of oscillation of a body from one extreme to its neutral position (peak), or to its opposite extreme (peak to peak). Amplitude is a measure of the severity of vibration, and can be expressed in three interrelated quantities:

> **Displacement** is the amplitude of the position changes measured in thousandths of an inch or mils (0.001" = 1 mil).
> **Velocity** is the amplitude of the speed changes measured in units of in/sec.
> **Acceleration** is the amplitude of the time rate of change of velocity measured in g's (1 g = 386 in/sec^2).

Which units are selected for vibration analysis depends on the type of vibration equipment used, type of analysis, as well as personal preferences. Acceleration, though used only occasionally, is most useful when dynamic forces are to be determined. Velocity is the most common unit of measure because it gives a direct indication of severity which accounts for both displacement and frequency. Displacement yields information on the extent of motion. Most vibration analyzers are capable of displaying any of the three units of measure.

Antinodes are positions of maximum movement on a structure when excited at a particular natural frequency.

Boundary Conditions are constraints applied to points on a structure that can be either fixed (no movement) or free (no external restraint). Boundary conditions for a point must be specified for each of the 6 kinds of displacements: 3 translations and 3 rotations. An important concept is that boundary conditions have a tremendous effect on natural frequencies. For this reason, the designer should make efforts to maximize fixation at roller and framework supports.

Additionally, play and clearance in a structure, often neglected in analysis, can have a deadly effect on vibration by reducing the natural frequencies and increasing the severity of vibration. Amplitudes are nearly guaranteed to exceed any clearances already present.

Critical Speed is the speed at which a roller will reach resonance and the resulting vibration may increase tenfold. However, the simple design equations used to predict critical speeds of rollers are nearly always inappropriate for real machinery as it assumes completely rigid support. Unfortunately, using these equations results in nonconservative values which greatly overpredict actual resonant speeds of machine systems. Finally, reduced system resonances due to framework and foundation compliance are often mistaken as half-criticals. Vibration problems due to half-criticals are rare.

Damping (or dampening) can be defined as the resistance to motion which dissipates vibratory energy. Damping is the mechanism that keeps machinery from literally flying apart when operating at resonance. There are a number of different types of damping, but the textbook case is viscous damping where the resistant force is proportional to the speed. Damping can be measured by observing the rate of vibration decay after a structure is struck. Most metal structures are very lightly damped.

Eccentricity is the distance between the center of mass of a roller and its axis of rotation. Unbalance force is proportional to eccentricity, mass and the rotational frequency squared.

FEM is an abbreviation for Finite Element Modeling. This is a numerical/computer technique that can be used to predict the dynamic behavior of web machines. Indeed, FEM is required to predict any system more complex than a single roller on simple stands.

FFT is an abbreviation for Fast Fourier Transform and the instrumentation that uses this mathematical principle. An FFT takes a amplitude versus time signal from a sensor and displays it as amplitude versus frequency. Spikes on this plot will usually correspond in frequency to an integral multiple of the rotational frequency of an offending roller.

Forced Vibration occurs when a structure is driven by a periodically varying force such as an unbalanced roller and will continue to oscillate as long as the varying force remains. Forced vibration due to rotating elements is the most common cause of web machine vibration problems.

Forcing Function is a broad term describing the cause of vibration in terms of amplitude, frequency, position and direction. A forcing function can be a displacement input such as a rotating cam shaft moving a valve or a force input such as roll imbalance. In many cases, winder vibration is driven by rotating roll unbalance with the energy being supplied by the electric drive motors and dissipated through the structure as heat, hysteresis and noise.

Free Vibration occurs when a body continues to oscillate at its natural frequencies with no outside forces to drive it, such as a bell after it is struck or a guitar string after it is plucked, and will die out after a time due to damping. The thump or rap test utilize free vibrations to determine natural frequencies.

Frequency can be defined as the rate at which a body moves from one extreme to the other and back again and is commonly measured in Hz (Hertz, CPS or cycles per second) or KCPM (kilocycles per minute) or radians/sec. 1 Hz = 0.06 KCPM = 2π rads/sec. The usual frequency range of interest for web machine vibration is 1-50 Hz and for noise is 20-20,000 Hz.

Roller **imbalance** force is a primary forcing function for machine vibration. Unbalance is proportional to eccentricity, mass and rotational speed squared.

Mode Shape (or Eigenvector) is the dynamic shape of a structure at a particular natural frequency. Each natural frequency has its own mode shape determined in a complex manner by the mass and stiffness distribution of the system. Mode shape information is invaluable because it tells us how we may increase the natural frequencies for a particular mode. This may be done by removing mass at or near antinodes and by increasing stiffness on components that experience large relative dynamic deflections.

Natural Frequency (or Eigenvalue) is the frequency at which a structure tend to vibrate. Examples of this are the pitch (frequency) at which a bell rings, or a guitar string vibrates when struck. Any structure possesses an infinite number of natural frequencies, each with its own characteristic mode shape. However as a practical matter, usually only the lowest half dozen modes or frequencies are of significance.

Nodes are the points on a structure which do not move when excited at a particular natural frequency.

Resonance is a condition where the frequency of the forcing function nearly equals one of the natural frequencies of a structure. Resonance can often be a very serious problem because the amplitude or severity can be magnified as much as 10 times when approaching this critical condition which is often called the critical speed. The severity depends upon many factors including:

1. Frequency ratio = frequency of forcing function/natural frequency. The closer this ratio gets to unity, the worse the vibration.
2. The severity increases somewhat linearly with increasing magnitude of the forcing function.
3. Severity increases with decreasing damping.

Unstable systems have responses that are not always predictable and may be divergent (ever increasing swings in amplitude). Machine examples include the flutter of airplane control surfaces at high speed and control systems set at excessive gain. However, there are at least two web machine systems that can be unstable. One is a slender shaft or roller that gets bent, and the other is a nipped rewound roll.

Balancing Terminology Reference

ANSI S2.7-1980 (R1986) (ASA 42). American National Standard Balancing Terminology.

A Simple Roller on Simple Mounts

The simplest roller system is one that can be modeled as a uniform cylinder (no heads, journals or nips) on infinitely rigid supports. It natural frequency can be calculated from classical beam theory as:

(11.1)

$$f = \frac{\pi}{2} \sqrt{\frac{g E I}{w l^4}}$$

where
f = natural frequency (Hz)
g = acceleration of gravity = 386 in/sec^2
E = Youn'gs Modulus (lb/in^2)
I = area moment of inertia (in4) = $\pi(D^4-d^4)/64$
D = outside diameter (in)
d = inside diameter (in)
w = weight per length (lb/in)
l = length; and where
Critical Speed (ft/min) = 15.71xfxD
[Natural frequency can also be calculated in a consistent set of metric units. However, see Appendix C for speed from frequency.]

However, it will seldom be prudent to use this formula as many do. Here is why. First, the formula does not consider journal flexibility. Second, the formula does not consider roller head weight. But most importantly, this formula is based on simple supports that are infinitely rigid. All of these errors are nonconservative and can over-predict natural frequencies by as much as a factor of two or more. This has led many to misinterpret system resonances as the 1/2 critical.

Real machinery does not have infinitely rigid roller mountings (framework and foundation). Quite commonly the stands upon which rollers are mounted are little more rigid than the rollers themselves. Even a stout framework will be flexible if it is tall. Similar reductions in real resonant frequencies will result from mass that is not considered such as framework and other rollers or components.

Framework flexibility is illustrated in Figure 11.1 for the slightly more complicated case of a roller mounted on stands. As seen here, modes occur in pairs with the lowest corresponding to the direction of stand flexibility.

Figure 11.1
Vibration Modes for Rollers on Stands

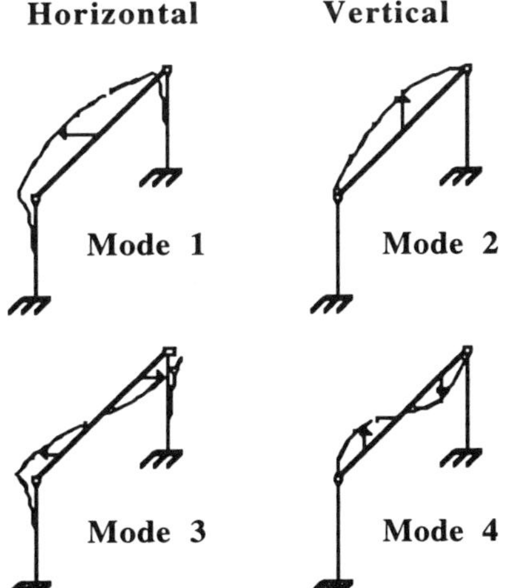

Closed form solutions for natural frequencies do exist as Rayleigh extensions to equation 11.2 for Mode 1 without the effects of heads and journals, and Mode 2 with journal flexibility but without stand flexibility.[1] However, they are usually too restrictive to be useful. Indeed, we need to consider the entire machine (and even flooring) as a system in order to get useful answers to dynamic behavior.[2,3,4] However, a simplistic critical speed calculation may have application merely to ensure ease of balancing (avoids flexible rotors).

The only sure methods to determine natural frequencies of even extremely simple web machines are either by FEM (Finite Element Modeling), or FFT (Fast Fourier Transform) measurement. Unfortunately, it is extremely rare for a web machine to be engineered or analyzed using these kinds of tools because of the cost and the skills required to execute and interpret.

The realities are that most wide web machinery running over a few hundred FPM is operated at or through resonances without notice or complaint. This illustrates that while you should avoid rough spots, running near a resonance need not be debilitating. In most cases, all one needs to avoid excessive vibration are rollers that are well balanced and true, and the framework that is quite tight and robust.

Measuring and Identifying Vibration

Certainly, a whole book could be written on the subject of web machine vibration. Here, however, we will present a simple procedure that will identify the most common sources of vibration. Figure 11.2 illustrates the technique.

First, ask the operating personnel what conditions (speed, grade, settings, etc.) and locations seem to give the most problem. Attempting to reproduce the severest conditions will make the problem(s) easier to identify. Knowing the problem location will reduce the number of positions that need to be checked.

Next, the vibration sensor is attached to a close but accessible part of the machine. In most cases, the sensor will be an accelerometer which is attached to a bearing housing by a magnet. It is important to orient the sensor in the direction which is estimated to be the most compliant (usually MD) because it will often yield the greatest vibration amplitudes. For completeness, other locations and orientations should be monitored as time permits.

The cable from the sensor is safely and securely routed to the FFT instrument. The FFT converts the amplitude versus time signal from the sensor to an amplitude versus frequency. The amplitude can be either displacement (movement), velocity (severity) or acceleration (force). However, the most useful information is to identify the diameter of the offending roller(s) from:

$$\text{(11.2)} \quad \text{Dia(in)} = \frac{1}{\text{integer}} \; \frac{\text{speed(ft/min)}}{15.71 \times \text{freq(Hz)}}$$

Roller unbalance will usually show up with a steady amplitude at 1x rotational frequency (integer=1). However, misalignment, looseness and bent shafts will show also show smaller spikes at 2X and higher multiples. Much higher multiples are seen with components such as bearings, gears, roller chains, timing belts and electrical drive motors. If the readings wander and are not repeatable, it can indicate either looseness or instability (such as on wound rolls). If any significant axial motion is detected, it can be a symptom of mechanical problems that should be obvious by other measures.

Figure 11.2
FFT Identification of the Offending Rollers

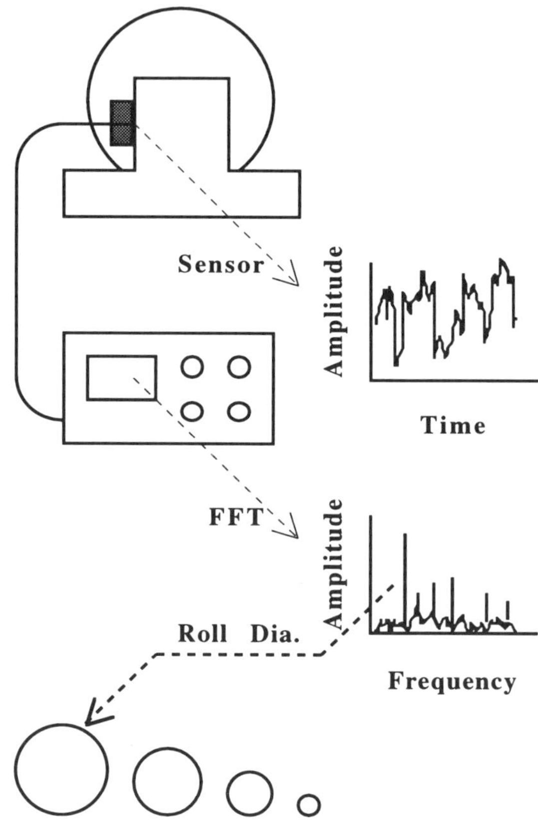

It is important to keep in mind the proximity effect because a sensor will weight and read the effects of the nearest (or heaviest) roller more than other sources. Thus, all suspect rollers should be checked for relative comparison or a neutral point on the framework can be used as an arbitrator.

The amplitude of frequency spikes may slowly oscillate due to beating. This is caused by nominally identical diameter rollers in proximity where the heavy spots go in and out of phase. Beating of different diameter rollers in proximity will show up as sum and difference frequencies. This is also normal and can be ignored.

This simple and powerful technique will usually identify the offending roller, and often the precise cause (such as roller unbalance). However, a more complete understanding can be achieved if resonances and mode shapes are diagnosed.

Resonances and Mode Shapes

Many vibrating situations can be improved merely by identifying the source and reducing its forcing function. The classic case is roller unbalance where the treatment is roller balancing. However, some moderate speed (>100 FPm) and most high speed (>1000 FPM) machines will operate through and at resonant conditions, especially if they are wide (>100 inch). Also, some components such as bowed spreader rollers, load cell rollers and unwinding/winding shafts and cores are often operated at and through resonance.

The difficulty with resonance is that the amplitudes will usually double and perhaps increase as much as 10X at critical speeds than at speeds only 10% away from the criticals. To understand the resonances, we want to know their speed, severity and shape. Knowing the resonant speed allows us to avoid the rough spots of operation. Knowing the severity will allow us to judge whether we can operate at those rough spots or should accel/decel to nearby smoother speeds. Knowing the mode shape will tell us how the machine would have to be rebuilt in order to move or eliminate the resonance.

One powerful technique for obtaining these kinds of answers is to model the machine with FEM. This is especially useful for visualizing mode shapes and testing alternative designs. Unfortunately, modelling even a very modest machine will take at least a month of skilled and dedicated labor. Also, the FEM model should always be verified and tuned by field measurement. Thus, it may be advantageous to pursue quicker experimental procedures.

The easiest way to find resonances is to use a *sweep test.* This is done by slowly increasing the speed over its operating range and watching for large and sudden increases in amplitude at resonance followed by a decrease as the resonant speed is passed. Similar information can be obtained from a *bump test* where the machine frame is struck in various locations and directions by a big hammer. Another way to excite the structure is to bump the foundation by dropping a heavy object from a small height.

For either the sweep or bump test, the FFT is set to peak mode and the resonant frequencies will show up as spikes. The frequency of these spikes can be converted to speeds through roll(er) diameter and equation (11.2). Special difficulties will arise with winders and unwinds where the roll frequency is continuously varying (even at constant machine speed), and the changing mass (diameter) will move resonances.

However, there are several conditions that must be met to excite the resonance sufficiently to become a problem. First, the roller must be run at (1X) the indicated speed. Second, the forcing function (e.g. from roller unbalance) must be oriented in the appropriate direction. Third, the forcing function must be close to an antinode of vibration. Finally, the forcing function must be large enough.

Determining the severity of the resonances, once found, is easy. All one has to do is to run at those speeds and observe the consequences.

Mode shape is a bit harder to get. The most elegant solution is to use vibration equipment that is equipped to do modal analysis. Unfortunately, the equipment and skills required to do this are quite specialized and not readily utilized.

There is, however, a poor man's modal analysis. This is done by running at the resonant speeds (remember that there may be many), and recording the amplitude of vibration at many locations and directions. These amplitudes can be scaled and sketched on top of a scaled drawing of the machine as a series of dots. Finally, the dots are connected with smooth curves much like a dot-to-dot picture. One difficulty is dealing with antisymmetric modes, such as the front of the machine moving in the +MD and the back moving in the -MD. Another is confusion from multiple sources and modes.

A modal picture tells us how to redesign the machine to change (usually increase) natural frequencies. The cross section is increased on beams or rollers which have large dynamic curvature. Alternatively, the mass is decreased near antinodes.

How Much Is Too Much

Defining an acceptable level of vibration is difficult because it depends on the type of machine, component or process, the operating speeds, and the web material. In many cases, the opinions and perceptions of the owners or operators of the machine will define acceptability. However, a neutral party with considerable experience may best arbitrate the occasional disputes between machine builders and their customers.

It is unusual for the direct connection between vibration and a problem to be intimately established. Exceptions include the need to avoid obvious resonance, some web breaks, nipped roll barring and shear slitter jumps. Sometimes it is not the vibrating machine itself that experiences the problem. Rather, for example, it might be operator discomfort or disturbances to sensitive instrumentation that is located nearby.

However, the first decision that is encountered is to select a unit of measure for vibration. The choices include displacement (movement), velocity (severity) or acceleration (forces). Fortunately, it is easy to convert from one set of units to another as long as frequency is known.[5]

The most common unit to evaluate vibration levels is velocity (in/sec in English units). Charts are widely circulated and distributed by instrument makers which are used to convert between the units. These charts will often classify vibration severity as given in Table 11.3.

Unfortunately, these universal charts do not really provide a good interpretation of acceptability for web processes. For example, wide high speed paper machinery will at best be "fair" by that scale, and will most often be in the "very rough" region. Furthermore, web flutter and vibration will be well off the scale. Yet these machines feel relatively smooth and seldom cause obvious problems at these levels.

I have taken a different tack in my interpretation of vibration severity. First, I will use displacement because it is more intuitive for mechanical people and is readily compared to other geometrical errors in web machinery.

Table 11.3
Vibration Severity

Condition	Vibration (in/sec peak)
Extremely Smooth	<0.0049
Very Smooth	<0.0098
Smooth	<0.0196
Very Good	<0.0392
Good	<0.0785
Fair	<0.1570
Slightly Rough	<0.314
Rough	<0.628
Very Rough	> 0.628

Second, my scale is based solely upon web machinery. Furthermore, the levels are set by the threshold for which resulting problems were identifiable or were perceived so by machine owners. Finally, they are based on economically achievable levels (because "smooth" is just not going to happen on wide high speed paper machinery). These levels are presented in Table 11.4.

Most wide (>100") high speed (>1000 FPM) machinery will easily be able to meet these values provided that the design is robust and the machine well maintained. Thus, we should also expect slower or narrow web processes to easily meet the criteria. Unfortunately, sometimes corners are cut when designing and building small converting machinery. Designers should consider that metal at only $1/lb and concrete at less than 1¢/lb can be a very wise investment. While rollers can be rebalanced with modest trouble, rebuilding framework is seldom economical after the fact.

Table 11.4
Guidelines for Vibration Maximums

Component	Vibration (mils. pk-pk)
Foundation	0.5
Slitters	2
Rollers (process)	1
Rollers (transport)	5
Rollers (lay-on)	10
Rolls (corechuck/shaft)	20
Rolls (wound)	50
Web (mid span)	100

Unbalance

The simplest form of unbalance (also called imbalance) is when the center of mass of a roller does not lie along the axis of rotation. As seen in Figure 11.3, the CG and axis are not coincident perhaps because the journals were not mounted centrically or they were bent. This unbalance cause may be easy to detect because it would result in TIR or runout. However, as seen in Figure 11.4, the shell of a roller could contain light and heavy spots in a casting which would still cause an unbalance, but would not show up as static TIR.

There are many other ways that rollers can be made unbalanced or later be put out of balance as given in Table 11.5. Obviously, it would be ideal to know the precise cause(s) so that they could be minimized in the future. However, in many cases it is just easier to rebalance the roller than trying to figure out how changing tolerances stack up.

In any case, the radial distance between the CG and axis for this unbalance is called eccentricity, and is shown as an "e." The product of roller mass and its eccentricity is the value of unbalance. Finally, the centrifugal force resulting from this unbalance can be calculated as

(11.3)

$$F\,(\text{lbf}) = 1.77 \text{ x } \left(\frac{\text{rpm}}{1000}\right)^2 \text{ x Unbalance (oz} - \text{in) or}$$

$$F\,(\text{N}) = 1.1 \text{ x } \left(\frac{\text{rpm}}{10}\right)^2 \text{ x Unbalance (kg} - \text{m)}$$

Note that most balance and vibration (and drive) people use rotational speed in rpm (revolutions per minute), whereas most web people use surface speed in FPM or MPM. However,

$$(11.4) \quad \text{rpm} = \frac{3.82 \text{ x Speed (ft/min)}}{\text{Diameter (in)}}$$

A few sample calculations will show how unbalance forces will soar at high speeds. This is especially critical in the paper industry[6,7,8] where rollers are also massive. In any case, unbalance causes premature failure of many machine components.

Figure 11.3
Unbalance with Runout

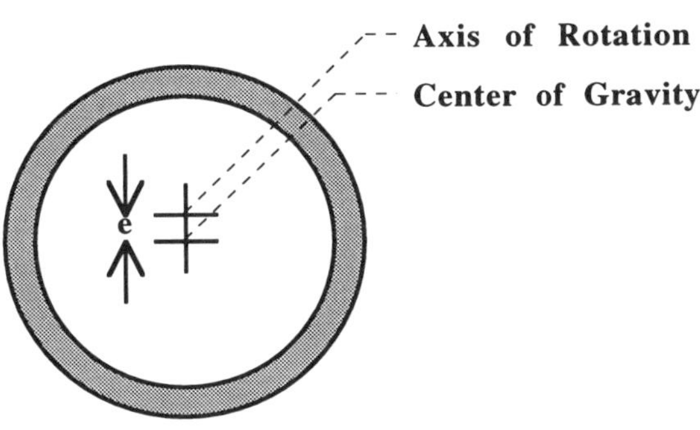

Figure 11.4
Unbalance without Runout

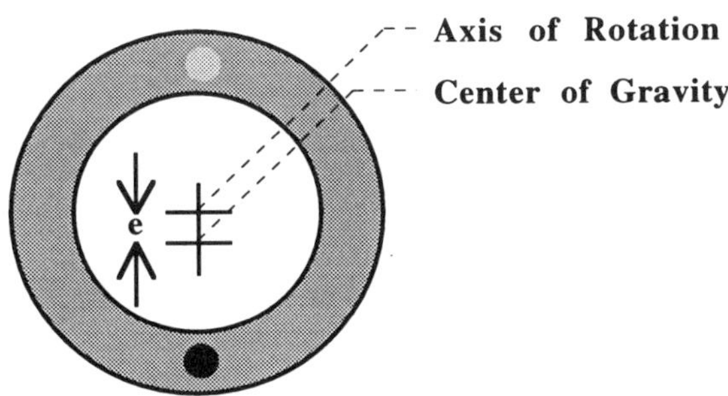

Table 11.5
Sources of Unbalance

<u>Original Unbalance</u>
Bearing eccentricity
Castings - unmachined portions
Homogeneity - lack of in raw materials
Manufacturing tolerances
Nonsymmetry of parts
Stress relieving

<u>Subsequent Unbalance</u>
Bearing wear
Cleaning
Changing orientation of assembly clearances
Changing, rearranging parts and fasteners
Shafts - bent due to overload
Shell wear
Thermal distortion

Simple Static Balancing

The static unbalance shown would cause the roller to tend to **pendulum** to the heavy side, particularly if the bearings were very free (coast for >10 seconds with a sharp hand spin). Indeed, the simplest field check of balance is to watch where the roll stops. If it consistently stops in the same location, the roller will exceed any reasonable specification for even the least demanding applications.

Any roller that pendulums can be field balanced without any instruments as an interim step until it can be more properly balanced. The procedure is to add trial weights at the top of the roller until it just begins to pendulum to the trial weight side. Then the weights are permanently fixed with half on each end (front and back) side of the roller as near as possible to the outside diameter of the roller.

This procedure has sensitivity limitations due to bearing friction. However, this can be improved upon in several simple ways. First, the roller mounting can be shook or tapped to reduce friction. Second, the (dead shaft) roller can be taken out and supported by the journals on knife edges or on low friction cam roller saddles.

Another field technique is a pencil marking method. As seen in Figure 11.5, the roller is turned at speed with a pencil held very steady in close proximity to the shell. The side that has the mark may be the heavy side. This can be extended to dynamic balance by doing the procedure on both ends independently. Unfortunately, this method is very dependent on the roller being quite true (very low static TIR), and that it is operated well below the first critical. Even worse is that while phase is known, unbalance mass is not (trial & errors).

The problem with static balance, even supplemented with instrumentation, is that it can correct for only two distinct cases of unbalance as seen in Figure 11.6. The first is where the rotor is short compared to its diameter (l/d < about 3), which is rarely the case for rollers. The second is where the unbalance distribution for a more slender roller is evenly distributed or balances about the centerline of the roller.

Figure 11.5
Pencil Method for Simple Balance

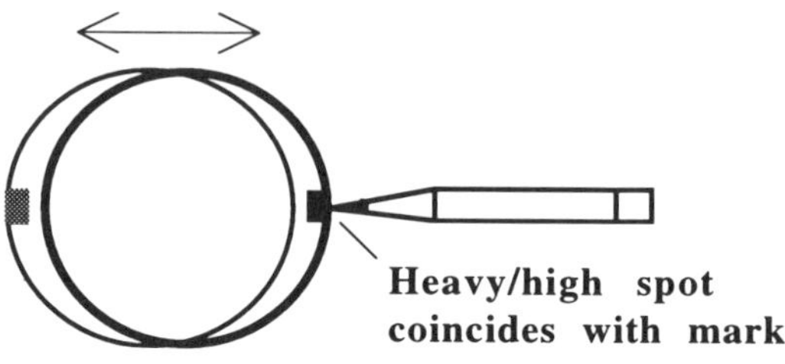

Figure 11.6
Static Unbalance Cases

Short Roller

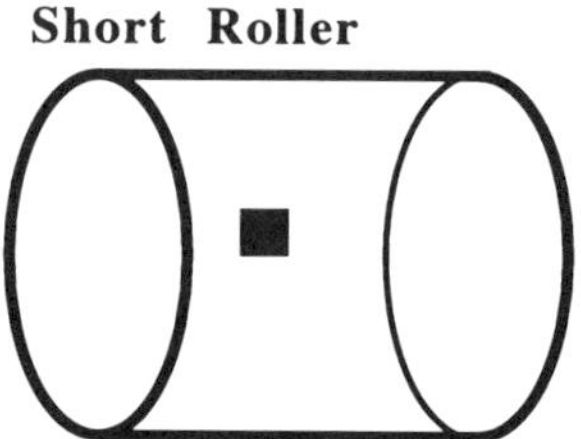

Uniformly Distributed Heavy/Light Spot

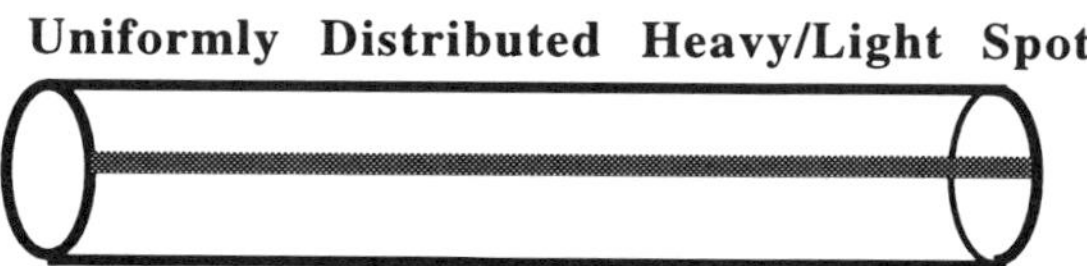

CG of Heavy/Light Spots at Roll CL and Principle Axes of Spots & Roll ||

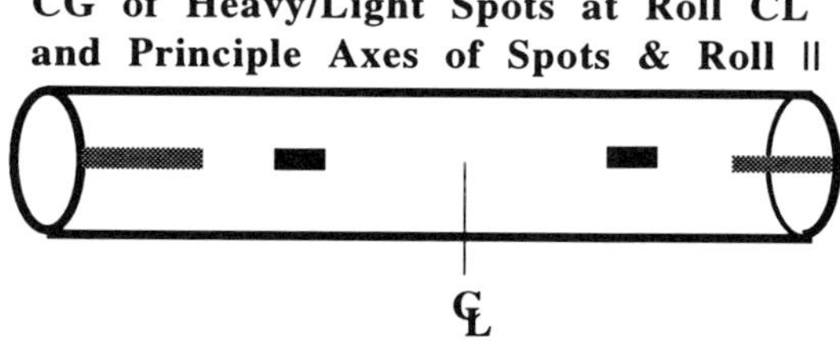

While a good static balance may be sufficient for slow speed operations (<100 FPM), there is no practical reason to use it except in a field emergency where no instrumentation is available. The reason for this is that with very modest instrumentation (in machine or in shop), a dynamic balance can be achieved with only minimal additional effort.[9] While static balance is a simple first step and will often reduce vibration considerably, a dynamic balance is needed to truly minimize unbalance forces transmitted by the roller through the bearings and into the support structure.

Balancing Weights

Before we proceed to dynamic balance, it would be appropriate to mention unbalance equivalency as well as how mass is redistributed during a shop or field balance of rollers.

The real mass distribution of any roller will be complex and indeterminate. However, Figure 11.7 will illustrate the concepts of net unbalance with two heavy spots and a light spot (relative to the average). The procedure to determine net imbalance is to use vector math, or the more basic tip to tail addition. The length of the individual unbalance will be the mass of the spot (m) times its distance to the axis (r). As discussed earlier, the centrifugal force produced would be $F = mr\omega^2$, where ω is the rotational speed in radians per second. The direction of the individual vector is the direction from the axis to the heavy spot.

In our example, the first vector is short and oriented at 0° (straight up). However, the void is slightly longer (greater radius), and oriented at 225° in the opposite direction (lack of mass). Finally, the heavy spot at 3 o'clock has the longest vector because of a large mass and radius and is oriented at 90°. The resultant unbalance is large and directed toward 4 o'clock. A weight with an opposite vector will bring the roller back into balance.

We can take advantage of unbalance equivalency to give some flexibility in the positioning of balance weights as given in Figure 11.8. They are all equivalent because the unbalance vector is the same length and orientation.

Often, balancing methods require the use of trial weights to calibrate the instrument and readings.[10] These weights are temporarily but securely attached to the roller (usually to the outside) by means such as tape, strapping and so on. The balancing instrument will calculate the required unbalance (magnitude and orientation) to move the roller back into balance. The technician will determine where (radius) to add/subtract the weights, and the mass is determined by the principle of equivalency.

Figure 11.7
Net Unbalance

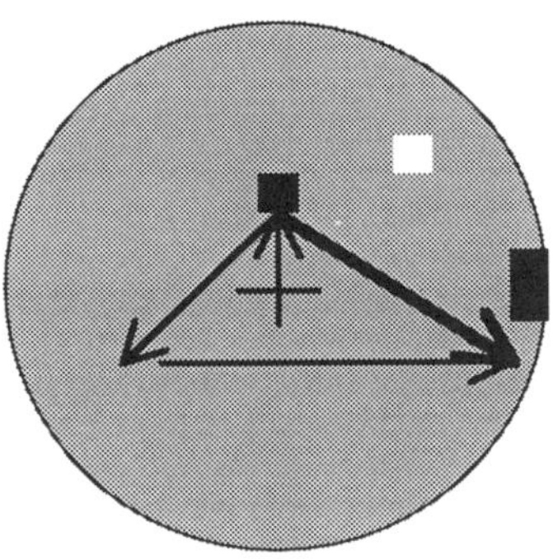

Figure 11.8
Balance Equivalency

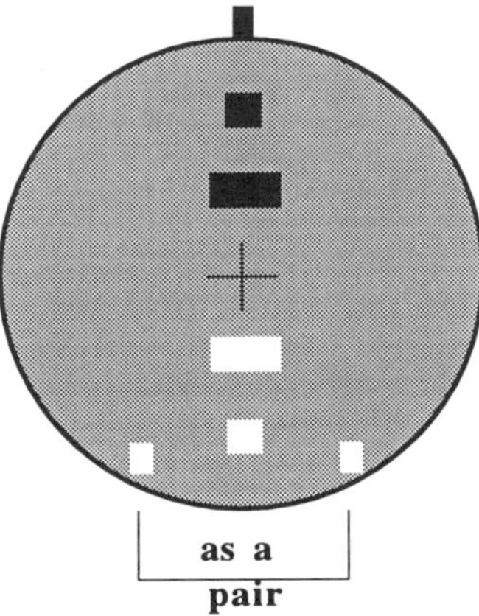

The most common methods of adding a weight to a roller is to drill and tap a stud which protrudes into the interior for small weights, or by welding/bolting weights to the head or shell interior if larger masses are required. More commonly, mass is removed by drilling out the heads, which may be covered later with a lightweight filler.

It is very important for maintenance people to remember that balance weights should not be disturbed, or if necessary returned exactly to their original location. Furthermore, good practice would indicate that all roller parts are returned exactly to their former position <u>and</u> orientation to preserve balance.

Couple Unbalance

The simple static balance, sometimes called single plane balance, described earlier is not adequate for most rollers. Rollers are generally much longer than their diameter so that a couple unbalance is also likely which will transmit vibration into the framework.

A simplified pure couple unbalance is shown in Figure 11.9. Here, two equal mass heavy spots are located 180° apart around the shell. Additionally, they are equidistant from the roll centerline (and have the same radii from the axis). In this example, the roller is already in static balance, so that the *net* centrifugal unbalance force is zero. This does not mean that the roller does not impart forces to the structure, however. Here, there will be a rotating unbalance force at each of the bearings that is equal in magnitude but 180° out of phase.

The magnitude of the bearing reactions generated by couple unbalance is easily calculated from statics if the distribution of masses is known. It has all of the same factors as does static balance such as mass, radius and rotational speed squared. Additionally, it has the ZD distance from the centerline of the roller to the masses. Thus, the more slender the roller (l/d), the larger we might expect the couple unbalance potential.

The principle inertial axis of a pure static balance is parallel to the rotational axis, but offset or eccentric. The principle inertial axis of a pure couple balance is skewed with respect to the rotational axis, but intersects it at the centerline of the roller.

Real rollers have a degree of both static and couple unbalance, each of which must be corrected in the balancing process. This combination is called dynamic balance. While static balance requires only a single plane of measurement and balance weight correction, dynamic balance requires two planes of measurement and at least two planes of balance weight correction. Static unbalance will show as equal magnitude in phase readings at each end. Couple unbalance will show as equal magnitude readings 180° apart. Dynamic unbalance will show as unequal magnitudes.

Figure 11.9
Reactions from a Pure Couple Unbalance

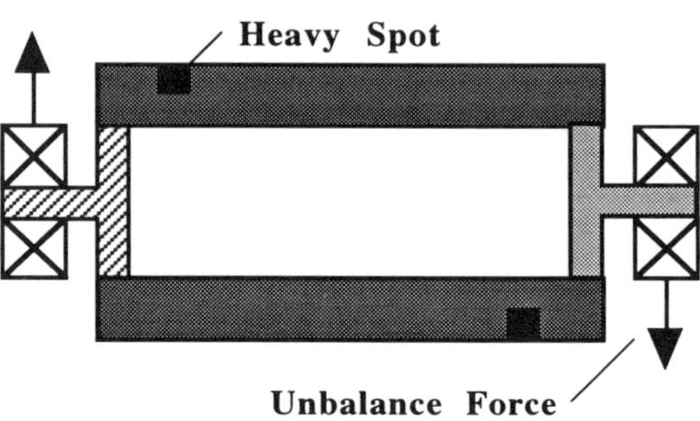

There is a couple equivalency, similar to static balance equivalency, that is used for flexibility in the placement of balance weights. Most of the sources of the couple unbalance may well be inboard of the heads. However, the correction may often take place at the heads.

As seen in Figure 11.10, the balance weights at heads will tend to be less than the couple unbalance sources because of the larger distance from the centerline. On the other hand, the weights will tend to be larger because the heads are at a smaller radius than the shell.

With the more typical dynamic unbalance (combination of static and couple), the weights will not be the same at each end, nor will they be 180° apart. Fortunately, modern balancing instrumentation does all of the vector mechanics and gives the exact position and mass required once the technician enters balance weight location preferences. However, those who are interested in the mathematical development, as well as other details and practice, can refer to Wowk's excellent book given at end of this chapter.

Figure 11.10
Couple Equivalency

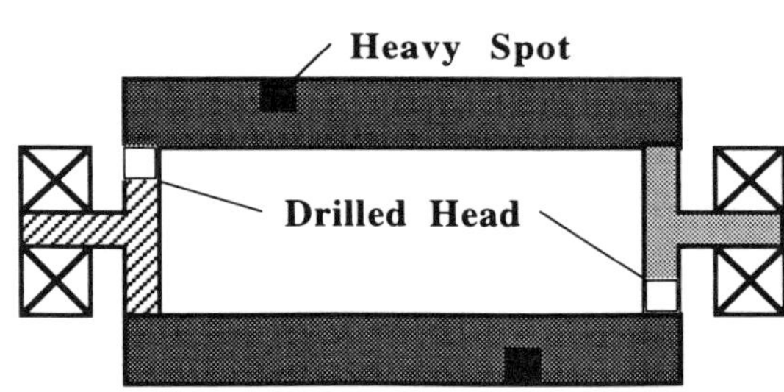

Balancers

Most balancers, in one fashion or another, are composed of the elements given in Table 11.6. The balancer types are given in Figure 11.11.

For **in situ** (field) balancing, the roller support will be the original framework.[11,12] Aside from not needing to remove the roller, the vibration (but not necessarily unbalance) reduction may be better when balancing in place. Unfortunately, the time to balance may likely to be longer. A **hard balance** machine will support the roller in very rigid mountings. This method has the advantage of not requiring calibration runs. In contrast, the **soft balance** machine supports the roller on constrained linear bearing slides. The principal advantage of this balancer is a higher sensitivity for most situations. Despite numerous differences in the details, all three balancer types are capable of achieving any reasonable roller balance specification.

A schematic of a typical balancer setup is shown in Figure 11.11. First, a sensor is needed to measure vibration. This can be any small probe capable of measuring displacement, velocity or acceleration provided it has sufficient sensitivity and frequency response. While the sensor is often an accelerometer, it is somewhat unimportant which units are measured because the instrument will be able to convert the signals to other related units. The probe may be permanently mounted in the saddles of hard or soft balance machines. Alternatively, the probe may be magnetically attached to the bearing housing, or even handheld to measure along the face of the roller.

An index mark needs to be affixed to the roller to define angular position. Often, the mark will be a piece of photoreflective tape. The position of the index mark is read by an angular sensor such as a photoeye. Finally, the instrument will need to return an angular position with respect to the index mark for purpose of defining the position for balance weights. Sometimes this is a strobe light which illuminates a sector position on the circumference (numbered tape) or roll end (protractor). In other cases, the instrument may merely display an angle for the front and back weight locations.

Table 11.6
Elements of Balancers

Roller Support
- In Situ
- Hard Balancer
- Soft Balancer

Roller Drive
- Drive Shaft
- Belt

Mass Alteration
- Removal (e.g., drilled holes)
- Addition (e.g., weights)

Instrumentation

Figure 11.11
Balancer Types

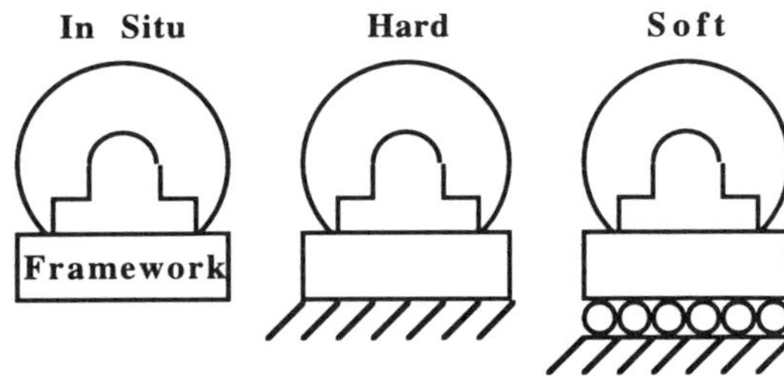

Figure 11.12
Balancer Instrumentation

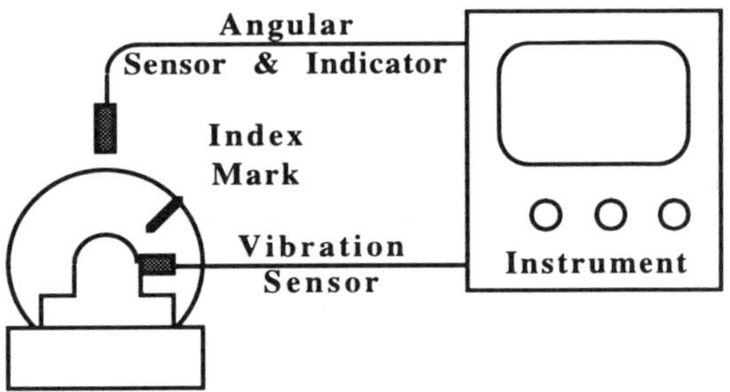

The modern balancing instrument is a digital computer based device. It will have provisions for operator inputs through switches, dials, etc., and/or through a touchpad or keyboard. The display could be as minimal as a meter, but more likely would be a CRT monitor. The computer based balancer is adept at digital filtering, vector math, and perhaps statics and FFT. Additionally, it will have the recipes for a variety of balancing methods.

Dynamic Balancing

Dynamic balancing means correcting both static and couple unbalance by changing weights in two or possibly three planes.[13] Here, we will briefly discuss the methods to accomplish this without describing the mathematics. Again, the interested reader can refer to Wowk for details.

Single plane balancing, is able to correct for cases of pure static imbalance. On rollers, this restrictive case would be confirmed by identical amplitude and phase of vibration at both ends. The single plane balance requires only a single run to calibrate the in-situ or soft balancer. The required balance weight would then be split in two and placed equidistant from the centerline of the roller. A final verification run is then performed to decide if a trim balance is required. Unfortunately, most rollers will exhibit some degree of couple unbalance that can't be treated with single plane balance.

The **four run w/o phase** is another single plane balancing procedure. It has the advantage that phase information is not required and thus could be performed by even the most primitive vibration equipment. In this method, amplitudes are measured in the original condition, and three additional runs where trial weights are added at various angular positions such as 0°, 120° and 240°. A simple graphical procedure using only a compass and ruler is sufficient to determine required balance weight and angular position.

The **two plane** method requires only three runs: original, trial weight at one end and trial weight at the other end, where amplitude and phase readings are taken at both ends. The drawback is the solution requires influence coefficient mathematics and thus requires a computer program for effective use. The primary complication is called cross-effect which means that a weight placed on one end of the roller will affect the amplitude and phase of vibration at both ends.

The **seven run w/o phase** is another two plane method.[14] Here, the tradeoff is reduced equipment requirements but additional runs. Again, the mathematics of any two plane method is complicated by cross-effect.

The **static-couple method** is another two-plane balance method. It has the advantages of being a simple single plane method extended to compensate for couple unbalance, and has a simple graphical procedure. The disadvantage is that weights must be placed/removed in three planes instead of two. The weight at the center compensates only for the static unbalance, while a pair of weights at the ends compensates only for the couple unbalance.

These methods described thus far are based on **rigid rotor** assumptions. That is, the roller will not change shape significantly with speed or the addition of weights along its face. The objective is merely to minimize the bearing reactions and thus framework vibration. It does not say anything about reducing the dynamic TIR.

However, the shape of a roller will change as it approaches and exceeds its first critical speed. It is the further objective of **flexible rotor** balancing to keep the roller cylindrical at and through its first critical and perhaps the second and third.[15,16] Flexible rotor balancing is sometimes referred to as **multi-plane** or **multi-speed** balancing.

Analytical solutions to multi-plane balance are quite complex, and thus not often used on rollers. However, the balance technician through experience and trial-and-error can reduce roller flexing. In simple terms, they add weights along the face of the roller to counteract the dynamic TIR as typically read by a handheld probe.

There is a practical limitation to flexible rotor balancing in a shop, however. That is, the effective stiffness of the balancer saddles can be much different than the stiffness of the framework to which the roller will be added. This difference in stiffness affects both the resonant speeds as well as their mode shapes. In most cases, the effective framework stiffness will be intermediate to that of soft and hard balancers, though rollers mounted low on very stiff framework and foundation will approach that of the hard balancer mounts.

Balance Grade

Most of us will not be involved in the details or mechanics of roller balancing. Also, we will not usually be involved with selecting balance methods, balancing equipment or defining how or where balance weights may be added/removed. More likely, however, we will need to specify acceptable unbalance.

In some cases, rotating elements may not require a formal balance. Situations that may not justify a balance step include: sufficiently tight control of manufacturing tolerances, very low speeds, very low mass, or where application loads far exceed unbalance loads. In these cases, merely controlling the cylindricity may be sufficient. Specific applications where the application loads are far greater than centrifugal forces might be some conveyor rollers or a few farm machinery components. An example of low mass is the fiber core which is central to most wound rolls.

However, most rollers which operate more than 30 MPM (90 FPM) or are longer than 1 meter (40") should be dynamically balanced to a specific criteria. One common standard is ISO 1940, or its updated American equivalents of ANSI S2.19-1989 and ASA STD 2-1975, or its Society of German Engineers VDI standards dated October 1963.

In all of these cases, the grade of balance quality is termed G where

(11.5) $G = e\omega$

where

G = balance quality grade (mm/sec)

e = eccentricity (mm)

ω = speed (rad/sec)

and where

$\omega = 2\pi f$ for f(Hz) or

$\omega = 2\pi n/60$ for n(RPM)

These balance grades are listed on charts which include example applications. I will interpret these examples for web machine rollers, as given in Table 11.8, because the original charts are not explicit. This interpretation would be typical of good machine design practices.

Table 11.7
Balance Quality Grades (from ISO 1940)

Grade	Applications
16	nonprecision agricultural components nonprecision mining components conveyors low speed (<30MPM) web rollers
6.3	most web machine rollers
2.5	>2000MPM transport rollers >1000 MPM nipped process rollers
1.0	high speed precision hard nips

Example Problem

Tell me about the unbalance at the outside of a G2.5 grade idler roller which is 7" in diameter, weighs 100lb and runs at 5000 FPM.

From (11.2), the frequency of rotation is

$f = 5000/(15.71 \times 7.00) = 45.5$ Hz or

$\omega = 2\pi(45.5) = 285$ rad/sec

From (11.5), the eccentricity of the roller is

$e = 2.5/285 = 0.00877$ mm $= 0.000345$ inch

This 0.3 mil cg eccentricity is somewhat less than typical static or dynamic TIR tolerances

A weight at the shell that would put a perfectly balanced idler out of the G2.5 spec would be

$w = 100$ lbx$(0.000345/3.5) = 0.00985$ lb

$w = 0.00985 \times 16 = 0.158$ oz

From (11.3), the centrifugal force produced by the maximum unbalance

$$F\,(\text{lbf}) = 1.77 \times \left(\frac{45.5 \times 60}{1000}\right)^2 \times (0.158 \times 3.5) = 7.29 \text{ lb}$$

Thus, the G2.5 balance grade will produce only modest dynamic forces with respect to roller weight, and is easy to achieve on modern balancers given sufficient balance time. The G6.3 grade would be quicker and cheaper, but would result in more significant dynamic loads.

Other Balance Specs

There are a few other common alternatives to the ISO balance specifications. One is the API (American Petroleum Institute) formula which is also an old **U.S. Navy standard**.[17] This formula calculates allowable unbalance as:

$$(11.6) \quad U = \frac{4\,w}{N}$$

where U = unbalance (oz-in)
N = rotational speed (rpm)
w = journal weight (lb)
(which is often roller wgt / 2)

In our previous example,

U = (4 x 100 / 2) / (45.5 x 60) = 0.0733 oz-in

which is considerably less than the

U = (100/2)x16oz x 0.000345 in = 0.2760 oz-in

of the G2.5 grade, and indeed lies between the G0.4 and G1.0 grade.

Another balance criteria is to limit dynamic (centrifugal unbalance) loads to some fraction of the static weight of the roller. A common criteria of 10% is equivalent to a 0.1g vibration. Even at this modest value, however, bearing life will be reduced 33%. In the previous example, the 10% criteria would fall between the ISO G2.5 and G6.3 balance grades.

Finally, field balancing and many flexible roller specifications use a value for maximum permissible dynamic TIR. **Dynamic TIR** is the synchronous (1 x rpm) displacement of the shell at some location (plane) and at some speed. It is the cumulative effect of static TIR, unbalance, amplification at resonance and many other factors.

All of the previous standards are based on rigid rotor assumptions where balance is specified in two planes (heads or bearings) only. No mention is made of vibration or other dynamic movement elsewhere. Thus, very demanding (high speeds and precisions) rollers which will commonly operate near resonances may be specified by a maximum dynamic TIR anywhere along the width and at any speed up to 110% of design.

In a crude fashion, the dynamic TIR can be related to ISO and other balance standards. This relationship begins by noting that the numerical value of the G grade is the velocity the roller would vibrate at (mm/sec) if it were suspended freely (soft balance). The dynamic displacement can be calculated from this grade as:

$$(11.7) \quad D = \frac{0.322\,V}{f}$$

where D = displacement (in pk-pk)
V = velocity (in/sec)
f = frequency (Hz)

Thus for our previous G2.5 example,

D = (0.322)x (2.5/25.4)/(45.5) x 1000 = 0.7 mils

which is in the range of typical precision roller dynamic TIR specifications.

Conversions and generalizations between various roller measures and standards can be risky because of the complexities. However, there is an underlying similarity. Indeed, we can coarsely define an overall class which covers static TIR, dynamic TIR, balance and vibration as given in Table 11.8.

Most well engineered web manufacturing machinery will have transport rollers which tend to be in the good class, with process rollers approaching the excellent class. Unfortunately, all too often web converting machinery falls in the poor class which causes many product/process compromises that vary from subtle to obvious. Wound rolls will tend to be in the poor class at their best, and may occasionally be much worse.

Table 11.8
Guidelines for Balance, Runout and Vibration

Class	Accel g peak	Velocity in/sec pk	Displacement mil pk-pk
Poor	1.0	1.0	10.0
Good	0.1	0.1	1.0
Excellent	0.01	0.01	0.1

Nipped Rollers

Obviously, nipped rollers will have some degree of unbalance that will contribute to mechanical vibration. Usually, however, much larger forces will result from out-of-round nip rollers. The effect of a roller with (static TIR) runout when riding against a circular roller will be much like a cam follower, which can generate large dynamic loads.

The magnitude of the pounding dynamic forces developed in a nip depends on a number of complex factors. First, the forces will increase with machine speed. This makes paper calendering and supercalendering, which is often run in excess of 2000 FPM, a very demanding application. Second, the forces will increase with the magnitude of the static TIR runout. While good metal nipped rollers may have runouts much less than 1 mil, covered rollers may be closer to 5 mils, and wound rolls will be much larger still. Third, the forces increase with the hardness of the nip. Metal-metal nips with a thin product will tend to pound much more than rubber-rubber nips with a thick product. Hard nips also tend to have less damping than soft nips. Fourth, the forces will depend in a very complex manner on the mass and stiffness distribution of the entire machine which includes foundation, framework and rollers. Fifth, the forces will increase frequency content of the runout, as will be discussed shortly. Finally, the roller-machine system will vibrate considerably more during resonance where the frequency of the out-of-round driving forces coincide with one of many system resonances.

One simple measure of nip roller out-of-roundness is the maximum static TIR as measured with a dial indicator. However, this does not in itself give a picture of the shape of the roller which also affects nip roller vibration. The non-circular shape of a roller can be resolved by a FFT into elemental components as:

(11.8)

$$R(\theta) = r_0 + r_1\cos\left(\theta+\theta_1\right) + r_2\cos\left(2\theta+\theta_2\right) + \ldots$$

$$R(\theta) = r_0 + \sum_{i=1}^{n} r_i\cos\left(i\theta+\theta_i\right)$$

where

$R(\theta)$ = radius at angular position θ

r_0 = average roll radius

θ = angular position

i = sidedness number

Figure 11.13, shows a polar plot of runout derived from several generated shapes. In practice, however, it would be the actual shape which would be described by the amplitude and frequency components of an FFT.[18]

As seen here, the roll(er) will have a basic circularity which is modified by several simultaneous components of sidedness. The one sided roller would be one with a circular shell which is slightly eccentric from the journal axis, but could be thought of as egg shaped. Higher order shapes will typically be of lower amplitude. However, it is the integral multiples of roller rpm from these higher order sidedness components which almost inevitably drives structural resonances at higher frequencies.

As one might imagine, this picture of roll shape is very simplistic.[19,20] We have not even touched on the subject of the interaction of two rollers which both have significant non-circularity.

Figure 11.13
Components of Roll Sidedness

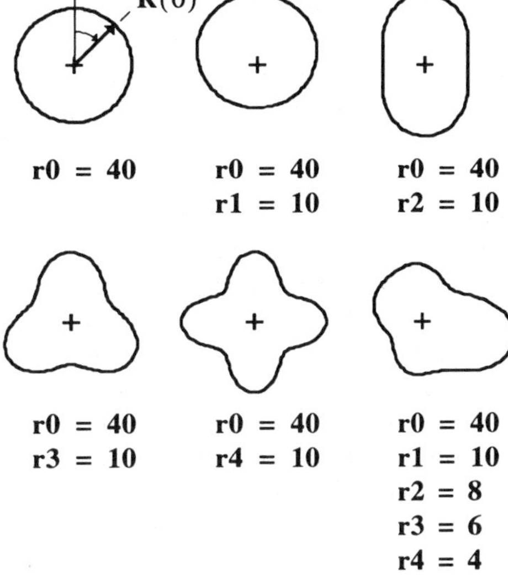

Wound Rolls

Wound rolls present special vibration challenges. First, the roll is not explicitly balanced and large centrifugal forces can be generated. Second, the roll is not anywhere near as cylindrical as even the worst roller.[21,22] Indeed, core eccentricities in excess of 1 inch have been observed. Third, winders run at least as fast as web manufacturing lines, if not considerably faster as in the case of many rewinders or roll-to-roll operations.[23] Furthermore, some elements such as unwinding cores are regularly run near or at resonant conditions, often with explosive results.[24,25,26,27]

However, a distinguishing feature of wound rolls is that they suffer permanent deformation in a bouncing nip.[28,29,30,31] As seen in Figure 11.14, an initial disturbance in the nip will cause it to bounce with a preferred frequency and thus wavelength. As the nip converges from the bounce, a new deformation is suffered by the roll. If the spacings are such that they coincide to an integral number times roll circumference, the deformation pattern is self reinforcing. This is analogous to the generation of washboard gravel roads due to the repeated passage of cars whose suspension is tuned similarly. However, this phenomenon is not limited to wound rolls. Some roller covers will suffer permanent compaction if repeatedly impacted in the same place. Even if these covers are reground to eliminate runout, the pattern of hardness variation may be retained to start the cycle anew.

Winder vibration may become even more violent if the frequency of bumps or sides on a wound roll also coincide with one of numerous natural frequencies. A simplfied schematic of the principle modes of winder vibration are given in Figure 11.15.

Another unique factor of winder vibration is that the driving frequencies are continually changing. The fundamental roll frequency is continually decreasing due to the ever increasing wound roll diameter during winding. If that weren't enough, the sidedness of the rolls are always changing as well. These factors combined with all of the other machine and nip vibration complexities make winder vibration an exceedingly difficult problem to diagnose.

Figure 11.14
Washboard Road Analogy for Roll Vibration

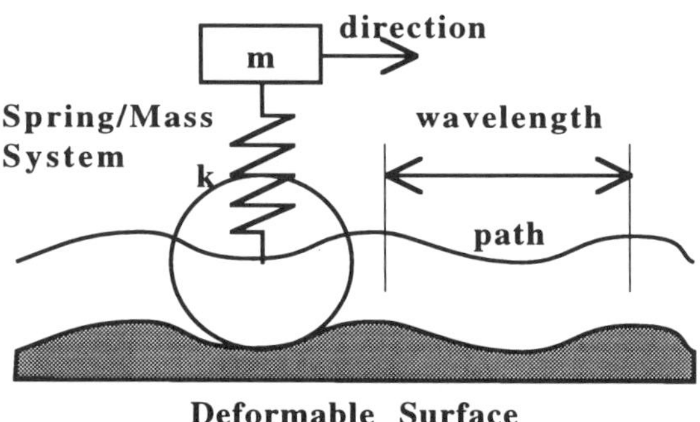

Figure 11.15
Rewound Roll Vibration Modes

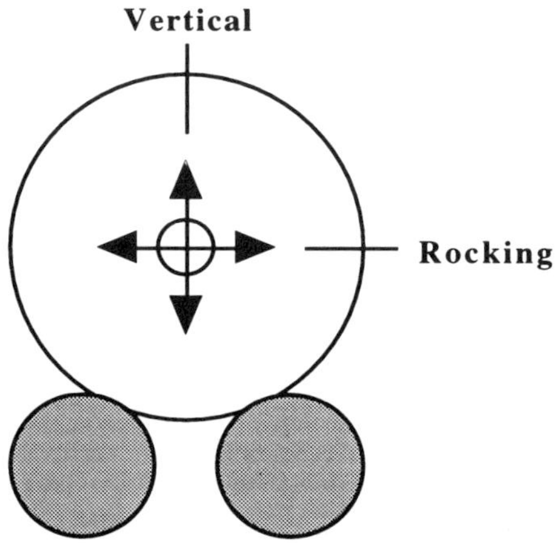

Single Drum Radial

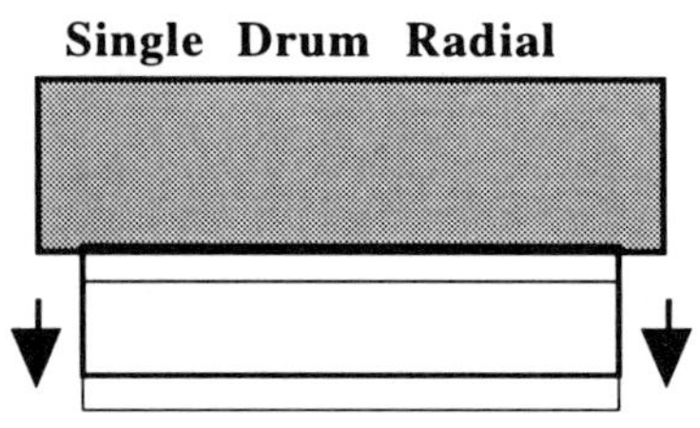

Single Drum Rocking

Wound Roll Instability

As mentioned in the last section, unwinding cores, such as found on high speed printing presses, can explode when run through resonance. However, wound rolls can suffer a similar fate.

One type of wound roll instability is the spiral throwout, which is found on some high speed shaftless two drum winders.[32] The sequence of events for this failure is shown in Figure 11.16. It usually starts with a rocking mode of the rolls in a set. When two rolls get out of phase, as they inevitably do because of small differences in diameter, the roll centers will be slightly offset. There is a relative surface velocity difference between the rolls, which when they are in contact, causes a separating force at right angles. This increases the offset, which in turn increases the force, and so on. If the initial offset and force is sufficient, the system becomes unstable in an outward spiral fashion. Eventually, one of the roll edges will be launched up and out. The surest solution is to put a shaft through the set because that ensures that there is no relative velocity difference, even if the rolls are extremely eccentric and out-of-round.

However, some shafted winders are also prone to violent instability. The most common is the turret center/surface winders sometimes used for small consumer rolls. When winding toilet tissue or kitchen towels, these shafted winders can exhibit a failure that is called a blowout when passing through resonances and other instabilities.

Unfortunately, there are no currently known solutions or even aids to reduce wound roll vibration on those applications where it is a problem. Rather, the only safe treatment is to change speeds, usually down. However, sometimes one may be able to speed up again after the roughness passes. The speed toggle helps prevent a reinforcement of the roll's out-of-round shape. One final caution is that changing speeds might move you out of one resonance but into another which could be even worse! The speed programs which can avoid vibration buildup during winding are given in Figure 11.17.

Figure 11.16
The Spiral Throwout

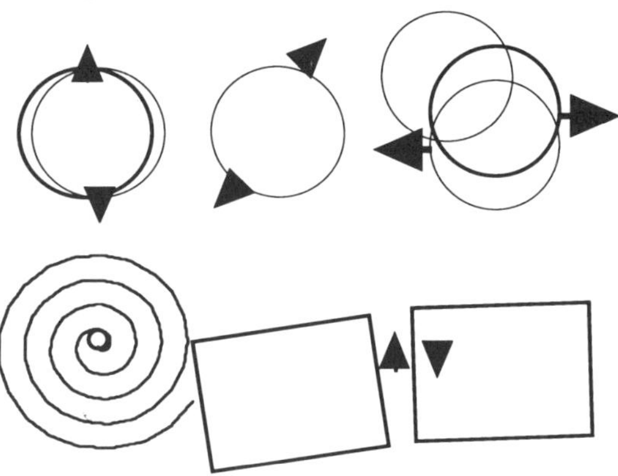

Figure 11.17
Speed Programs to Avoid Vibration

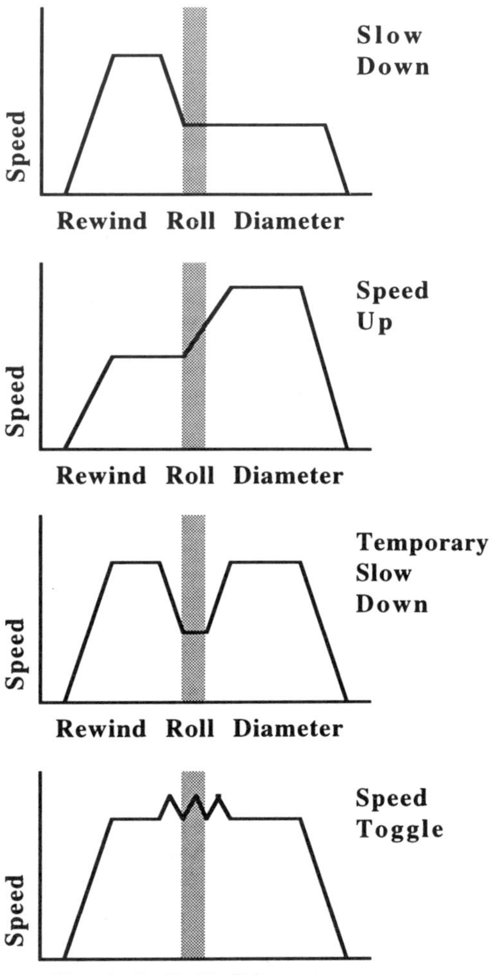

Foundations

Many web machine problems could have been prevented had the machine been mounted on a suitable foundation. Inadequate foundations can result in the need for frequent realignment and unnecessarily high levels of vibration.

The most common problem with small converting machinery is the practice of merely setting the machine on a concrete slab floor. In some cases the erector will take the time to anchorman the machine down, but this only keeps it from walking around (due to vibration or getting bumped) or tipping over (earthquakes). One problem with this penny-wise pound-foolish approach is that slab flooring is not dimensionally stable because of swelling of the earth beneath due to seasonal changes in moisture. The result is the need for frequent realignment which will far exceed the costs of an appropriate foundation.

Small and slow web machines should be mounted on a footing that is at least 12" wide by 12" deep at every attachment point. This is easy to do with an existing floor by first sawing a strip out of it, trenching the earth at least 12" down, and then filling with concrete. Studs or anchormans can be used for small and slow web machines, but baseplates that are leveled and epoxy grouted in place will be sturdier and provide an easier surface to align the framework from.

A problem that is endemic in the paper industry is the improper mounting of heavy rotating equipment on the second or higher floors. The common mistake is to use a stress based design (the usual civil engineering approach) to make sure that the flooring can safely handle the loads. However, as any good web machine designer will know, *static* deflection is a much more demanding criteria for most structural elements than is stresses. Those designers who are the rare exception will recognize that minimizing *dynamic* deflection is even more demanding.

Figure 11.18 schematically illustrates modes and compliances that are common to elevated flooring. If the floor moves a little, the machine will obviously move more. The solution to these modes is two continuous walls underneath the frame/baseplates extending the complete length of the machine. Indeed, some have suggested using 3-10X as much concrete as the mass of the machine for demanding applications. While a closer answer might be found with FEM, the engineering costs will likely exceed even the most extravagant use of concrete.

Figure 11.18
Modes of Flooring Vibration

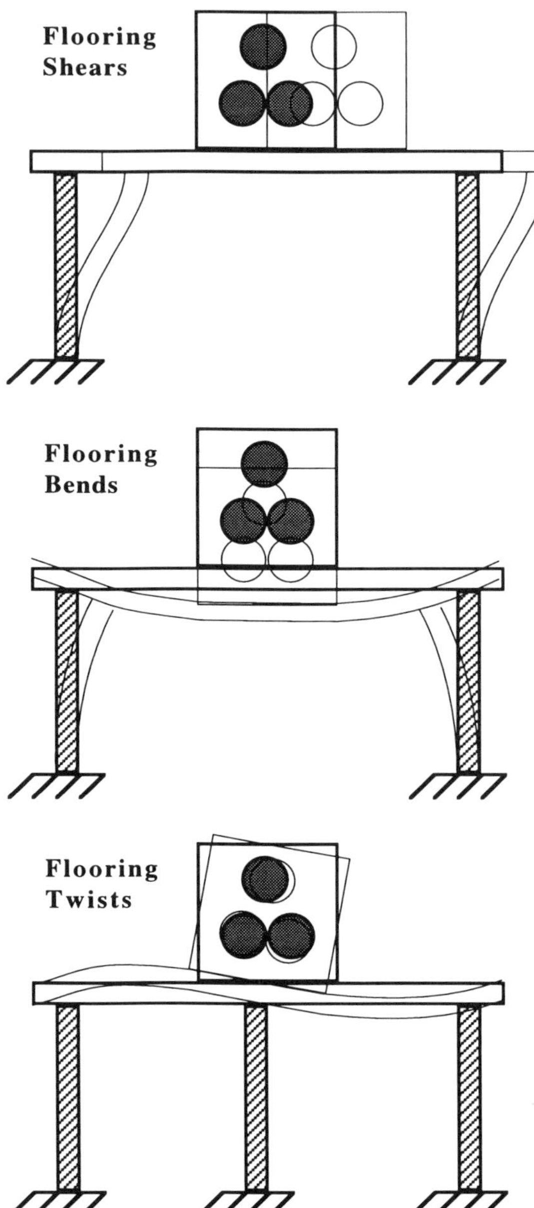

Framework

Vibration, deflection and many other problems can be reduced greatly by proper framework design. Web machine frames are seldom engineered in the strict sense as are airplanes and automobiles. Rather, it is usually guided by experience and judgment. Fortunately, good frames can be designed without expensive engineering provided that the goal of a maximum dynamic deflection is kept in mind.

Good framework design begins by reducing the overall height of the machine as much as possible. Of particular importance is to keep vibrating sources such as nips, heavy rollers and high rpm rollers as low and as close to the floor as possible. While it might be an operator convenience to have overhead web runs, it may well compromise the operation of a machine. Rather, design the process and machine as robust as possible and the need for regular operator access will be minimal.

A second common design deficiency is the tendency to cantilever mountings which span between a roller and the main framework. If unavoidable, the cantilever can be stiffened by triangular (usually plate) gussets as seen in Figure 11.19. Related shortcomings include lack of shear stiffness and using open beam sections to carry torsion.

Another common error is to attach (weldment or bolts) one member to unsupported plate section of another. A better approach is to mount near a corner to pick up additional stiffness in at least one rotational plane.

However, perhaps the most common error of all is to use framework that is not stiff enough. The worst offenders are often converting machinery builders where cost conciousness is at a premium and appreciation of structures at a minimum. However, the cost of metal for framework, at about $1/lb, is far smaller than engineering, controls, drives, purchased components and so on. Indeed, the entire machine costs are usually a very small factor in the costs to manufacture or convert webs (raw materials and labor are much larger).

Figure 11.19
Poor Framework Design Practices

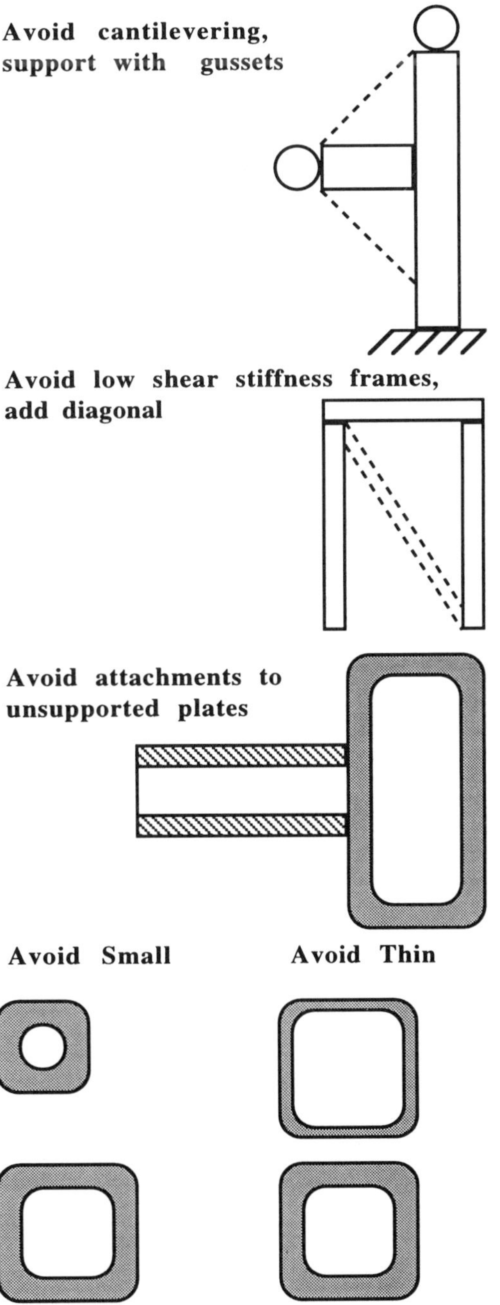

However, it is not only the lack of stiffness that compromises the machine performance. It is also the lack of mass which slows and minimizes dynamic motion. Some builders even fill their frames with concrete![33]

Vibration Summary

Ensuring a smooth running web machine always begins in the earliest part of the design stage. It means laying out the rollers so that they are mounted as close to the floor as possible. Even if operator and maintenance access becomes more limited, the compromise may be justified as operators should not often be crawling around operating machinery, and the machine should be running far more often than it is down for maintenance. Also very early in the design stage, the flooring and foundation details need to be worked out between the machine builder, contractor and customer. Always keep in mind that concrete at 1¢/lb and steel at $1/lb will be an investment that should pay for itself many times over in the life of a machine.

Next, specifications must be selected for the unbalance of all rollers because that is the source of much machine vibration as well as some wear-and-tear. Certainly, a G6.3 should be considered for even the most undemanding applications, and a tighter 2.5 or even 1.0 might be requested for precision applications.

Similarly, roller roundness should be closely specified. While transport rollers may only require a 1 mil TIR, rollers in nip should be much tighter to reduce the dynamics of lobed rollers beating against each other. While wound roll roundness can't be easily controlled, it should be monitored if the application is prone to vibration.

Once a machine is installed, however, the efforts to minimize vibration must still be kept up. Maintenance should regularly measure the cylindricity of process rollers, but not overlook the occasional check of transport rollers as well. Rollers which were perfectly balanced when new may be put seriously out of balance if they are removed for maintenance, have fouled internally, or worn externally, and even due to normal bearing wear. Vibration monitoring might provide a means to detect increasing unbalance without actually removing a roller from its frame. Finally, rooting out any unnecessary clearances and excessive misalignment should be done on a regular basis.

Despite all of these efforts, machine vibration problems may still crop up. To determine the cause and cures will require skilled diagnostics from someone who has considerable vibration experience with very similar machinery. Finding such a person can be a problem. Inhouse personnel usually have too many and too varied duties to maintain, much less develop, the specialized skills. Contracted vibration experts, on the other hand, may have the skill but not the familiarity with the particular equipment.

Diagnostics will almost inevitably involve a vibration instrument with FFT capabilities. The first step would be to determine the source(s) and condition of each vibration mode. The FFT is indispensable here because it will quickly identify the diameter of the offending roller, if not its exact position. This roller could be considered for a rework if it has lost its balance, or its cylindricity if in nip.

If the machine has always vibrated since startup, despite the best maintenance efforts, one of two likely possibilities exist. First, the machine design could be compromised as described in the foundation and framework sections. This would be a very unfortunate occurrence because it is usually very difficult to economically redesign and rework a machine after the fact.

However, the second possibility is even worse. That is the application is prone to an instability. The classic example is vibration caused by the winding of punky and deformable materials. In these situations, it can be immensely difficult to achieve even a noticeable much less modest improvement. Fortunately, these unstable applications are rare.

Selected Bibliography

Wouk, Victor. *Machinery Vibration: Measurement and Analysis.* McGraw-Hill.

Wouk, Victor. *Machinery Vibration: Balancing.* McGraw-Hill, 1995.

Bibliography

1. Shelton, John J. *Deflection and Critical Velocity of Rollers.* 3rd Int'l Conf. on Web Handling, Oklahoma State Univ., June 18-21, 1995.

2. Roisum, David R. *The Winder Vibration Book.* Beloit Corporation internal publication, 1986.

3. Roisum, David R. *Winder Vibration.* TAPPI Finishing and Converting Conf. Proc., October 1987.

4. Roisum, David R. *Winder Vibration Can Reduce Operating Efficiency and Increase Maintenance.* Tappi J., vol 71, no 1, pp 87-96, January 1988.

5. Kaufman, A.B. *Monitor Acceleration, Velocity or Displacement.* Instruments and Control Systems, October 1975.

6. Borbas, James and Constello, Donald. *Roll Grinding Technology for the Paper Industry.* TAPPI Finishing and Converting Conf. Proc., 1981.

7. Rothenbacher P. and Vomhoff E. *Mass Centering of Cast-Iron Rolls.* TAPPI Finishing and Converting Conf. Proc., pp 31-34, October 1984.

8. Schminke, K.H. *The Importance of Roll Balancing on High Speed Paper Machines.* Papier, July 1991.

9. Lang, George F. *Balance with Your Real Time Analyzer.* Nicolet Scientific Application Note 11, October 1977. Note, similar instructions can be found with most other manufacturers of vibration instrumentation.

10. Wowk, Victor. *Developing Safe Trial Weights for Balancing.* Vibration, vol 7, no 3, pp 3-9, September 1991.

11. ANSI S2.38-1982 (R 1990) (ASA 44). *American National Standard Field Balancing Equipment - Description and Evaluation.* (same as ISO 2371).

12. Baur, Paul. *Field Balancing of Rotating Machinery.* Power, October 1983.

13. Senger, W.I. *Specifying Dynamic Balance,* Parts I-IV. Machine Design, November 1944 through February 1945.

14. Everett, Louis J. *Two-Plane Balancing of a Rotor System Without Phase Response Measurements.* ASME Trans., J of Vibration, Acoustics, Stress, and Reliability in Design, vol 109, April 1987.

15. ANSI S2.42-1982 (R 1990) (ASA 46). *American National Standard Procedures for Balancing Flexible Rotors.* (same as ISO 5406).

16. ANSI S2.43-1984 (R 1990) (ASA 54). *American National Standard Procedures for Evaluating Flexible Rotor Balance.* (same as ISO 5343).

17. MIL-STD-1 (Ships). *Mechanical Vibrations of Shipboard Equipment.* May 1, 1974.

18. Hussain, Mirza S. *Problems Associated with Rollers in Nip.* Tappi J., vol 76, no 1, pp 246-248, January 1993.

19. Nayak, P. Ranganath. *Contact Vibrations.* J. of Sound and Vibration, vol 22, no 3, pp 297-322, 1972.

20. Gray, G.G. and Johnson, K.L. *The Dynamic Response of Elastic Bodies in Rolling Contact .* J. of Sound and Vibration, vol 22, no 3, pp 323-342, 1972.

21. Ryti, N. et al. *A Method to Measure the Structure of Newsprint Rolls.* TAPPI Finishing and Converting Conf. Proc., 1972.

22. Eriksson, Leif G. *Deformations in Paper Rolls.* Winding Technology Conf. Proc., Swedish

Newsprint Research Center (TFL), Stockholm, March 16-17,1986.

23. Mohle, H. and Buschmann, G. and Muller, G. *Vibration on Paper Winders.* Papier, vol 24, no 11, pp 845-850.

24. Caccase, Vincent. *Dynamic Behavior of Steel Core in the Unwinding Process.* TAPPI Finishing and Converting Conf. Proc., Nashville, pp 283-296, October 18-21, 1992.

25. Caccase, Vincent. *Dynamic Behavior of Steel Core in the Unwinding Process.* Tappi J., vol 76, no 10, pp 51-61, October 1993.

26. Gerhardt, Terry D. and Staples, John F. and Lucas, Robert G. *Vibrational Characteristics of Wound Paper Rolls: Experiment and Theory.* TAPPI Finishing and Converting Conf. Proc., Nashville, pp 263-282, October 18-21, 1992.

27. Gerhardt, Terry D. and Staples, John F. and Lucas, Robert G. *Vibrational Characteristics of Wound Paper Rolls: Experiment and Theory.* Tappi J., vol 76, no 6, pp 121-128, June 1993.

28. Daly, David A. *How Paper Rolls on a Winder Generated Vibration and Bouncing.* Paper Trade J., December 11, 1967.

29. Crouse, Jere W. *Winder Noise Its Generation and Current Methods of Reduction.* TAPPI Finishing and Converting Conf. Proc., 1978.

30. Crouse, Jere W. *Winder Noise Generation and Current Methods of Reduction.* Tappi J., vol 62, no 5, pp 69-71, May 1979.

31. Yi, S. *Ultrasonic Application in Measuring of the Radius of an Eccentric Roll.* OK. State Univ., WHRC, M.S. Thesis, November 1985.

32. Olshansky, Alexis. *Roll Bouncing.* Beloit Lenox Internal Report, 1992.

33. Tuomisto, M. *Kl-1000 Wartsila Winder Features Stoneheart Structure to Dampen Vibration.* Paper Age, vol 16, no 3, pp 42-43, March 1980.

Chapter 12

Alignment

In this chapter we show that alignment is used to reduce web stress variations as well as to improve machine component life. While alignment can be checked with simple hand tools, alignment to tolerance can usually only be performed by instruments such as the optical transit.

Why Align

Alignment is the precision re-orientation of a roller in the level and square directions. While it is possible to build in near perfect alignment, it will not stay that way for very long. (Precision alignment is usually lost in the mere transport and handling of a new machine from the builder to the plant floor.) Thus, alignment is a maintenance procedure that needs to be done upon machine installation, then checked again about a year after startup, and then again periodically every 1-5 years. Realignment is needed because foundations settle, frameworks creep and elements are moved during maintenance. There are at least three reasons to align web machine components.

First, a misaligned roller will increase stress variations in the process. If web stress is not uniform across the width, the high tensioned areas may yield or break while the low tensioned areas can be floppy. Floppy lanes may not feed through nips and may slide over rollers erratically causing a loss of edge position control. One symptom, though not unique, of machine misalignment is a touchiness of web tension settings. A robust process is often able to run a 2:1 ratio of tensions on a single grade with few noticeable effects.

Second, misaligned roller is the tendency to wrinkle light webs. The fingerprint of this problem is a walking diagonal wrinkle. While a moderate case of troughing in an open web span may not cause obvious defects, a trough crossing a roller can ruin a web product. A wrinkle on a process roller is usually a disaster.

Finally, roller and other web machine component misalignment can cause premature part failure. For example, a pair of pulleys that are not coplanar (in the same plane and parallel in two directions) will tear up belts. Similarly, a motor misaligned with a drive shaft or roller can tear up bearings and couplings.

There is really no satisfying substitute for alignment. True, you can sometimes reduce wrinkles by changing tensions (usually up), but this constrains the process and makes it touchy. True, you can avoid running light grades, but this constrains the breadth of products. Finally, you can increase spans (lengths between rollers), but this may require rebuilding a machine. Similarly, component misalignment will either require expensive self aligning versions to extend life or frequent replacement otherwise.

It is interesting to note the cultural differences across the industries. In paper manufacturing, all mills know about alignment, and for the most part practice it. Alignments on the dry end are typically within the thickness of a human hair over spans up to 10 meters. Film is in transition where many have begun alignment programs. On the other hand, alignment is not frequently practiced in narrow web converting, nonwovens or textiles.

The reasons for these differences are to some extent pragmatic. The paper industry often has span ratios (length/width) of as little as 0.1, which increases the sensitivity to misalignment immensely. On the other hand, narrow web converting often has span ratios closer to 10. There is also a sensitivity to the modulus material property. Many grades of films and nonwovens are fairly extensible and will stretch or yield rather than break.

However, all industries are tending toward thinner materials, as well as higher qualities, speeds and efficiencies. This will only increase the need for better alignment and other web machine precisions.

Inplane Misalignment and Bending

The inplane bending stresses due to roller misalignment were first derived by Shelton of the Fife Corp. and Oklahoma State University.[1,2,3] This foundation of lateral behavior was initially applied to the statics and dynamics of web steering guides. However, we will apply the results here to one of the failure criterion for misalignment.

Figure 12.1 is a schematic of a pair of rollers where the downstream roller is mildly misaligned in the plane of the incoming web. Note how the web obeys the normal entry and exit laws on these rollers which we assume are in traction. The details of what happens before the first roller and after the second in terms of wrap, spans and so on is unimportant as long as they are both in traction. Perhaps ironically, the tension or stress distribution is uniform on the misaligned downstream roller while the upstream roller sees a bending stress which skews the tension. However, this is an artifact of the figure as the web has no sense of absolute alignment, rather only of relative alignment between two adjacent rollers.

The failure criteria of this misalignment mode is quite simple. We want to keep the stress deviation bounded within some quality limit. While you are free to choose any value, we might suggest a maximum deviation of 10% of the average setpoint tension as a starting point for most rollers. Exceptions to this value needs to be made in the case of guides where for practical reasons the deviation needs to be opened up to allow the guide to work effectively. On light webs, we should allow no more than a 50% deviation (factor of safety of 2 on a slack web). On thick webs (e.g., roofing) or strong webs (metals), the criteria could be only that the maximum stress is <1/4-1/2 of the yield.

This case can be modeled like a beam cantilevered on the upstream roller which carries a lateral force on the downstream roller sufficient to induce an angle equal to the misalignment angle. Knowing that the web acts like a beam allows us to use beam bending equations. However, we will neglect shear stresses and deformations to simplify the analysis even though almost all web spans are short beams.

Figure 12.1
Inplane Bending at a Misaligned Roller

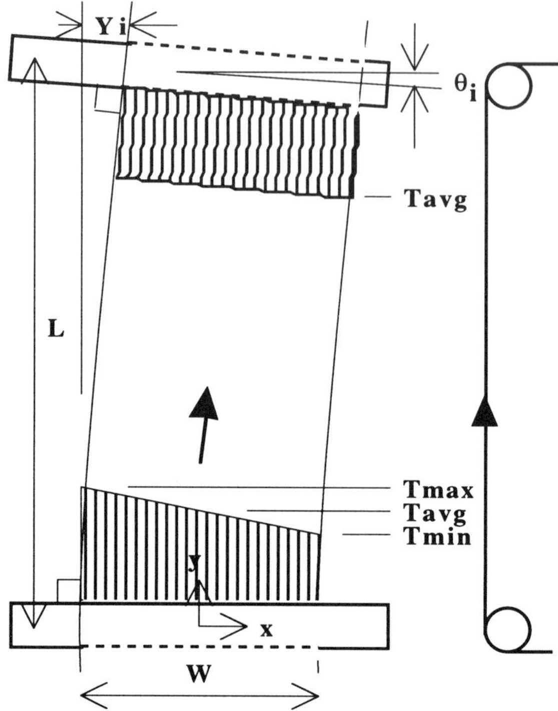

Rather than deriving this case from scratch, however, we will use the very convenient non-dimensionalization developed by Shelton. The interested reader should consult his or many subsequent works for the derivation and details.

The example problem solved in the next section is typical of many web applications. That is, we want to know the effect of inplane misalignment on web stresses to attempt to quantify a necessary alignment tolerance. The question to answer is whether sufficient accuracy of alignment can be obtained by hand tools or is optical transits necessary. The motivation is purely economic.

The first important parameter in this problem is the length/width ratio selected to be near the 1 typical of many spans. Longer spans, of course, would be more tolerant and vice versa. The second important parameters are the modulus and strength which is selected to be approximately that of paper and stiff films. Obviously a nonwoven or tissue would be more tolerant of misalignment.

Inplane Example Problem

A 30" long roller with a 40" long span is known to be misaligned 10 mils in the plane of the incoming web. What is the resulting tension deviation and, if excessive, how close must the roller be aligned? Also needed are

L = length = 40 inches
W = width = 30 inches
σy = yield strength = 4,000 lb/in^2
c = caliper = 0.001"
E1 = MD modulus = 5E5 lb/in^2
and
Tension minimum = 10% of yield
Tension deviation maximum = 10% of setpoint

First, we will calculate the minimum setpoint tension based on yield strengths.

(12.1)

$$\text{Tension} = (\text{Fraction of Yield})(\text{Caliper})(\sigma_y)$$

$$T = (0.1)(0.001\text{ in})\left(4000\ \frac{\text{lb}}{\text{in}^2}\right) = 0.4\text{ lb / in}$$

Next, we will calculate the average strain based on the setpoint tension.

$$(12.2)\quad \varepsilon_{1avg} = \frac{T}{c\,E_1}$$

$$\varepsilon_{1avg} = \frac{0.4\ \frac{\text{lb}}{\text{in}}}{(0.001\text{ in})\left(5E5\ \frac{\text{lb}}{\text{in}^2}\right)} = 0.000800\ \frac{\text{in}}{\text{in}}$$

Now we will calculate a very useful nondimensionalization factor, KL, where

$$(12.3)\quad K = \sqrt{\frac{T}{EI}}$$

$$(12.4)\quad KL = \left(\frac{L}{W}\right)\sqrt{12\,\varepsilon_{1avg}}$$

$$KL = \left(\frac{40\text{in}}{30\text{in}}\right)\sqrt{(12)(0.0008)} = 0.1306$$

The purpose of calculating a KL factor for this exercise is to use some very convenient approximations. If KL < 0.5, which it often is except for very extensible webs or very long spans, then we can proceed with the following.

The inplane misalignment angle in radians is

$$(12.5)\quad \theta_i = \frac{\text{inplane misalignment}}{\text{width}}$$

$$\theta_i = \frac{0.010\text{ in}}{30\text{ in}} = 0.000333\text{ radians}$$

The stress ratio can be calculated as

$$(12.6)\quad \frac{T_{max}}{T_{avg}} = 1 + \frac{\theta_i}{\varepsilon_{1avg}\left(\frac{L}{W}\right)}$$

Thus, the second term is really our stress riser fraction which for our problem is

$$\%\ \text{TensRise} = \frac{(100\%))\ (0.000333\text{ rad})}{\left(0.000800\ \frac{\text{in}}{\text{in}}\right)\left(\frac{40\text{ in}}{30\text{ in}}\right)} = 31\%$$

Since the problem is linear, we would need to decrease the misalignment to about 3 mils [(10%/31%)10mils] to reduce the tension rise on the tight side to only 10% more than the average. This would also make the slacker side 10% less than the average.

While we were somewhat conservative in this calculation by using a minimum tension and a tight quality standard, it does illustrate a point. Quality cross web tension control on stiff webs would require alignment to about half the thickness of a human hair across 30" (0.75m). This is far tighter than can be obtained by the best hand tools, thus requiring optical alignment.

Finally, the lateral offset of the web on the downstream roller can be calculated.

$$(12.7)\quad Y_i = \frac{2}{3}\,L\,\theta_i$$

$$Y_i = \frac{2}{3}\ 40\text{ in}\ \ 0.000333\text{ radians} = 8.88\text{ mils}$$

High Inplane Misalignments

We may, however, enter another regime if the misalignment becomes excessive, or on narrow webs which may be difficult to align and on many guides if the upstream span is long. This regime is where the web breaks loose on part of the width of the roll.

From the previous section, we showed how the tensions could be easily and significantly reduced on the slacker side of the web. This also would adversely affect the ability of the web to track there. This could cause the web to break loose there, particularly if the roll were lightly wrapped, the traction coefficient were low, or the web entrained significant air or fluids.

The net effect of a loss of traction on the slack side of the upstream roller could result in moment transfer upstream as shown in Figure 12.2. One problem with this is that slight changes in tension or traction can toggle between normal entry and moment transfer, causing sudden shifts in edge position. Another problem is that guides will become unstable in this regime which can be avoided by making L2 > L1.

If there is sufficient traction, a highly misaligned roller can create a slack edge as shown in Figure 12.3. The critical angle where the minimum web tension on the slack side goes to zero is

$$(12.8) \quad \theta_{icrit} = \varepsilon_{lavg}\left(\frac{L}{W}\right)$$

which for the previous problem yields

$$\theta_{icrit} = 0.000800\frac{\text{in}}{\text{in}}\left(\frac{40}{30}\right) = 0.001067 \text{ radians}$$

or about 3.2 times the current 10 mil over 30 inch misalignment. Thus, is takes very little misalignment to start a slack region in the web which grows in length and width with increasing misalignment angle. Note also how the web tension soars on the tight side. Obviously, this is a situation that is highly undesirable.

Figure 12.2
Moment Transfer on Upstream Roller

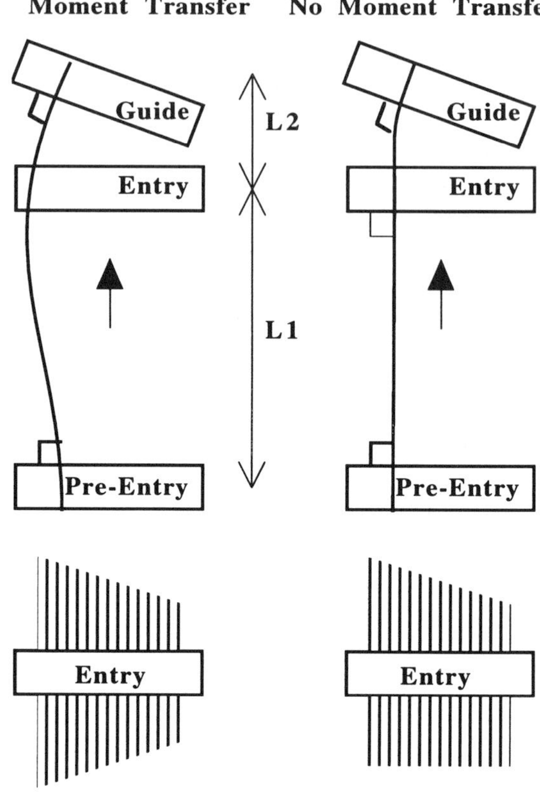

Figure 12.3
Bending Stresses Beyond Critical Angle

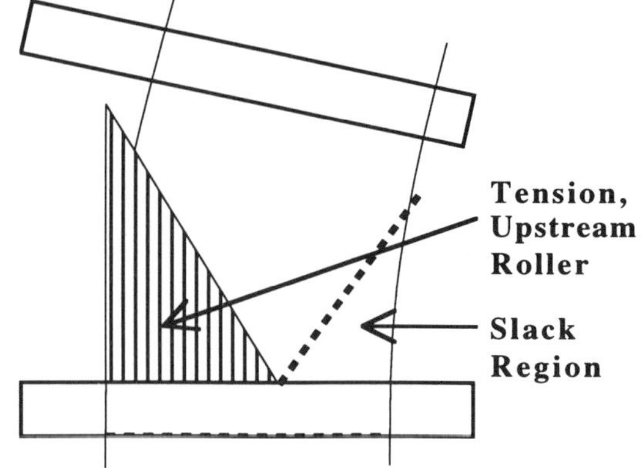

Out-of-Plane Twisting

The bending caused by inplane roller misalignment discussed in the last section can be quite damaging to web processes. The other component of alignment is out-of-plane which causes web twisting. This component is not nearly as serious but still must be controlled.

Figure 12.3 shows a roller pair where one (it matters not which) is tipped solely out of the plane of the web. Note how the longer paths at the ends versus the center of the web creates a parabolic tension distribution throughout that span. This stress riser created by this type of misalignment is easy to derive as illustrated by the following example.

A 30" long roller with a 40" long span is known to be misaligned 10 mils out of the plane of the incoming web. What is the resulting tension deviation and, if excessive, how close must the roller be aligned? Also needed are

Zo = out-of-plane misalingment = 0.010"
L = length = 40 inches
W = width = 30 inches
σy = yield strength = 4,000 lb/in^2
c = caliper = 0.001"
E1 = MD modulus = 5E5 lb/in^2
and
Tension minimum = 10% of yield
Tension deviation maximum = 10% of setpoint

From the previous example, we calculated the average MD strain to be 0.000800 in/in.

The strain (stress and tension) riser due to the path length difference can be calculated from trigonometry using the small angle approximation as:

$$(12.9) \quad \varepsilon_{omisalign} = \frac{1}{8}\left(\frac{Z_o}{L}\right)^2$$

$$\varepsilon_{omisalign} = \frac{1}{8}\left(\frac{0.010}{40}\right)^2 = 7.81E-9$$

$$(12.10) \quad \frac{T_{max}}{T_{avg}} = 1 + \frac{\varepsilon_{omisalign}}{\varepsilon_{1avg}}$$

Figure 12.4
Out-of-Plane Misalignment and Twisting

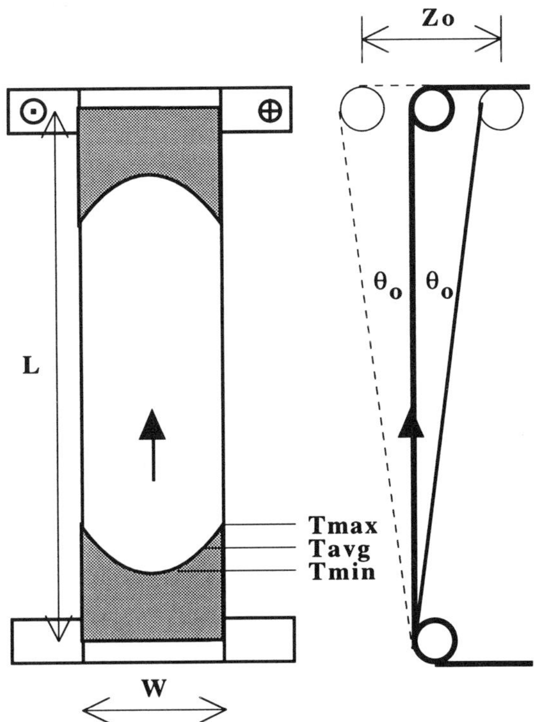

Thus, the second term is really out stress riser fraction which for our problem.

$$\% \text{ TensRise} = \frac{(100\%))\left(7.81E-9 \frac{in}{in}\right)}{\left(0.000800 \frac{in}{in}\right)} = 0.0009\%$$

Note how tiny the stresses are due to this type of misalignment. The reader can verify that the roller have 1" misalignment out of the plane and still be just under a stress rise of 10% over the setpoint average.

The approach used for both inplane bending and out-of-plane twisting can also be used to size **spreaders**. For example, bowed roller spreaders oriented into the sheet run further than the nominal will cause longer paths at the center. Also, the twisting stresses which are tiny compared to bending shows why the folding spreaders (dual bowed roller and Pos-z) are gentler on the web than fan spreaders (bowed roll etc.) Finally, this approach can be used to size guides, skewing and squaring rollers as well.

Diagonal Shear Wrinkles

The last failure mode we will explore here is diagonal shear wrinkles from inplane misalignment. All wrinkles are a compressive buckling of the web.[4] In the case of a shear wrinkle, the buckling is driven by shear stresses, which through Mohr's circle will result in compression at a slight angle to the CD.[5]

The misaligned roller is the source of the shear as the web is bent as a cantilever to achieve normal entry. The fingerprint of this classic problem is a walking diagonal wrinkle as seen in Figure 12.5. If the misalignment is moderate, the web may only be troughed in the free web span and may not be a problem in itself. As the misalignment angle increases, however, the wrinkle may cross the roller as a bulge, foldover or crease which is most destructive.[6]

Pioneering research by Gehlbach *et. al.* has modeled and experimentally verified the misalignment shear buckling failure.[7] Though the resulting equations can be used as a quantitative misalignment design criteria, we will only present an overview here.

As seen in Figure 12.6, a graph of tension versus misalignment angle can be divided into three regions. Region I at small misalignment angles is where the web is flat (but not necessarily evenly stressed). Region II is where the web is wrinkled and its boundary with Region I has been modeled. The boundary shows that increasing the span ratio (length/width) and especially increasing the caliper makes the system much more tolerant to misalignment. In Region III, the edge of the web is slack (out-of-plane) but is not troughed or does not cross the roller as a bulge or wrinkle. The boundary between Regions II and II illustrate that decreasing the traction can make the system more tolerant.

The beauty of this work is that it explains why, for example, narrow web converting and heavy gauge materials can sometimes get away without alignment without visible effects. It also explains the paradox where sometimes wrinkles can be reduced by increasing tension and at other times are better reduced by decreasing tension.

Figure 12.5
Diagonal Shear Wrinkle

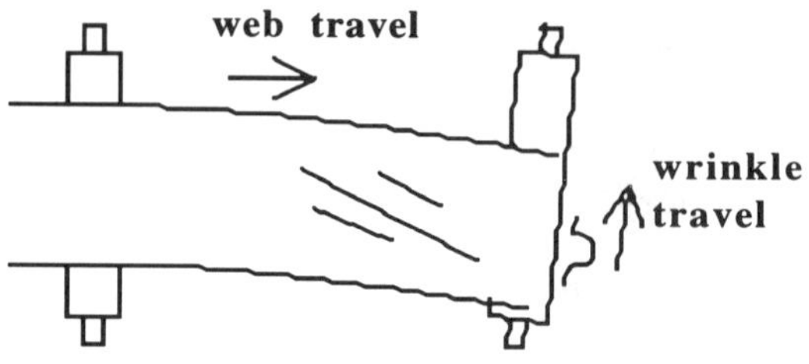

Figure 12.6
Wrinkles and Inplane Misalignment

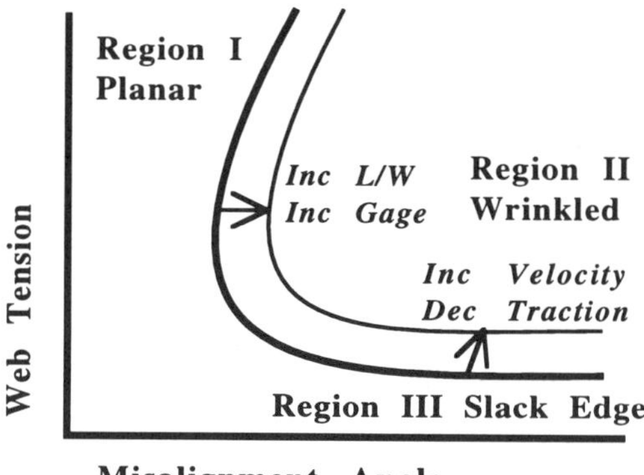

However, the misalignments that would move a system into the wrinkled or slack edge regions are truly gross. Long before this ultimate insult is reached, the web stress variation would be far higher than advocated in this book, indicating that this approach is not sufficiently conservative. Similarly, even the slightest hint of a walking diagonal wrinkle should be cause for grave alarm.

Obviously, the best approach to eliminate the diagonal shear wrinkle is to optically align the web machine. Against my better judgment, however, I will describe how to make the system more tolerant of wrinkles without re-alignment. First, reduce the web tension as much as possible and still maintain web control. If this does not eliminate the wrinkles, then the traction on the roller can be reduced. This may involve increasing the diameter, unwrapping the roller, stopping it from turning and/or polishing the surface very smooth.

Manual Alignment Checks

While re-alignment can seldom be done adequately without an optical transit, an alignment check can be made with simple tools. The first thing to do is to carefully observe every web span with incident lighting to look for diagonal shear wrinkles which persist in a span. As given in the last section, even a hint of wrinkles indicates severe misalignment.

The next thing that might be done is to check for non-parallelism between rollers, which is the direction of maximum sensitivity. Customarily, this has been done by comparing flat (Pi) tape readings on both ends of a roller pair. However, this method can be an order of magnitude too crude unless exceptional care is taken. An alternative method is by observing the travel of a ribbon or tape as shown in Figure 12.7, but requires turning a roller and high roller trueness. Thus, the dual dial indicator method of Figure 12.8 may be easier and more accurate, but suffers from accumulated errors.

While the web cares only about inplane and out-of-plane misalignment, it is often more convenient to work in level and square. Roller level can be checked using a machinist's level, which is only good to 1 mil per foot (about 5X too crude), or better yet a master level. In any case, however, there are two cautions when using a level. First, the bubble tube must be checked by reversing the level and making sure the reading is the same. Second, the level is very sensitive to imperfections of the roller surface so that many readings must be made across the width and an average used.

Square is a bit harder to check, but I will pass on one technique that got me by in a pinch. First, a pair of threads weighted with plumb bobs are set up to define the centerline of the machine. Next, a laser is aligned so that the shadow of the first string falls upon the second (this takes some doing). Now, a third string is placed in front of the target roller and exactly in line with the other two and the laser realigned to pass through the front two strings and hit the target roller. Finally, an optical quality mirror is placed on the target roller and the beam is redirected back toward the laser. Square errors will show as a reflection off to the side of the laser lens.

Figure 12.7
Parallelism Check Using the Tape Method

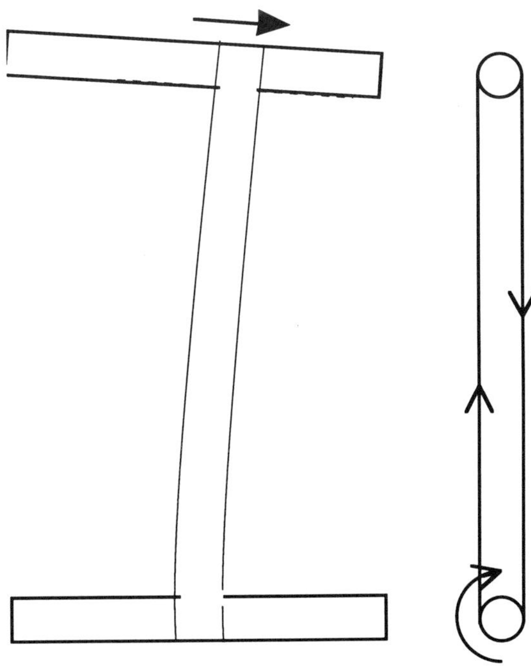

Figure 12.8
Parallelism With Dual Dial Indicators

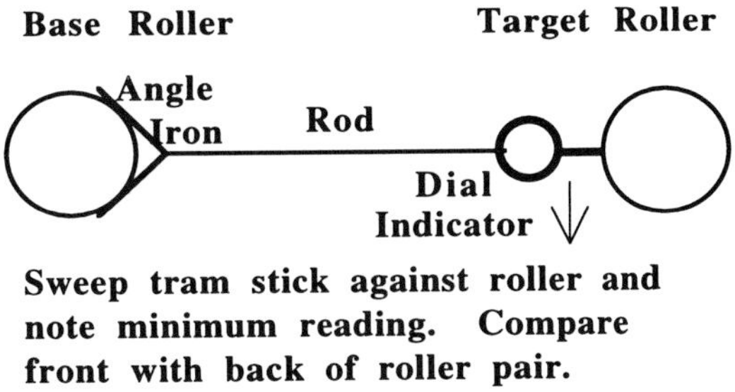

Sweep tram stick against roller and note minimum reading. Compare front with back of roller pair.

All of these techniques should only be considered disaster checks. If a valid discrepancy is measured, the misalignment is quite gross. However, these methods are crude enough that mild misalignments would not be picked up.

Finally, I would advise against using these methods to align a single roller. The reason being is that moving a roller will very likely just move a wrinkle to the next span.

Alignment and the Machine Builder

You can tell a lot about the expertise, knowledge and craftsmanship of a web machine builder by how they handle alignment. On the quality end, typified by wide paper and printing machinery, alignment is considered as important as any other machine parameter. Alignment specifications giving **tolerances** and maintenance intervals similar to Table 12.1 are written into their instruction manuals. They may also indicate how to prepare for an alignment as given in the next section.

Another way that one can judge the expertise of the builder is by looking at their machines. If the builder understands the web's alignment needs, they will make provision for easy re-alignment as in the example roller mounting of Figure 12.9. Here, a machined L-shaped **shear ledge** provides a precision mounting surface upon which to mount a bearing housing. Level and square are obtained by shimming between the bearing housing and the shear ledge. The reason for the ledge, for example, is to allow a squaring operation without the roller dropping and losing the previous level (alignment is a two step serial process). While there are techniques to align nearly all elements, a system like this makes it much quicker. Thus, if one sees shear ledges (or the like), brass shims and dowel pins, one can be assured that the builder is expert in this area.

Intermediate quality builders take great care to build the machine square and level by employing tight tolerance machining of frameworks such as by NC machine. What the builder (and the customer) do not understand and appreciate is that this initial alignment will be lost immediately during transport and installation, and subsequently due to framework creep (especially if it has not been thermally or vibratory stress relaxed) and foundation shifting.[8] The problem is that many of these machines are difficult to align because there was no provision to do so designed in. Thus, one has to drill out body bound bolt holes and so on to allow components to be moved. This would be similar to having to cut and splice a line to drain your automobile oil when it needed changing. It is not arrogance which motivates a builder's claims their machine is built square, it is simply their lack of understanding.

Table 12.1
Alignment Specification Example

Tolerance (for each class of roller)

0.002" per 100" of machine width, level and square with respect to a datum, not to exceed 0.005" regardless of width.

Optical Alignment Schedule (suggested)

During erection (to make sure parts fit)
Upon installation
After the first year (framework creep)
Check every 2-5 years thereafter or
Whenever a roller is added or replaced

On the low end of the quality scale is the builder who has no quantitative design understanding of how intolerant webs are to inplane misalignment. This is typified by some builders of narrow web, nonwoven and textile machines. Their instruction manuals may refer to alignment using levels and Pi tapes, if any mention is even given. On the plus side, these machines are often easy to re-align because they are often built on open framework. On the minus side, these machines are often too crude, flimsy or compromised to achieve or sustain a quality alignment.

Tip: You can use keys to preserve alignment of rollers that are frequently removed or interchanged.

Figure 12.9
Roller Re-Alignment System Example

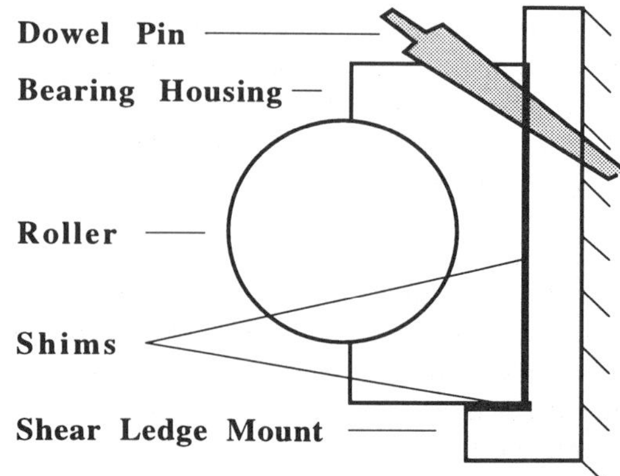

Alignment and Plant Maintenance

Optical alignment of web machinery is a very expensive process. At the very least, it will require two alignment personnel and two erectors to move the rollers. In many cases, however, additional crewing or other trades may be required if the job is complex. The cost of these people could exceed $5K/day if they were all hired from the outside.

The amount of time to realign a machine depends primarily on the number of rollers in the machine (or section). As a very ballpark number, it takes about one hour per roller. The actual amount depends largely on how easy the rollers (or components) are to move, and the ease of access for an optical line of sight setup. Obviously, there are many other factors affecting the alignment rate that need to be discussed when budgeting for an alignment. However, the cost of the alignment crew is not the only factor, the cost of machine downtime must also be considered.

Since alignment is so expensive, you want to make sure that you get your money's worth as discussed in this section covering maintenance and the next section covering the alignment crew.

There is a lot of preparation work needed before an alignment. The first step is to remove all excess or **redundant rollers**, because the number of rollers is the largest factor in alignment cost. This may involve upward of 10% or even 50% of all of the transport rollers in many converting lines, and sometimes excess drying or chilling process rollers. This streamlining of equipment is headed by a web handling knowledgeable process engineer who first tries bypassing rollers where easy, and physically removing others where a bypass thread route can't be found. After successful operation for a month or so, the redundant rollers can be permanently removed.

Next, rollers with **excessive deflection** must be replaced with stouter ones. Excessive deflection causes small variations in the CD placement of the optical target and tension variations to affect the ultimate alignment quality.

Next, the roller surfaces must be trued up if they are out of spec. For example, a roller with 0.005" **runout or diametral variation** will be difficult to align any closer than that value because the optical target is placed on the imperfect roller surface. Similarly, play and clearances must be minimized to a sliding fit.

Other machine components that need to be checked are chains or wire ropes used to raise devices such as festoons or rider rollers. If chains are worn or stretched, the effective pitch changes. Thus, the device could be level in the down position but out of level in the up position. Chain wear is best checked as the radius of curvature of the chain when bent in the stiff direction when lying on the floor. An analogous problem results if the center-center distance of pivots are not equal on the front and back sides of the machine.

During the alignment process, the maintenance crew will have some influence on how quickly the alignment will proceed. First, neatness counts. If the mountings are not scraped flat and brushed clean, the alignment may chase moving contaminants. Second, old shim packs should be removed, and new ones installed that have no more than three shims. Finally, it is imperative that all bolts are checked for tightness and the mountings **doweled** before the machine is started to preserve that hard fought alignment.

Lastly, the maintenance crew must be very conscious of preserving alignment when working around the machine. For example, a roller could be bumped out of alignment. Also, the shims and dowels must be marked and remain with a roller if it is removed for maintenance. A more difficult problem arises if a roller is regularly interchanged with other spares. Hopefully, the builder would have a system to make sure all rollers moved into that position would be in alignment. If not, you could check with your contract alignment crew for ideas on how to handle this situation, such as by using keys.

Alignment Service

There are three sources for optical alignment services. The first is the machine builder who, if large and knowledgeable, will often provide initial alignment as part of the equipment sale to make sure it goes in right, and later as a chargeable service. The second is a contract engineering or construction firm. The concern here is to make sure that the crew has a lot of familiarity with the demanding needs of web machinery, which is higher than many other types of machines.[9] The third source of alignment is your own company if there is enough work to justify it. While the equipment is not especially expensive and the training not particularly demanding, experience counts and must be developed. While anyone can be taught to read a transit, experience helps overcome real world difficulties by knowing the tricks of the trade.

Since one does not actually look through the transit to verify the quality of the alignment work, the selection of a crew is often based on the trust and reputation earned by its company. Only rarely has one the opportunity to verify work by having a second crew spot check the first crew's work or by comparing readings between two different alignment periods.

However, there are indicators of professionalism to look for. First, the estimator should be able to answer all of your questions to your satisfaction (except, ironically, how long it will take). Then when the crew shows up, one will get some idea how smoothly things are set up. Look to make sure that the crew uses landmarks, or offers to install them if they have not already been placed. What may not be a reliable indicator, however, is the appearance of the crew or their demeanor. I've seen some pretty scruffy and surly erectors perform flawless alignment work.

During the alignment process, one can expect that most rollers will be brought into level and square with one move in each direction. For example, if the front side of the roller is shimmed up, we would expect that the shims on the front would not need to be redone when the alignment is rechecked.

Inclined surfaces can pose problems if the erector doesn't know trigonometry (have a process engineer help here). Also, one should watch how the alignment crew handles pivots and slides. In all cases, the alignment needs to be set on the most important end of the stroke and checked on the other.

Finally, one should expect prompt and thorough documentation of the alignment. As shown in Figure 12.10, this should minimally consist of circle drawing with web run upon which four numbers are placed alongside each roller or component.[10] While not universal, the top pair is often the before and after level where a plus indicates the back side is higher than the front. The second pair is the before and after square where a plus indicates the back side is in the +MD of the front. Sometimes arrows are used instead of plus and minus signs. The report should also contain recommendations.

The utility of this report is such that one should expect it as part of the alignment service. The after readings will show the current state of the equipment. This will also include compromises in alignment that were made for timing or practical reasons and serves notice to keep an eye on the area. The difference in the before and after readings will show two things. First, accidents or careless maintenance where a roll is put out of alignment will stand out from normal slow machine movements. Second, one can anticipate the timing of the next alignment based on the rate of the settling of that machine.

Figure 12.10
Alignment Report Example

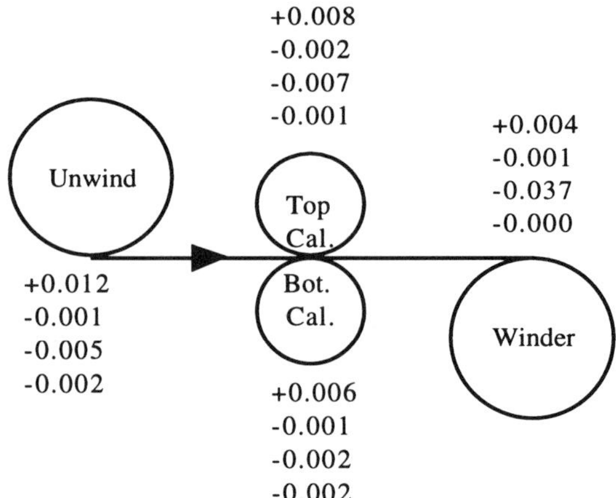

Alignment Process

The alignment begins with a clear understanding of the objective (which rollers), timing (how long can be taken), priorities (if time is running out) and required resources or needs. In some cases the goal is merely to survey the condition of key rollers to determine if a formal re-alignment is needed. In other cases, only a suspect roller is corrected.

The crew sets up two transits to perform the alignment along a **datum** line defined by two or three landmarks (monuments, plugs). A transit is a small telescope on a tripod similar in appearance to what is used by surveyors of lands and roads. A **landmark** is often a small brass plug epoxied into the floor with a prick mark on it to define the datum. As seen in Figure 12.11, the datum is often conveniently offset in the aisleway.[11]

The rollers in a section are usually leveled first, which only requires a single transit. This transit must be able to clearly see a target (small ruler with a bubble level) either on the top of the bottom of the roller. The roller is shimmed until the front and back sides read the same.

The roller is squared by setting up a transit on a datum and then rotating the head a precise 90 degrees. Often a second reference transit is set up for ease of moving from roll to roll. It is helpful if the working transit can see a target on one of the sides of the roller.

There are many problems that can be encountered that may require a judgment call. For example, what if the entire machine settled evenly 0.1" on the back side? One could shim the entire back framework up, or one could shim all of the individual rollers. Another strategy is just to leave it because the web does not know about absolute alignment, only relative differences between rollers. Judgments may also need to be made with imperfect slides, pivots, chains or bolt bound mountings. This is where a process engineer must stay with the job or be immediately available when a judgment is needed.

Figure 12.11
Optical Alignment System

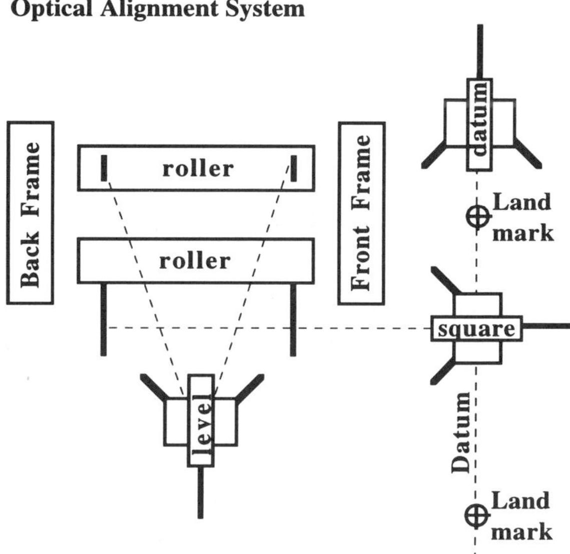

What can you expect after aligning the machine? If one is practicing regular alignment and this was only a mild correction, the results may be subtle. If this is the first alignment for a machine, one might see an entire change of character for the better. The process will run smoother, and with fewer wrinkles, breaks and other defects. The process should also run more consistently in terms of edge position and registration control.

The machine will also require less **maintenance** because bearings, couplings and other elements will last much longer. Indeed, it has been estimated that 50-70% of all rotating machinery failures can be directly attributed to shaft misalignment.[12] Finally, troubleshooting will become much simpler if all the rollers are true, level and square because that leaves machine design, process settings and web properties as the only major sources of problems left left.

Conclusion: Web machines should be regularly aligned with optical tooling because there is plenty of problems that can result if this important machine quality parameter is ignored.[13,14,15,16]

Bibliography

1. Shelton, John J. *Lateral Dynamics of a Moving Web.* Ph.D thesis at the Oklahoma State Univ., July 1968.

2. Shelton, John J. and Reid, Karl N. *Lateral Dynamics of a Real Moving Web.* Transactions of the ASME, October 1971.

3. Shelton, John J. and Reid, Karl N. *Lateral Dynamics of an Idealized Moving Web.* J. of Dynamic Systems Measurement and Control,Transactions of the ASME, September 1971.

4. Roisum, David R. *Why is My Web Wrinkled?* Converting Magazine, Web Works column, pp 16, February 1994.

5. Roisum, David R. *What Causes Diagonal Shear Wrinkles?* Converting Magazine, Web Works column, pp 22, June 1994.

6. Friedrich, Craig Richard. *Stability Sensitivity of Web Wrinkles on Rollers.* Tappi J., vol 72, no 2, pp 161-165, February 1989.

7. Gehlbach, Lars S. and Good, J.K and Kedl, Douglas M. *Prediction of Shear Wrinkles in Web Spans.* Tappi J., vol 72, no. 8, pp 129-134, August 1989.

8. Weiss, Herbert. *Yearly Machinery Alignment Controls Quality Loss, Cost.* Paper, Film & Foil Converter.

9. Dunn, Jim M. and Pepin, S. and Allen, D. and Leonard, W. and Carter, M. *Precision Alignment of Winders.* TAPPI Finishing and Converting Conf. Proc., San Diego, CA, pp 119-124, October 15-18, 1995.

10. LeBel, Gary A. *Optical Diagnosis for Misalignment.* Converting Magazine, pp 46, 48,50, November 1993.

11. Allen, George Jr. *Optical Tooling for Paper Machines.* TAPPI Engineering Conf. Proc., pp 595-599, November 1994.

12. Anon. Title Unknown. Engineer's Digest, August 1989.

13. Masse, Raymond L. *Realignment Program Keeps Quality In Check.* Paper Film & Foil CONVERTER, April 1990.

14. Boggs, R.E. *Rotating Machine Misalignment:* The Silent Disease but not the Symptom Tappi J., vol 73, no 12, December 1990.

15. LeBel, Gary A. *Improving Dryer Section Runnability through Optical Alignment.* Tappi J., vol 77, no 5, pp 285-288, May 1994.

16. LeBel, Gary A. *Paper Machine Alignment.* Papermaker, pp 12-14, December 1994.

Chapter 13

Mechanical Drives

In this chapter we cover mechanical means of producing torques and of transmitting these torques to a roller. Torque can be generated by pneumatic band, shoe or disk brakes; eddy current, electro-magnetic or magnetic particle brakes and clutches; and tendency drives. Torque can be transmitted via belts, chains, inshafts and couplings. Torque and speed can be varied via variable transmissions and pulleys.

Mechanical Motivation

Until perpetual motion is perfected, all web machines will require drive points. A drive point is wherever torque is externally supplied to a roller. **Motoring** is a positive torque which will tend to decrease the web's tension as it goes over that point as well as counteracting roller drag and acceleration inertia. **Braking** is a negative torque which will tend to increase the web's tension as it goes over that roller, and assist overcoming deceleration inertia. An exception to this sign convention is on an unwind where braking increases or provides a starting web tension.

As discussed in Chapter 4, these drive points may be used to provide tension at the ends of a line, counteract an unacceptable tension change at a roller due to drag or inertia, to intentionally step tension up or down at a specific location, or to maintain traction over lightly wrapped or slippery rollers. The number of drive points may be as few as one to perhaps a hundred or more. Whether a roller should be driven is merely a calculation once one defines an acceptable tension range. The nominal setpoint tension should be typically controllable within a range of 10-25% of the yield strengths of all products to be run. Allowable tension variations should typically not exceed 10% of the setpoint at a load cell with respect time or 25% within a drive zone (bounded by driven rollers).

While at least one of the drive points in a web line will be an electric motor, the other required points could be mechanical brakes or clutches, or hybrid electromechanical devices. The designer must weigh cost, performance, maintenance, size and other issues when deciding which of several eligible devices will be used. A drive device is eligible when it is capable of the required torque range, holds a command torque within the required precision, has sufficient power, and does not exceed speed, RPM or other limitations.

There are at least two strong motivations for choosing mechanical drives over electrical motors. First, mechanical technology is usually simpler to understand and maintain. Mechanical drives can be serviced by anyone with a mechanical aptitude who will take a moment to read the accompanying maintenance manual. In contrast, electrical drives require considerable specialized training and expertise. Troubleshooting and tuning an electrical drive often requires a visit by the vendor, particularly in the smaller plants. Second, mechanical drives are often cheaper than the electric motors and controllers of the same torque and power range.

However, there are at least two disadvantages to mechanical drives. First, they tend to require more frequent maintenance. This is because there are more parts which can wear out. In contrast, modern electric drives require very little attention after proper setup and tuning, particularly those which are digital instead of analog. Second, mechanical drives can suffer from cruder control precision and flexibility. This translates to poorer control of web tension.

However, mechanical components will be important on even the most electrically oriented drives. That is because electric motor torque is usually transmitted to the roller via mechanical components such as belts, couplings and shafts.

Belts

Belts are used to transmit power from one location to the next. This could be, for example, from a motor, brake or shaft to a roller. Belts could also be used to cascade or slave several rollers together. Belts can also provide a speed or torque ratio between two points. For example, a drive motor may be more efficient (higher power) at RPMs different than appropriate for its roller's speed. Here, a belt may be used to gear the motor down so that roller and motor both turn a appropriate speeds.

All but a few percent of the power supplied by the driver is transmitted to the driven pulley. The remainder is frictional losses. However, both rotational speed and torque will vary as the ratio of pitch diameters as shown in Figure 13.1.

$$(13.1) \quad N_2 = N_1 \frac{D_1}{D_2}$$

$$(13.2) \quad Torq_2 = Torq_1 \frac{D_2}{D_1}$$

If a roller were connected to the driven pulley, its surface speed given its RPM in the English system and where V is the roller's surface speed in FPM, N_2 is the RPM of the roller (pulley), and D is the OD of the roller in inches would be:

$$(13.3a) \quad V = (0.2618)(N_2)(D_{roller})$$

or in the Metric system and where V is the roller's surface speed in MPM, N_2 is the RPM of the roller (pulley), and D is the OD of the roller in centimeters would be:

$$(13.3b) \quad V = (0.03142)(N_2)(D_{roller})$$

There are three principal types of belts used on web machine drives as shown in Figure 13.2.

Figure 13.1
Schematic of a Pulley System

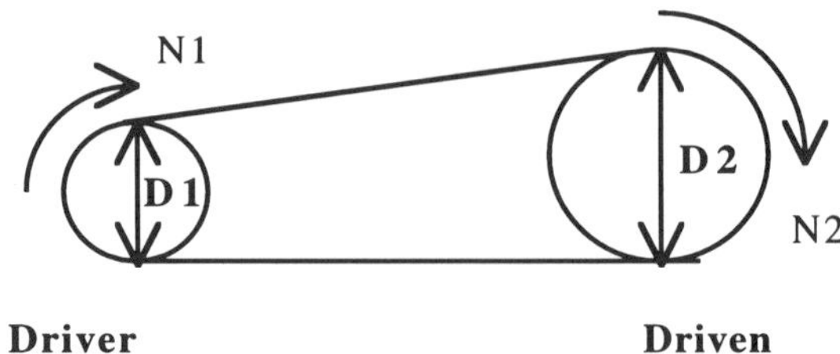

Figure 13.2
Drive Belt Types

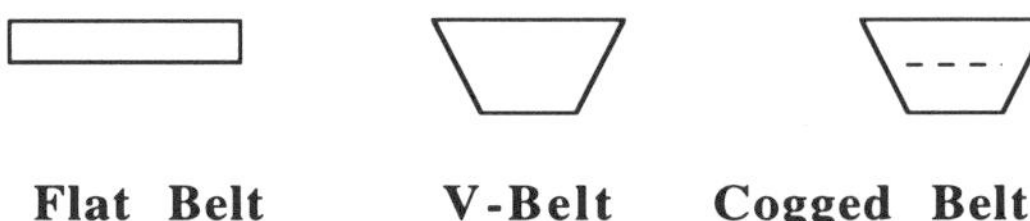

Figure 13.3
Cone Pulleys for Speed Ratio Adjustment

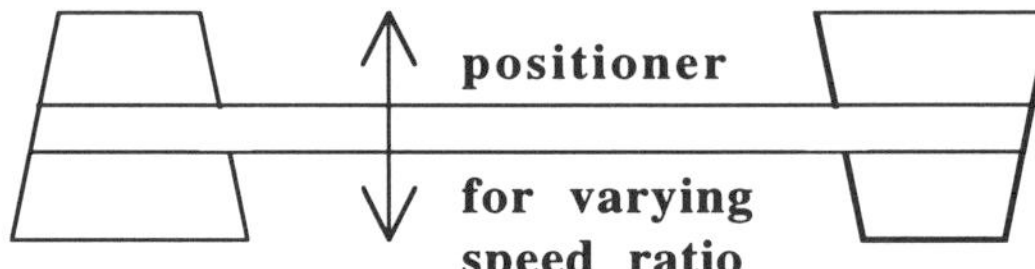

Compared to the V and cogged belts, a **flat belt** can transmit very little torque without slipping on the pulley. The maximum tension difference across a pulley is determined by the band-brake equation given in Chapter 4. This difference times the effective pitch radius determines the maximum torque. Another disadvantage of the flat belt is that it requires some type of guide to keep it on the pulleys. Various guide methods include edges on the pulleys or tightener, crowned pulleys, side rollers and pulley tilting.

One important application for flat belts is on cone pulleys for adjusting speed ratios as shown in Figure 13.3. Here, a positioner moves the belt so that it rides on a different pitch of the driver and driven pulleys. Similar methods are used to adjust speed ratios of specially designed V-belts and chain belts.

V-belts can carry more torque than a similarly sized flat belt because the sides are wedged into the pulley and thus permit a higher normal load for the same belt tension. V-belts should not, however, be used on speed controlled sections because the effective pitch ratio is not very precise. Variations in drive torque, belt tension, wear, temperature, belt modulus and other factors conspire to make the ratio ever varying.

If speed control is desired, **cogged belts** (often called by the tradename Timing Belt) should be used. Cogged belts have a couple of disadvantages, however. First, they can be much noisier than other mechanical drive components. Second, the cogs in the belts and pulleys can get fouled with debris easier.

Roller chains can carry much more load than belts and are much more durable. However, **chains** should never be used on a roller drive train. The reason is that the output speed of the sprocket will cyclically vary several percent about the mean as the teeth come into engagement.

Belts must be checked against several design criteria. They are speed, speed ratio, torque capacity, power and minimum pulley size.

Tension is a very critical setup item for all styles of belts and chains. If the tension is loose, the belt may slip, skip or vibrate. If the tension is tight, the loads on the shafts and bearings increase unnecessarily. Belt tension is controlled by adjusting the position of one of the pulleys (such as moving the base of a motor), or by an adjustable or spring loaded idler on the low tension side. Belt tension is usually measured as a deflection of the belt with a specified point load in midspan. Belt tension needs to be checked frequently.

Another very important item is pulley **alignment**. If pulleys do not lie exactly in the same plane, belt life will be short and belts can even jump the sheaves. The three types of alignment errors include offset (CD), out-of-plumb and out-of-square

Belts that are under distress will make noise, vibrate, have glazed surfaces, or leave black wear particulate underneath.

Gears

Like belts and chains, gears are used to transmit torque and power from one point to the next, often with a unity speed ratio. The advantages of gearing include a higher torque capacity and longer service life. However, gearing is much more expensive, requires more elaborate lubrication and enclosure, and can be noisy. Also, the center-center distance is fixed, very critical for life, and some backlash may be inevitable. Gearing is most commonly found where two nipped rollers must be timed together, such as on rotary die-cutters and some embossers. Gearing is also found on some older paper machine dryer sections.

There are three principal gear styles used for parallel shaft applications. The **spur** gear is the least expensive, but has more limiting pitch line velocities. The **helical** gear is more suitable for high speeds, but is expensive and produces large thrust loads. The **herringbone** gear has side by side helical teeth with opposite leads so that thrust is canceled out. Any of these gear styles may be split down the middle where one half might be rotated with respect to the other. This adjustment allows some control over backlash and can accommodate minor differences in center-to-center distance.

After the designer selects the nominal gear style, the gear pitch diameter will be determined primarily by the tangential force that must be carried to transmit the necessary torque. Other sizing tasks are recipes similar to sizing bearings, and can be found in many machine design textbooks. Since two nipped rollers are friction limited, the maximum (internal recirculating) torque produced can be calculated from the coefficient of web-roll or roll-roll friction, the unit nip load, roller width and roller radius as:

$$\text{Torq} = \mu \, N \, w \, r \qquad (13.4)$$

There is a common misconception that if you match the diameters of the rollers well, that they will not fight each other. As we have seen, however, torque loads will be present and independent of the amount of roller surface velocity mismatch. The motivation to match velocities as close as possible is merely to extend roller surface life by reducing scrubbing.

Couplings

Couplings are used to connect two shafts together, such as between an electric motor and a live shaft journal. Couplings are connected at assembly, but may be disconnected for occasional maintenance or repair. Applications which require periodic disengagement, such as on unwind shafts, call for the use of clutches.

Aside from assembly and disassembly, couplings serve to accommodate inevitable misalignment between shafts. If the shafts were welded together, for example, tiny misalignments could create bending stresses large enough to break the joint or damage adjacent bearings. Aside from torque capacity, the other very important selection and maintenance criteria for couplings is the type and amount of misalignment present. The two types of misalignment are parallel offset and angular misalignment, which each have components in the horizontal and vertical directions. While shafts can be aligned by a dial indicator method, optical transits may be convenient if they are already set up during major downs for machine repair.

There are two principal classes of couplings. The rigid coupling is used with shafts with have excellent collinear alignment. **Rigid couplings** are usually cheaper, smaller, more durable, and are torsionally much stiffer than flexible couplings. If alignment is not excellent, however, flexible couplings must be used. **Flexible couplings** are torsionally very soft, which might require detuning of drive controls to avoid resonance. However, torsionally soft couplings may be advantageous to reduce shock loading produced by unusual applications.

If misalignment is very large or highly variable, a **U-joint** (aka universal joint, Hooke's joint or Cardan's joint) may be used. However, constant velocity U-joints must be employed to ensure that the roller's surface velocity does not vary cyclically as it rotates. A constant speed U-joint is a pair of U-joints where the outermost yokes are in the same plane.

The reader should consult machine design textbooks and vendor catalogues for more information on the numerous coupling types that are available.

Shafts

Shafts oriented in the CD which are used to transmit torque to a roller from a motor/brake which must be placed some distance away are often called **inshafts**. Shafts oriented in the MD which transmit torque to several machine sections through right angle gearboxes are called **lineshaft** drives. Small shafts, similar in size to live shaft journals, are usually solid while very large shafts may be hollow tubes.

The first requirement of a shaft is torsional strength. The maximum shear stress on a shaft in torsion is:

$$(13.5) \qquad \tau = k \, \frac{\text{Torq}\, D}{2\, J}$$

where J is the polar moment of inertia which for a hollow shaft is:

$$(13.6) \qquad J = \frac{\pi}{32}\left(D^4 - d^4\right)$$

The diameter, D, is the minimum outside diameter of a possibly stepped shaft, and d is the inner diameter. The stress concentration factor, k, for a stepped shaft depends on the radius ratio and the fillet radius and can be calculated or looked up in tables. Other possible stress concentrations which must be investigated include keyways, holes, pressfits and so on.

The next requirement of the shaft is to avoid rotational resonance and can be calculated similarly to that of a roller as given in Chapter 11. Finally, the shaft and associated couplings must be torsionally stiff enough to avoid torsional resonance which would require detuning of drive response. While this type of resonance is complex and beyond the scope of this text, it depends on the torsional stiffness of the shaft which can be calculated from the angle of twist as:

$$(13.7) \qquad \phi = \frac{\text{Torq}\, L}{J\, G}$$

where ϕ is the angle of twist is radians due to a torque on a shaft of length L, and where G is the shear modulus of the shaft material (usually about 1/3 of the tensile modulus).

Tendency Drives

All rollers have some amount of bearing drag which tends to increase the tension as the web passes over idler rollers. If necessary, this drag can be counteracted by using a tendency drive. As seen in Figure 13.4, a tendency drive is composed of a drive rotating faster than web speed connected to a live shaft journal through a bearing. This figure is schematic in that most often a single motor will turn a series of tendency idlers via a common belt. Also, both sides of the rollers are typically driven.

On tendency drives, the **bearing drag** torque which tends to slow it and add web tension is counteracted by the tendency bearing which tends to overspeed and subtract web tension. If the two bearings are identical, the torques will tend to cancel. This principle is similar to motor/clutch drives, though the bearing transmits much smaller torques. Tendency drives are a primitive version of a helper drive, but without the ability for adjustment or inertial compensation.

Though some converting machines have tendency drives, few are truly justified. A tendency drive is indicated only when the tension rise due to bearing drag is unacceptably high and at the same time the inertial tensions are negligible. As was indicated in Chapter 2, a 10% tension rise through a drive section should be acceptable for most applications. Chapter 4 showed how to measure or calculate bearing torque. Unless the web is weak and the bearings brutish, the web should be able to turn the idlers without problem or even compromise.

Figure 13.4
Tendency Drive Schematic

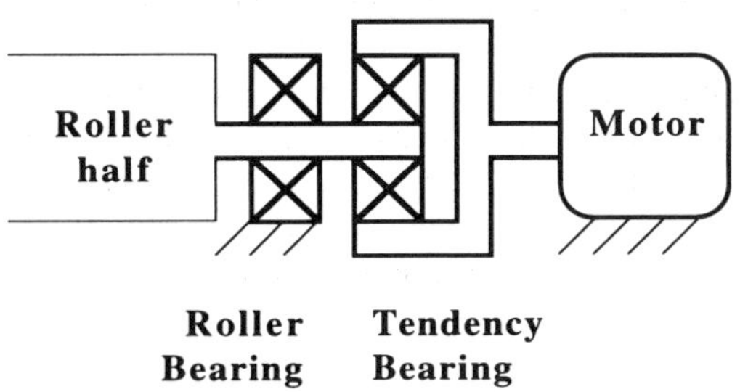

Variable Speed Transmissions

In speed or registration control, one may want to adjust the speed of one section versus another as discussed in Chapter 2. On most modern machines, draws are set via an operator panel which are input to a sectional electric drive controller. However, draws on older machines and some specialized processes are varied mechanically via variable speed transmissions. An example given earlier was flat belts on a cone pulley. Other means include **PIV'**s™, hydraulic pump/motor systems, harmonic drives and so on. However, in this section we will focus on sizing rather than describing the means.

First, the variable speed transmission must have a sufficient torque capacity. The required torque could be estimated from Chapter 4, but will certainly be less than the maximum torque output of the electric motor or other prime mover. Next, the range of speed ratio adjustment must be adequate. For simple systems the speed ratio may be estimated as 1 ± material yield strain / 4. Hygro-thermal or solvent processes will need an additional amount to compensate for anticipated elastic expansion or contraction. Finally, drawing processes which intend to permanently stretch the web (such as on tissue, some films and some metals) must accommodate the expected range based on prior product development or experience.

Next, the drive must have suitable RPM ratings. Lastly, one should check the **accuracy, resolution** and repeatability of the transmission. For simple systems, the speed ratio should be repeatable to better than the minimum material yield strain divided by 100.

For example, an inextensible material such as paper might have a yield strain of only 1%. The nominal setpoint draw between sections might need to be as little as 10% of that (to prevent web breaks or damage), or in other words a draw of 0.1%. For even the crudest control, we would expect that a setpoint be held within ±10%, or a speed control consistency of 0.01%. Unfortunately, few mechanical devices can achieve that kind of resolution so that they are generally suitable only for extremely extensible materials.

Pneumatic Brakes and Clutches

Pneumatic brakes are common on small converting machines because they are simpler and cheaper than electric drives. Brakes are used to provide unwind tension, to step up tension on an intermediate section and to assist the stopping of driven and undriven rollers. Motor/clutch systems are similar to tendency drives in arrangement, but are intended for tension control rather than bearing drag compensation.

Brakes and clutches are activated from a short-stroke air cylinder or from an air bag. Air pressure is preferably **closed-loop** controlled manually via a regulator or automatically via an E/P (electrical to pressure transmitter) from a load cell tension or dancer controller. However, **open-loop** systems using ultrasonic or other methods of diameter calculation can be used. Air pressure is used on both caliper or band/shoe brakes.

Horsepower, which is a function of thermal dissipation, is largely determined by physical size. Horsepower can be increased without increasing component temperatures (to avoid pad fading, fire streaking and alteration of metallurgy) by forced air cooling. Aggressive cooling can be achieved by circulating water through a drum brake. However, water cooled drums tend to sweat and maintenance is higher.

Torque is a function of the cylinder pressure, cylinder area, disk or drum radius and the coefficient of pad/disk or shoe/drum friction. Some vendors provide high and low coefficient pads to move the torque range of their devices. Horsepower, torque, speed and other specifications can usually be found in the vendor's catalogue.

Perhaps the most common web application for pneumatic brakes is on unwinds to provide web tension. As seen in Figure 13.5, the required unwind torque or nominal **brake pressure** drops off proportionally to the unwind diameter decrease. However, during accelerations the pressure will decrease from the nominal because the unwind roll inertia will provide some of the tension and vice versa during deceleration. Note how the sizes of the inertial steps gets much smaller as the unwind expires.

Figure 13.5
Unwind Brake Torque and Pressure

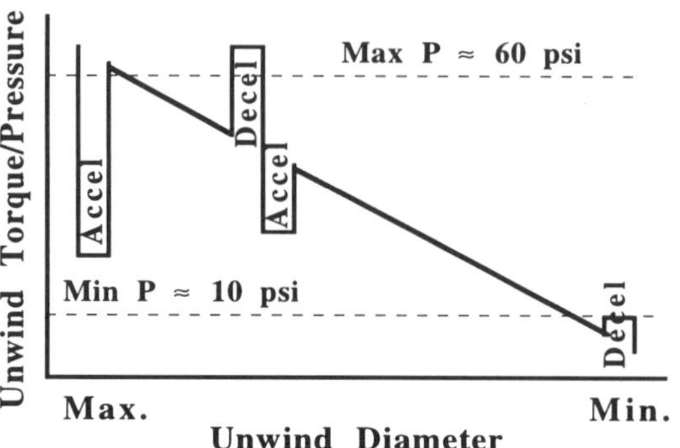

There are two very common unwind **design** mistakes. The first is to control brake pressure manually with a regulator. Even if the operator could and would follow a chart to decrease pressure manually with diameter, the results would be poor. First, the corrections would be in steps instead of smoothly. Second, the operator could not respond appropriately during speed changes. Third, any changes in brake friction (fading) or loading would result in a tension change with this open loop scheme. Unwinds require closed loop tension or dancer control on almost all applications.

The second mistake is to size brakes inappropriately for horsepower or torque. Note from the figure how the required brake pressure increased beyond 60 psi (the most common minimum supply) during the deceleration for a splice. More commonly, however, the required brake pressure may drop below an acceptable minimum at the end of the roll. For example, a common regulator is only consistent to about 1 psi. Thus, even at a 10 psi, the system may yield marginal performance (10% tension variation due just to one source).

In almost all unwinds, several **calipers** need to be turned off as the unwind diameter expires to ensure sufficient brake pressure. This will happen whenever the max/min diametral ratio times the max/min grade tension ratio exceeds 6:1. The best way to turn off calipers is automatically via a shuttle valve. Modern brake control systems can also sequence the cylinders to ensure even friction pad wear.

Hydraulic Brakes and Drives

Hydraulic motors are occasionally used on winders and rollers. The configuration is a pump followed by a valve/regulator followed by the motor (often a vane pump running backward). Hydraulic motors have the advantage of a much greater power to size ratio than electric motors.

Hydraulic clutches are occasionally used to connect and control torque from an electric motor to a driven device, such as on textile machinery. The hydraulic clutch has a smooth engagement and absorbs shock loads well.

Hydraulic brakes are occasionally used on very heavy unwinds. The brake may be a wet disk brake where fluid shear absorbs power which is carried away to a heat exchanger. The amount of torque is controlled by axial loading of the disk pack.

While hydraulic drives are strong, smooth and usually economical, they have several disadvantages that make them unsuitable for most web applications. First, hydraulics are shunned in many plants because of the ever present threat of leaks and contaminations. More importantly, the control quality of hydraulic devices is much poorer than electric or other drive counterparts. Hydraulics are more sluggish and may suffer from a large parasitic drag (minimum torque). Finally, it seems that at time passes there are fewer maintenance people who are comfortable with analog hydraulic controls because they are becoming somewhat uncommon.

Electromechanical Brakes & Clutches

Tendency drives and motor/clutches are two examples already given of electromechanical brakes and clutches. In other designs, an electromagnet can be used instead of a pneumatic cylinder to pull brake pads against a disk. In fail safe brake designs, a spring loaded pad is unloaded via an electromagnet when the brake is turned off.

Another example of an electromechanical device includes the magnetic particle clutch or brake. Here, current flowing through the brake causes iron particles to bridge between and cause increased friction between a rotor and stator. Finally, eddy current clutches and brakes are occasionally found on light equipment.

Unwind Clutches

Unwind and windup shafts must be changed frequently throughout the day. Thus, a shaft must be easily coupled and uncoupled from its motor or brake. Though there are scores of ways of achieving this function, most have problems with vibration, backlash, torque capacity, speed, life and so on.

One clean arrangement is a clutch head that is moved in the CD on a splined drive shaft with a pneumatic cylinder for engaging and disengaging. The clutch head is a star gear that engages another spline cut into the wind/unwind spool or mandrel. Though splines are expensive, they can carry a lot of load and are smaller, more durable and have less backlash that square block clutches, keys or pins in holes.

An alternative to the spline coupling between the drive shaft and spool is an air expandable clutch. In the external version, a bladder (much like a tire with friction pads on the outside) is air expanded to contact the inside of a shroud on the wind/unwind mandrel. The advantage of this type of clutch is that it can be engaged even when there is a speed mismatch between the drive shaft and wind/unwind mandrel.

Band Brakes on Rolls and Rollers

It is not uncommon to find band or strap brakes used to set web tension on an unwinding roll. As seen in Figure 13.6, this brake is composed of a strap wrapping the unwind roll about 90 degrees which is fixed on one end and weighted on the other. The principle of this self energizing brake is described by the band brake equation given in Chapter 4.

$$(13.8) \quad F_2 = F_1 e^{\beta \mu_k}$$

where F1 is the weight, β is the wrap angle in radians and μ is the coefficient of band/roll friction. The lineal web tension can then be calculated as

$$(13.9) \quad \text{Web Tension} \left(\frac{\text{force}}{\text{width}}\right) = \frac{F_2 - F_1}{\text{Web Width}}$$

Despite the crudeness of the band or strap brake unwind, it will often control tension better than a pneumatic brake with simple open-loop regulator. The reason being is that tension is only mildly dependent on unwind diameter (through wrap angle) for the surface brake, whereas the torque requirement is strongly dependent on diameter for center brakes.

Band or strap brakes can be made of brass strip, leather, belting or old aprons. The width is usually greater that one foot so that the bearing load is reduced (to protect web and band) and so that there is a larger wear surface (longer band life). Often the lower end of the strap brake will have a pocket in which weights are added or removed to adjust tension. These weights may be lead blocks, or more commonly, wrenches and other tools. Seldom is the band brake tension control sophisticated enough in these applications to include pneumatic cylinder loading or closed loop control, though there is no functional reason why it could not.

Figure 13.6
Unwind Strap Brake

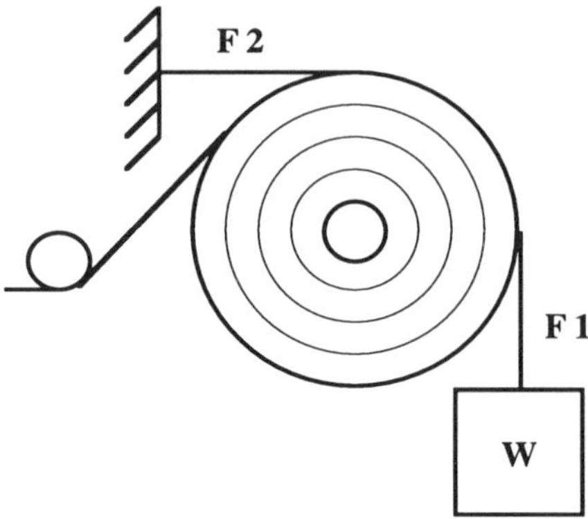

The band brake does have several severe disadvantages. First, tension is adjustable only to the smallest sized weight. Second, tension will change if the grade (friction) changes. Third, the rubbing may mark the web product. Fourth, the brake has very limited thermal capacity. Fifth, the web shear between the braked and unbraked part of the unwind may cause diagonal wrinkling.

Band brakes are occasionally used on rollers as well. Sometimes it is used to provide a mild web tension increase. More often, however, the band brake is used to stop a roller quickly so that an operator can gain safe access or for emergency stops. Obviously, the band must bear against a wear resistant material like steel instead of aluminum.

Small converting rollers are seldom braked because the inertias and speeds are small. At higher speeds and inertias, rollers can be brought to a stop quickly using surface band, pad or prony brakes. Very large diameter idler rollers, on the other hand, would be center braked through the live shaft journal.

Chapter 14

Electric Drives

This chapter is a layman's introduction to components and concepts. Covered are scalar and vector AC, as well as brush and brushless DC drives.

Electric Motivation

Electric drives are a clean and mechanically simple means of providing power to rollers. They are in many ways more versatile than mechanical drives because they can operate over a wider control range and they usually can motor as well as brake. Electric drives also offer, in general, tighter control of tension, torque or speed than their mechanical counterparts.

However, electric drives trade a good deal of control complexity for a modest mechanical simplification. While any mechanical engineer should be able to size, select, install and tune a mechanical drive, only a fraction of control engineers are suitably experienced to work on electric drives. Thus, we can and usually must rely on the drive manufacturers for these functions. Indeed, even most of the machine builders will not get intimately involved in electric drive selection and installation.

The choice of drive type is largely determined by how closely and how responsive control reactions must be. In general, the tightest control is a servo drive which is generally only needed for pick and place types of registration where the motor must follow rapidly changing speed and position setpoints. In the upper middle of control finesse is vector AC and brushless DC. The lower middle is the brush DC drive, which is by far the most common web drive type. At the low end of the scale is the scalar AC invertor without tachometer feedback. One measure of control accuracy is static speed regulation. While a scalar AC drive may hold speeds only to a few percent, very tight drives can hold speeds to within 0.01%. Responsiveness of a drive system may vary from more than a second to less than 1 millisecond.

The required tightness of control is in part determined by the stretch (yield or failure strain) of the material and the length of the web span. Stiff materials such as dry paper and many films are very demanding. In contrast, tissue or wet paper, stretch film and some nonwovens are easy. The tightness of control is also quite dependent on the control method. For example, load cell control is more demanding than dancer control because the only buffer or storage with load cells is in the stretch of the web. Speed control is very demanding on stiff webs or for registration, while draw (permanent stretching) or loop control are generally easy. Also, drive sections can be more difficult to tune when strip tensions are low compared to mechanical loads (inertia, friction or rolling resistance). Finally, we must also consider responsiveness. For example, it is more difficult to accelerate a winder at a snappy 150 ft/min/sec than it is to accelerate a coater at a sluggish 10 ft/min/sec.

In addition to the demands of control finesse, flexibility and responsiveness, we also must consider costs. The cost of a drive system is primarily determined by the number of drive points, power (tension x speed), and the demands for tightness or complexity of the control. This may result in a number of different drive means used on a single line. Using a rewinder, for example, an unwind might be purely mechanical, or have a regenerative brush DC motor with a disc brake for normal or E-stop assist, or be a vector AC. The slitters would probably be an AC inverter or belted off the winder drums. Finally, the drums would probably be a brush DC or vector AC.

With all of these options and control-speak, however, we must never lose sight of who is the customer of the drive. The customer, of course, is the web rather than the company or machine. Web tensions (or strains in the case of true drawing) must be held within a given tolerance at all times and at all locations. This is covered in Chapter 2 and elsewhere.

AC Inverter Drive

The AC motor is the simplest, cheapest and most rugged of all electrical motors. However, with few exceptions (see AC vector in the next section), the AC scalar drive is not a good choice for regulating tension, torque or speed in most web handling applications. Thus, it is relegated to auxiliary functions such as fans, pumps and slitter knives where speeds can vary by a few percent without compromise.

The reason that conventional AC scalar drives have such loose regulation is that speed varies with load. Thus, changes in load will result in speed variations. Variable loads are present due to tension surges, bearing friction, windage, nip rolling resistance and of course, roller inertia during acceleration and deceleration. The two types of AC motors most common with variable frequency (speed) drives systems are the squirrel-cage and synchronous motor.

The squirrel cage or induction motor has a **rotor** consisting of a number of axial conductors shorted at the ends. The **stator** consists of a number of windings arranged to make an even number of poles. The stator field is in motion by virtue of the AC input. The stator field moving past the slower rotor induces an EMF in the rotor which provides torque.

The **speed** of the AC motor is governed by the frequency of the input (60 Hz in the US, 50 Hz most elsewhere) and slip (typically around 2-5%) for squirrel cage induction motors) is:

$$N_{rotor}\,(\text{rpm}) = \frac{F\,(\text{freq. Hz})\;120}{P\,(\text{poles})} \times \text{Slip} \tag{14.1}$$

where slip is

$$\text{Slip} = \frac{N_{synchronous} - N_{rotor}}{N_{synchronous}} \tag{14.2}$$

The rotor of the synchronous AC motor has a series of windings which take power through brushes and collector rings, while the field of the synchronous AC is similar to the induction AC. However, the synchronous AC always travels at synchronous speed.

There are a number of ways of changing the speed of an AC motor including varying the number of poles, changing resistance in series to the rotating field of an induction motor and so on. However, the primary method for web machine auxiliary components is a variable frequency inverter drive. The inverter generates an output waveform with a variable frequency relative to the line input. This synthesized AC wave directly determines synchronous AC motor speed and largely determines induction AC motor speed.

The AC static inverter or static frequency drive converts constant voltage constant frequency AC power into variable voltage variable frequency power using solid state electronics. As seen in Figure 14.1, the inverter consists of a converter (power) bridge and an inverter (chopper) bridge. The converter bridge rectifies the AC line supply into a fixed voltage DC. The inverter bridge takes this and generates a PWM (pulse width modulated) output form of alternating current at a controllable frequency. Thus, the motor speed is indirectly controlled by varying voltage and frequency.

Figure 14.1
AC Static Inverter Drive

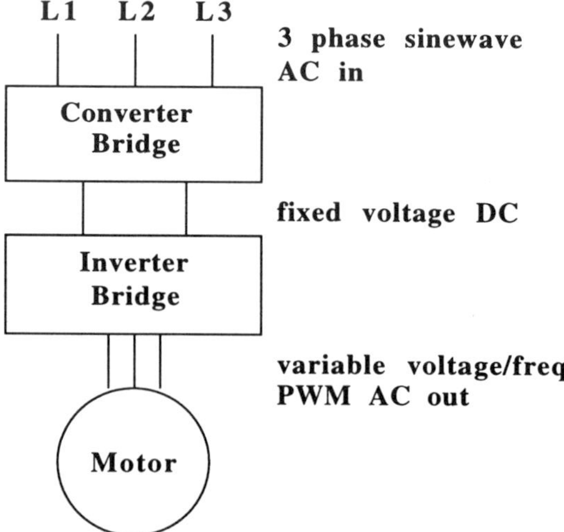

AC Vector Drive

There are several similarities between the scalar AC drive of the last section and the vector AC drive of this section. First, the motors are squirrel cage induction motors. These simple and rugged motors do not have the mechanical maintenance associated with the brushes commutators on DC motors.

Second, the heat produced on the AC motor is primarily on the stator which is located on the outside of the motor instead of the armature/rotor of the DC motor. Removing heat from the outside stator is more easily accomplished because of the large surface area of the (sometimes ribbed) casing. Cooling can be enhanced by blowing air over the outside of the totally enclosed and protected motor. However, even at high torques and low speeds which may require forced internal ventilation, the AC motor does not have the corrosion and fouling issues associated with DC brushes and commutators.

Finally, the AC drives have only a single power supply, network and motor component (stator) compared with the two of DC drives (field and armature). While at first sight this may seem advantageous, the parts count and complexity is not all that different.

There is a primary distinguishing difference between vector and scalar AC drives. That is, the vector AC drive knows the angular position of the rotating stator field vector (torque producing) with respect to the rotor field vector flux (magnetizing). However, it is the fundamental component of current rather than harmonics that produces the usable motor torque.

One advantage of the vector control is that speed can be controlled to within about 0.01%, compared to about 0.1% for a scalar AC with tachometer feedback and about 1% for a scalar AC without feedback. Also, the dynamic response of the system is quite high so that the drive is quite responsive. Finally, the vector AC can control speeds or torques more accurately near/at zero speed without overheating or during torque reversal than a brush type DC motor.

The torque and power curves of a vector AC drive and brush DC drive are similar and are given in Figure 14.2. Note how the torque is constant to the base speed and then drops off above the base speed.

The motor torque and current (at constant supply voltage and frequency) as a function of slip or speed for an induction motor is shown in Figure 14.3. Note how the torque and current rise sharply as the motor is forced away from synchronous speed.

Figure 14.2
Power and Torque for a Motor

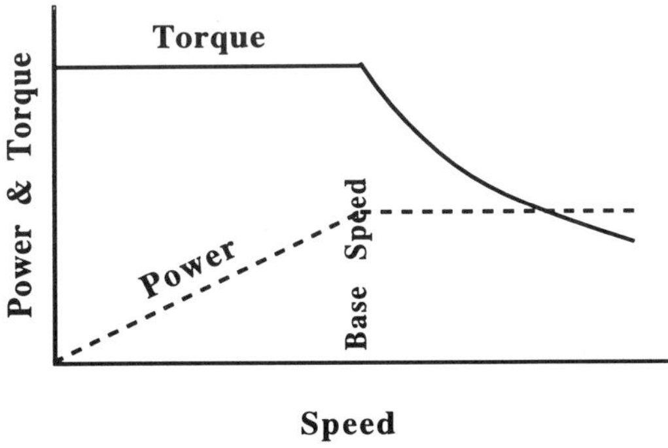

Figure 14.3
Induction Motor Torque and Current (constant voltage and frequency)

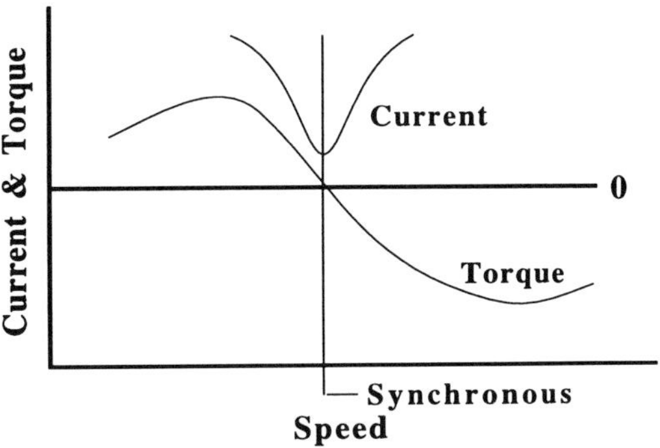

DC Drives

The DC drive in its many variations serves as the workhorse of the web industry. DC drives outnumber all other electric drive means combined. This drive is quite versatile and is capable of smooth control of torque and speed over ranges exceeding 100:1. The DC motor is composed of the motor itself, power conversion (covered in a later section) and a control section (control schemes were introduced in Chapter 2).

The rotating part of the DC motor is called the **armature** and consists of a number of coils of wire or windings. The DC current is fed through stationary carbon **brushes** to a series of segmented copper bars called **commutators** on the back end of the armature shaft. The brushes and commutator are what distinguish the DC motor from other motors, add to the costs, and are typically high maintenance items. The outer casing of the motor holds the stationary coil of the motor which is called a **stator** or **field**. DC power is supplied to both the armature and field.

The **speed** of a DC motor is determined by:

$$(14.3) \quad N = K \frac{E_a - I_a R_a}{\Phi}$$

where

N = motor speed (RPM)
K = motor constant
Ea = armature voltage
Ia = armature current
Ra = armature resistance
Φa = field strength

From this equation we can see that a motor could be sped up by increasing armature voltage, or by decreasing the field strength. The maximum speed of the DC motor at maximum field strength is called the base speed. The most common base speed is 1750 RPM; however, other discrete base speeds are available from less than 850 to more than 3600 RPM. Thus, one way of modulating speed is to use **field weakening**. The downside to this approach, aside from complexity, is that the torque is proportional to field strength. This means that there is less ability to accelerate rollers, pull tension, overcome friction and so on. Motors designed for field weakening are designated by a slant, such as 1750/2500 RPM.

When choosing a motor, it is important to try to choose a base speed that will give a surface speed of the driven roller just in excess of the design speed of the machine. Field weakening can be used to get slightly higher speeds, but the torque drops off. If much lower speeds are needed, a gearbox, pulley set or other fixed transmission will be needed. The largest sizing difficulty will be on unwinders or rewinders, particularly for those which have widely varying diameters, widths or tensions of the products.

The torque of a DC motor is given by:

$$(14.4) \quad T = K \times \Phi \times I_a$$

From this formula we see that torque is proportional to armature current if the field strength remains constant. This is why we can in many cases use an (armature) ammeter and simple formula as a poor man's load cell. From this formula we can also see that the DC motor, particularly the shunt motor, is able to output 100% of rated torque at very low speeds. The DC motor's torque and horsepower as a function of speed for a field weakened system is summarized in Figure 14.4.

One problem with low speeds is that forced air cooling may be needed to carry away the heat from losses in the armature coil. Other protections for the motor include heaters or other thermal protection devices, field loss relays and other means of preventing inadvertent overspeed. The DC motor may be overloaded for short periods of time within specified limits, which is useful for accelerations.

Figure 14.4
Torque and Power for a DC Motor

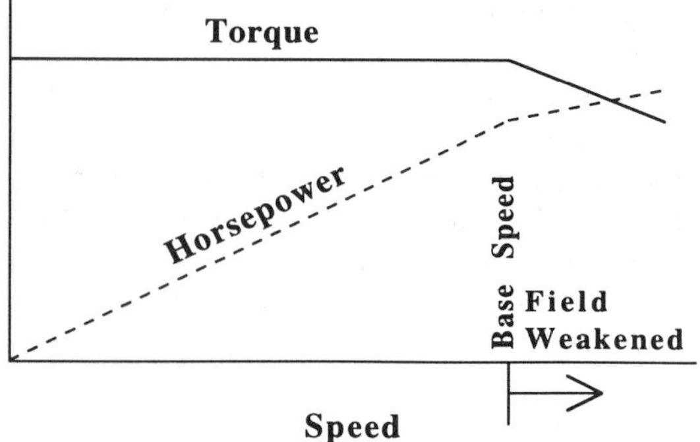

Despite the immense variety of DC motors and applications, the direct and immediate control of a DC motor is by varying the amount of armature current (and occasionally by field weakening). However, neither the driven roller nor the web knows about current. The question is how do we translate the conventional control schemes (of Chapter 2) into motor control?

The simple block diagram of Figure 14.5 shows three nested control loops for the motor. Of course, the inner loop is armature current or torque loop which is converted to shaft torque by the motor. The current is controlled by the firing angle of the SCR thyristors. The next is the speed loop which compares the motor tachometer speed reading with a setpoint for draw or speed control. Any error here is sent as a vernier modifier to the inner current loop. The outer loop is the tension or dancer feedback loop. Here, the tension reading is compared with a setpoint and the error modifies the speed reference loop. Also note how much slower the response is for the outer loops than the inner loops. Unfortunately, without the slow outer tension loop, the drive is open loop on the web.

A DC motor can be configured as a **helper** drive (compensate for constant drag and inertial upsets) by using a speed control loop with very low gain or by torque control. Alternatively, a DC motor can be configured as a **follower** (electronic lineshaft) by tight control of registration, position or speed.

The aggressiveness of the drive in responding to a sensed error is called **gain**. The gain for a tension system, for example, may be expressed as ft–lb of torque per lb/in of tension error. This setting is extremely important because, if it is too low, the drive will be sluggish in dealing with errors. Alternatively, if it is too high the system will overshoot and may become unstable. One important parameter in determining the optimum gain is the inertia of the driven element. If the inertia is high, the drive must have an aggressive gain to not only pull an extra tension but to actually change the velocity of the high inertia roll(er). On winding and unwinding rolls, any drive should be equipped with **inertia compensation** so that the gain is automatically increased for changing roll diameters.

Figure 14.5
DC Motor Control Loops

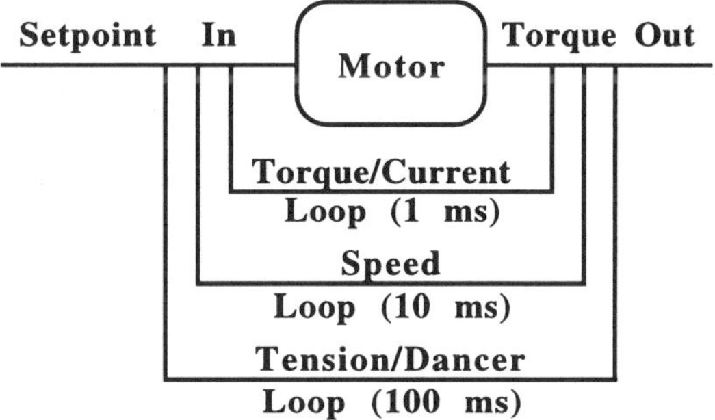

Occasionally, a drive must provide braking (minus) as well as motoring (plus). For example, an unwind motor will be braking except during the occasional acceleration. The three common methods for turning a DC motor into a controllable brake are **MG sets**, dynamic braking and regenerative braking. Regenerative braking feeds most of the power back into the lines instead of dissipating it as heat. While regeneration will be credited on the electricity bill (because electric company meters can rotate backward), the quality of the power is not very clean (some ripple superposed on top of the nominal sine wave). **Power factor** is a measure of the current efficiency of an actual AC sine wave and a pure sine wave.

The MG or motor generator set automatically provides regeneration. When slowing down or creating unwind tension, the motor becomes a generator which changes the generator into a motor which drives the AC motor which then becomes an alternator feeding current back into the lines. **Dynamic braking** of DC motors routes the armature current via diodes through a set of resistors (coils of heating wire). Lastly, the **regenerative** DC drive uses a second SCR bridge to route power back through the lines in a manner very similar to what was used during motoring. Since regenerative drives are costly, they should only be used where braking is necessary and mechanical brake assist is not sufficient or suitable. Power conversion is covered in the next section.

AC to DC Power Conversion

Since AC (Alternating Current) is supplied to a plant, it must be converted to DC (Direct Current) for use by a large DC motor. This power conversion can be accomplished by rotating generator or by static electronic means. The motor-generator, or MG for short, was the standard until the advent of electronic means commercialized in the mid 1960's. Despite a few niche advantages of the MG set, most have been replaced with the more capable electronic drive so that spare parts and service are now becoming difficult to obtain.

The **MG** set, as seen in Figure 14.6, is composed of an AC motor which propels a DC generator which supplies the electricity for the DC motor. The output voltage of the DC generator, which is delivered to the DC motor armature, is controlled by varying the strength of its field. The speed of the DC motor is a function of both its DC motor armature voltage and its field strength.

Modern DC drive power conversion is accomplished via **SCR's** (silicon controlled rectifier) also known as thyristors. A thyristor is a semiconductor which only allows current to flow in one direction, and then only when the gate is open. The **firing angle** of the gate, as seen in Figure 14.7, is controlled by accurate timing circuits. The amount of power, shown shaded in the illustration, that is passed to the motor is thus controlled by the firing angle. Also note that the AC has now been converted to a bumpy (varying voltage) DC.

The next major element in the SCR drive is a **bridge** circuit that appropriately combines both halves of the AC input from all three phases to power the drive motor. The bridge of SCRs serves two main purposes. First, it makes use of more of the AC input. Second, the combined DC output has less ripple. A schematic of a 3-phase 6-leg SCR bridge is shown in Figure 14.8.

Finally, DC drives will have an **isolation transformer** before the motor controller to protect it from incoming AC line spikes, reduce the noise transmitted from the drive back into the line, and change nominal voltage levels.

Figure 14.6
MG Set for AC to DC Power Conversion

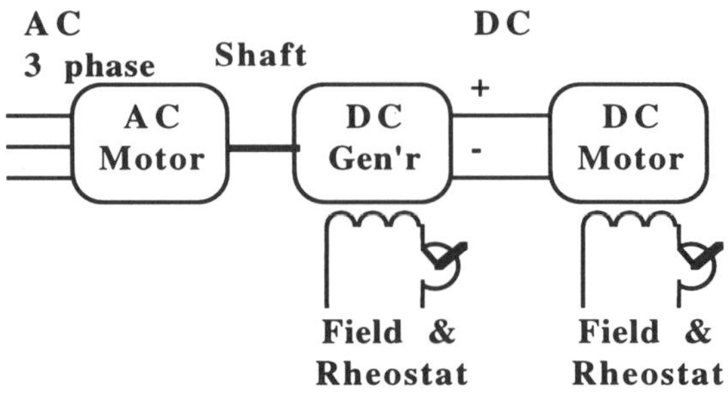

Figure 14.7
Thyristor Firing Angle

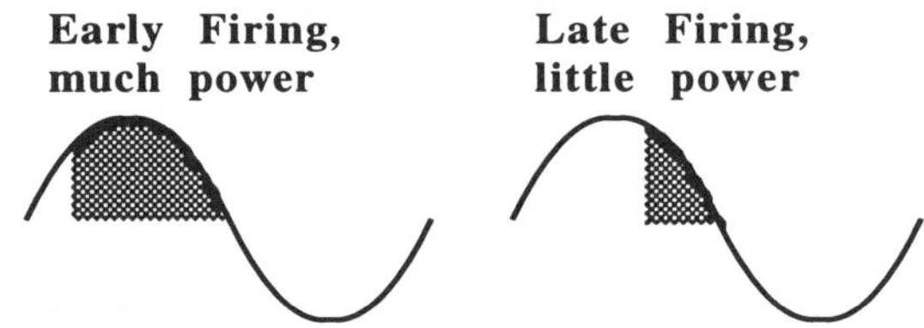

Figure 14.8
SCR Bridge for AC to DC Power Conversion

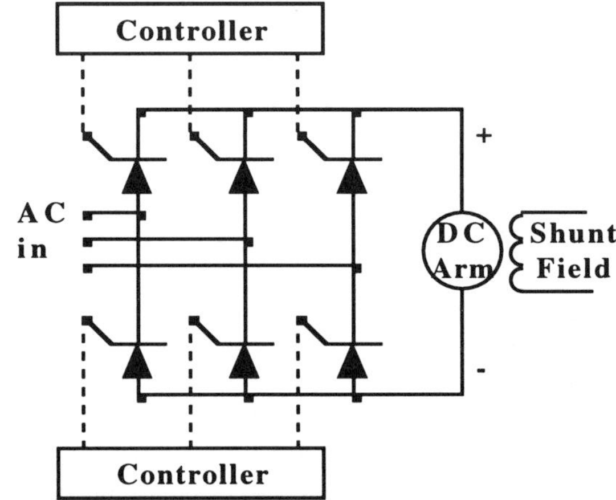

In addition to these major elements, DC drives may have a variety of other devices including contactors, relays, ammeters, voltmeters, fuses, chokes, microprocessors and so on. Typically, the power devices (for larger drives) will be in refrigerator sized metal enclosures located in the **drive room**. The drive room serves to protect the sensitive equipment from an adverse environment, serves as a single point for multiple drive troubleshooting and reduces unauthorized entry into hazardous areas.

Brushless DC

Brush type DC drives, though they are by far the most common electric drive for web applications, have several limitations. First, a brush DC motor controller can alter torque (current, speed, etc.) only at distinct time intervals just prior to the arrival of an input sine wave of the AC line. Thus, the torque output is rippled and controllable at a maximum response of 720 Hz for 3 phase power. This response, while adequate for most applications, is not fast enough for many pick and place registration or positioning operations involving complex speed versus time cycles. Second, the brush DC motor is self commutating which means that power is applied inefficiently. Third, the DC motor has difficulty holding highly accurate speeds even with a tachometer feedback. This is because the motor output and speed both vary with load. Indeed, it can be difficult to tune a DC motor to stop while holding a load (**stall tension**), particularly with the old analog drives.

The brushless DC motor, on the other hand, can alter torque (current, speed, etc.) at a rate that is about an order of magnitude higher than brush DC. This is because rectified input AC power is stored in intermediate buss capacitors from which DC power can be drawn off at will. Also, this configuration will allow the motor to run at low speeds and high torques while drawing much less current from the AC line compared with brush DC. Finally, the speed response of the brushless DC is better because the speed/direction/position encoders are typically much more capable and the speed versus torque curve is flat.

The brushless DC motor is wound like an AC induction motor, except that the rotor uses permanent magnets instead of shorted rotor bars. Thus, the lack of brushes eliminates one of the high maintenance areas of traditional brush DC motors. The stator consists of six windings, perhaps in a single wye connection. A simplified schematic of a two pole brushless DC motor is given in Figure 14.9, though most will be 4 or 8 pole motors.

Figure 14.9
Brushless DC Motor Schematic

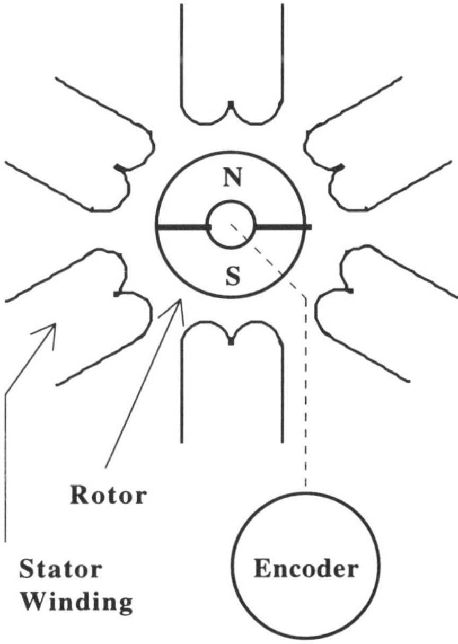

The astute reader will note the similarities of a brushless DC motor to the very capable and responsive stepper motor. The primary differences are that the stepper contains more windings, and that the stepper is powered by DC rather than AC. However, steppers are fractional horsepower devices and have a very ripply output torque, and are thus better suited to positioning rather than web handling.

The basic power section of a brushless DC consists of a diode bridge, followed by buss capacitors, followed by power output transistors and is contrasted with brush DC in two primary ways. First, the buss capacitor stores and releases power at will instead of during discrete intervals corresponding to the input AC sine wave of a brush DC. Second, the power transistor can vary both voltage and timing instead of the mere on/off timing of brush DC.

In conclusion, the brushless DC is more efficient than many other electric drives and is more capable than the traditional brush DC. However, the controller technology is more complex than brush DC and is available only from a few manufacturers. Indeed, capability and complexity are characteristics shared by both brushless DC as well as AC vector drives.

Web Drive Process Health Evaluation

Control of web tension everywhere and at all times is key to web handling process health. While the control engineer is likely to think in terms such as the tightness of speed regulation, armature current or inertia compensation, the web only knows about forces imparted to it by driven and undriven rollers. Indeed, the web does not know whether the roller is driven by a squirrel cage induction motor or a squirrel running in a cage.

Thus, the best measure of web drive system health is to monitor tension via sensitive and responsive load cells. For typical applications, the tension should be held to within 5% of the setpoint during steady state and to within 10% during transients such as speed changes.

The few exceptions to this application of load cells to evaluate drive system health include a true drawing (permanent stretching) operation or where tensions are impractically low to measure (such as on the wet end of a paper machine or the span after a film extruder). Here, we would want the speed regulation to be such that **caliper/weight control** is not compromised. For example, if a drawing operation needs the total weight variance to not exceed 1%, then the speed regulation needs to be much tighter than that.

In a speed control (but not drawing) application, the speed must be held tight enough so that the indirect control of tension is not compromised. As given in Chapter 2, a web may be tensioned as little as 10% of its yield strain. If we wanted to hold that tension to within 10% of its nominal, then true surface speed must be held very close on stiff materials. For example, a stiff material might have a yield of only 1%, and thus may be run at a strain of only 0.1%, and thus require a 0.01% speed regulation to not allow the indirect control of tension to vary by more than 10%.

We may also make an inference of the quality of the drive health by looking at load readings. We should expect the ammeters (or equivalent) to remain quite steady during normal running. During transients, the gauges should slew smoothly, synchronously and quickly without overshoot.

However, the quality of a perfect drive system may be compromised by factors outside of the driven roller such as poor load cell or dancer sensors, undriven roller inertias, roller slippage, variable drag, vibration and so on. Thus, we should look to the web for clues of compromised system control in locations where we do not have sensors (load cells) or actuators (drives).

If tension control is grossly compromised, you may see the **web sag** in certain spans or at certain times. Conversely, the web may tighten up to the point of a **web break**. If the problem occurs consistently on a span, there will likely be a control or mechanical oversight there. If the web behaves poorly during a certain event, such as during a speed change, the controls may not be responding appropriately.

While observation of webs sagging on the floor or snapping off is a sure sign of gross control problems, more subtle problems will be difficult to pick up by feeling the tautness of a web for example. Also, the portable and **handheld tension meters** are often so unreliable that they can't be trusted.

However, there is a way to determine if more subtle tension problems exist by observing either **web edge position** or **web width**. If a tension variation is accompanied by misalignment and partial tracking on a roller/bar, the centerline position of the web will dither. If a tension variation occurs on a material with a high stretch or Poisson ratio, the width may change noticeably.

The combination of centerline position changes and width changes will both show up as a change in the edge position of a web. This is most easy to observe on the edges of **wound roll** which records these insults much like tree rings. However, a (portable) edge sensor hooked up to a recorder will also pick up these responses to tension changes. The only limitation of the edge position technique is that it usually requires a slit edge.

Finally, remember that tension control is an electro-mechanical system issue. For example, it would be silly to have a tight drive and then load the system with an unnecessarily large number of misaligned idlers with high drag bearings.

Chapter 15

Temperature Control

In this chapter we look at various methods of heating and cooling a web, as well as a brief discussion of temperature measurement, and heat transfer. Covered are air flotation bars, electrically heated rollers, steam dryer cans, circulating liquid rollers and chill rollers.

Temperature Dependence

Webs are commonly heated via rollers in many manufacturing and converting operations. The water in paper manufacturing, for example, must be dried after the press section on large steam heated dryer cans. In the converting of many paper and non-paper grades, water or solvents from printing ink and other coatings are commonly driven off via heated rollers.

Webs may also be heated in calendering, embossing, laminating and other related processes. The higher web temperatures increase the plasticity of the web materials so that these processes can be more effective. In fact, most web properties are affected to one degree or another by temperature changes.

After heating, the web may have to be cooled by chill rollers so that it can withstand downstream web handling and converting operations without damage. For example, hot coatings are scratch sensitive and wet inks may smear merely going over idler rollers.

Another reason to cool a roller may be to improve the life and load range of some roller cover materials. For example, nip hysteresis will heat a covered roller which may make it more tender or even thermally degrade it. Water cooling is also used to extend the thermal horsepower capacity of drum brakes. For example, mechanical brakes may heat up sufficiently to distort metal if not cooled more aggressively than is possible merely by convection of free or forced air.

Despite the variety of applications, our most important thermal concern should be to maintain web temperature at a given setpoint within some required tolerance. This is especially true at elevated temperatures where the properties of the web and process become more sensitive.

However, it is difficult to measure and maintain web temperature directly. Rather, we rely on temperature control of roller surfaces to ensure control of web temperatures. In other words, we assume that the web will reach a temperature that is similar to that of the surface of the roller. But, if there is not sufficient dwell time (wrap angle, radius, 1/speed), the web will not reach roller surface temperature. This is an important concern in the drying operation where high weight (basis, moisture, solvent) lanes in the web will not reach as high a temperature as the low weight area. This further compounds material profile variation so that relatively wet areas become proportionally even more wet.

It is also difficult to measure and maintain roller surface temperature directly. Rather, we rely on circulating fluid temperatures, sensors embedded in the rollers or maybe only heat power settings to maintain process control. If conduction or convection varies, so too will web roller surface temperatures.

Thus, our control of web temperatures is very indirect in almost all applications. This need not be compromising if the web or process is quite tolerant. However, tighter temperature control will require a more thorough knowledge of the mechanics of the thermal system which is comprised of web, rollers and other elements. Though modeling is not difficult, obtaining certain inputs for the model can be challenging. In lieu of modeling we may opt for a more empirical approach. However, as we will see in the next section, merely measuring temperature is more complicated than might be initially suspected.

How To Measure Temperature

If we are to control temperature in even the broadest sense of the word, we must be able to measure it. Measurements may be desired for the ambient air, circulating fluid, roller shell, and the web itself. Temperature can be measured by a variety of sensors. However, repeatability, accuracy, signal processing and other factors must be considered when selecting a sensing system.[1,2]

The most inexpensive temperature sensor is the **bimetallic thermometer**. This type of thermometer has a metal of high expansion rate, such as brass or other alloy, bonded to one of near zero expansion, such as Invar which is a nickel steel. An increase of temperature causes the laminate to bend toward the low coefficient side which is read as a pointer deflection. The disadvantages of the bimetallic thermometer is that there is no capability for remote readout and it has an unimpressive accuracy.

Liquid in glass thermometers can be very accurate, often to well under 0.5°F. However, they are quite tender, have a slow response, and are also not well suited for remote readout. The most common liquid used is **mercury**, except for temperatures below its freezing point (-38°F) or for very high temperatures (>1000°F). Glass thermometers are specifically designed for either partial or total immersion.

Thermocouples are formed by soldering or pressing two dissimilar metals together. A temperature change will cause a very small voltage to develop across the junction which can then be read and calibrated as a temperature. Thermocouples are well suited for remote temperature readout if proper care is taken for the tiny signals. Thermocouples are potentially accurate to as little as about 1°F under calibrated laboratory conditions. However, process measurement accuracies will be considerably worse. For example, the handheld **contacting pyrometer** commonly used for measuring roller shell temperature is only repeatable to about 5°F.[3] Part of the additional measurement noise here comes from variable friction heating on the front and convection losses on the back of the thermocouple strip.

There are two basic types of **electrical resistance** sensors. The first are conductive sensors which make use of the fact that the electrical resistance of metals increase slightly with temperature. The other is semiconductor sensors, more commonly known as thermistors. Here, the electrical resistance of a semiconductor decreases strongly with temperature. The tiny resistance changes of these sensors are usually measured with a Wheatstone bridge circuit and must be corrected for nonlinearities.

Finally, noncontacting **optical pyrometers** (such as infrared cameras) are becoming more popular because of their convenience.[4] These handheld devices need only be pointed toward their target to get a reading. They work on a principle that all bodies emit radiation by virtue of their temperature. As temperature increases, the power and frequency of the peak of the broadband radiation curve increases. However, there are an unusually large number of complications that can render the readings worse than useless unless the engineer is very careful. The biggest problem is that the device makes some assumptions about the emissivity for the graybody radiation. This means that changes in the surface chemistry only a few molecules deep, such as by oxides, contaminants or coatings, can cause large inaccuracies. While a reference target can be placed on a roller to aid in calibration, it does not really address variance.

Finally, the measurement of **fluid flow** is also important for forced air or circulating liquid processes. While there are a number of means, the pitot tube principle is simple and widely used. Here, the difference between static (side) and dynamic (front) pressure for a given density represents a fluid velocity. Flow rate can then be calculated based on cross sectional area.

The process engineer would be best served by a healthy distrust of the readings given by any sensor, whether it be for temperature or other process parameter. In fact, I always keep in mind that the only sensor that truly reads what it advertises to is the ruler. All others sense a related behavior and convert it the unit of interest based on a number of assumptions and conversions.

Where to Measure Temperature

Ironically, the more important the location for temperature control, the more difficult it is to measure temperature. Obviously, the **web**'s temperature is most important, but is seldom measured because of practical difficulties. Many light webs will not tolerate the contact required with most temperature sensors. Also as mentioned earlier, noncontacting optical pyrometers do not perform well unless emissivity variations are small. Only when the web is in a wound roll or stack form is web temperature easy to obtain using conventional techniques.

The next most important location is the **shell** of a roller, particularly at the surface which will contact the web. At the surface, temperature can be measured during operation to about 5°F accuracy using a contacting pyrometer. If the roller is stopped, almost any temperature sensor can be used on the surface of the shell. However, the measurement must be made quickly before the shell temperature changes. Heat sink putty may be employed to improve heat transfer from the shell to the thermometer.

There are three areas of concern with shell surface temperature. First, is whether the average surface temperature is appropriate for the process. This is the easiest to work with because the accuracy demands are reduced. Second, is whether the surface **temperature profile** is tight enough for the process. This is harder because it requires a considerable knowledge of the behavior of the material and process, and the ability to measure temperatures closely. Third, is to determine why temperatures are varying such as due to variations in material, doctor or nip loading; end losses, bearing temperature, ambient air currents and so on.

In most cases, a constant temperature profile is desired across the width of the roller in contact with the web. However, there are niche applications where temperatures are intentionally varied locally to expand the roller (close up a gap in the nip) or for other reasons. Temperature can be profiled via refrigerated air or steam nozzles, or by zone adjustment of induction, infrared or resistance heaters.

It is also possible to measure temperatures inside the shell during operation using thermocouples or thermal resistance sensors. This is only infrequently done because of the expense and the complications of signal handling. Also bear in mind that the shell's interior temperature will be higher than its surface which is yet higher than the web's.

In most cases temperature is only measured at the **inlet and exit** for forced air and circulating fluid systems. Often the temperatures and/or flows are closed-loop controlled. The advantages for inlet/exit temperature measurement are simplicity and low capital cost. However, these inlet/exit temperatures may be far different than the web's.

Lastly, one may be concerned with the ambient air temperature and flow rate, particularly for forced air oven systems enclosed in hoods. Again the most common measurement is at the inlet and exit but may include other fixed locations. However, temperature and flow may need to be profiled when investigating certain problems. **Air temperature** profiles can be taken with long telescoping probes passed through holes in the sides of the dryers. **Air velocity** profiles can be taken using hot wire or hot air anemometers.

Often overlooked in the design and diagnostic process is the effect of a **idler** roller following a dryer or other heating element. In one case I observed, a hot calender stack was holding nip pressures and temperatures as close as could be measured, yet the product had a smile profile. Almost grasping at straws, we measured the temperature profiles of the idler roller which followed. The shell surface temperature was quite hot across the middle, but dropped about 50°F toward the edges of the web. A reasonably effective solution for this was to insulate the roller outside of the web run to improve temperature uniformity. This example illustrates how important it is to consider the temperature profiles in the entire area as opposed to a few points that may have been provided by the builder.

Chill Rolls

Chill rolls, as their name implies, are actively cooled below web temperature and often below room temperature. Chill rollers are more capable than forced air or refrigerated air systems. Most often the motivation for cooling is to prevent handling damage for webs or coatings that are thermoplastic or otherwise tender at high temperatures. High temperature damage can occur on rollers (marking) or wound rolls (blocking). Chill rolls are used on a variety of processes in most of the web industries. In the paper industry, for example, a chill roll may be used to bring a coated web temperature down after coming out of a hot dryer. In the film industry, chill rollers will usually be the first element after an extrusion die. In nonwovens, the chill roller may follow an air laid process or hot calender to set the web.

The chill roller itself is not usually as elaborate as a heated roller might be. In the simplest case, tap water is input into the hollow large diameter roller on one end through a rotary union and merely sewered out the other end. As the demands of the process increase, the chill roller will be composed of two annular shells with water circulating between. The outer shell may be thin to improve heat transfer while the inner is thick to resist deflection. The fluid path between the shells may not be straight through. Rather, the intershell zone may be baffled or spiraled to increase fluid velocity to generate enough turbulence for improved heat transfer but without excessive back pressure.

The chill roll cooling systems may also be more elaborate for demanding processes. For example, inlet temperature may be held by thermostatically controlled water chillers or mixing tanks. Also, flow rate may be controlled by a maximum setpoint temperature differential between the inlet and outlet.

Chill rolls can pose a number of different nuisances. The first is the maintenance associated with handling of any fluid, namely the potential for leaks. The **rotary joint** is particularly troublesome in this regard. Joint life of the (often carbon) seals may be weeks to years.

Second, chill rolls have tendency to condense humidity from the air. Condensation, more commonly known as **sweat**, occurs when the shell temperature is below the dew point of the ambient air, which can be as high as 100°F in some plants. The water dripping off the shell (particularly from the ends when stopped) can damage the web or equipment below. The most effective treatment is to only chill the water as much as necessary. After that, dehumidified air can be blown on the area to further reduce condensation.

Another problem with chill rollers is the formation of **scale** inside the roller due to contaminants in the water. Scale will decrease the heat transfer and fluid flow efficiencies of the system. Scale can be prevented on closed systems by using softened water. Scale can be removed mechanically with some difficulty if the roller can be disassembled or chemically if not. A related chill roll problem is corrosion which can be minimized by plating the internals of the roller, treatment of the water, and/or cleaning.

Typically, process rollers are not supplied in the excess that transport rollers commonly are. However, chill rollers are an exception. First, you must determine specifically why the web must be cooled immediately instead of naturally. (The web will cool by itself due to air convection and contact with other elements.) Second, you must determine what maximum web temperature is allowable at the next element downstream of the chill roll to avoid the problem/defect specified above. Third, you must use only as much chill roll as necessary to get the temperature below the defect limit.

In one plant I visited, all of their lines were equipped with a single forced air cooler followed by three tandem chill rolls. However, by merely feeling web temperature by hand it was determined that the web was brought below room temperature after only a single chill roll. This one minute exercise (in lieu of calculation) could have saved the plant over $100K in rollers, annual maintenance and utilities, as well as freed floor space up for better access in their tight plant. The need for chill rollers can be quickly determined for existing equipment by turning off the water and <u>objectively</u> observing the results.

Electrically Heated Rollers

Electrically heated rollers are clean and simple They are also very flexible because it is relatively easy to incorporate zone temperature control. Electrical heat is also quicker during startup than fluid heat. The only significant maintenance item on some designs is the slip rings or brushes on the journals which pass electricity to the elements. However, electrical heat may not be the best choice for larger rollers or for high speeds where power demands are considerable. In these applications, hot fluids are the norm.

One type of electrically heated roller is the **resistance heater**, which is more commonly known as a **cartridge heater**. The heat is provided by passing a large electrical current through a resistor material which comprises the cartridges. As many as a dozen of these long cartridges are inserted into axial holes drilled into the shell. The fit between the cartridge and shell must be close to provide good heat transfer.

The other major method of electrically heating a roller is through **induction heat**, also known as eddy current heat. Induction heating is also used to swell bearings which are to be pressfit over a journal. Induction heat uses a very high frequency AC electricity passing through a coil which generates an alternating flux (magnetic field). This alternating magnetic field will heat up any metal in proximity. It does so by molecular friction as atomic alignment tries to follow the reversing field.

There are two types of induction heater designs. In the internal design, a stationary coil is located inside the rotating shell which is heated by induction. The stationary coil has the advantage of a simple maintenance-free electrical connection. In many cases, the jacket within the roller is filled by a heat transfer fluid or by a vapor/liquid heat pipe.[5] In the external design, coils are mounted on a very stout beam parallel to the roller. The external design is used primarily for zone heating.

Heat Pipes

The temperature uniformity of most any roller can be improved significantly by incorporating heat pipes into the shell. Thus, while other rollers may be able to achieve 10°F or perhaps 5°F uniformity, a heat pipe design will often hold surface temperatures to within 1°F. Also, heat pipes can extend the temperature range of the roller to upward of 750°F. However, this considerable improvement in process uniformity will come at an increase in capital cost.

The heat pipe is an axially drilled cavity in the roller's shell. The cavity is filled with a thermal fluid, such as water, and then a vacuum is applied and the cavity is sealed. This process causes the thermal fluid to be partly in a liquid and partly in a vapor state. In other words, at the verge of boiling.

The principle of operation is based on the vaporization of the fluid at the higher temperature locations and condensation at the lower temperature locations. At the higher temperature location, the fluid will vaporize or boil which quickly and efficiently removes heat. Conversely, the low temperature area condenses the vapor which quickly releases heat. It is important to note that the heat pipe does not maintain a specific temperature, because that is governed by global heat transfer. Rather, it only ensures that the temperature distribution is very uniform.

There are two consumer toys which are based on the heat pipe principle. The first is the hand boiler. This dumbbell shaped glass toy is filled with a colored liquid (alcohol) which will boil when one end is enclosed in one's hand. While the assembly does get warmer, the temperature is uniform because the boiling moves heat from the held end to the other end where the fluid condenses. The other consumer toy which uses a heat pipe is the bobbing or drinking bird. Here, heat is extracted from water evaporation on the beak which causes fluid to move up the neck and imbalance the head down for another drink.

Heat pipes are also used in a variety of industrial applications.

Steam Heated Rollers

Steam is used to heat large dryer rollers, or dryer cans as they are commonly called. Steam heat is particularly common in the paper industry where steam is readily available. In most paper machines, scores of dryer cans will follow the press section. The most impressive of all steam heated rollers is the yankee dryer used on tissue which can be well over 20 feet in diameter and/or 10 feet wide and rotates at speeds in excess of 5000 FPM.

In converting, a steam heated roll might be constructed using a jacketed design as shown in Figure 15.1. Here, the steam is supplied through a rotary joint to a central shaft which is connected to the jacket via a rotating flexible hose. The condensate return out the other end is similar except the inlet should be placed very close to the ID of the outer shell to remove as much water as possible. In the paper industry, the much stouter steam heated dryers will not be jacketed and will have both supply and condensate lines on the back side. Both rotary and stationary **siphons** are employed.[6,7,8,9]

The inevitable **condensate** inside a steam heated roll is not generally desirable because it can reduce heat transfer efficiency and causes an increase in weight and inertia, and consumes drive horsepower.[10] The water condensate will be in one of three states given in Figure 15.2.[11,12] The state can be calculated from surface speed and diameter of the roller, or sometimes inferred from changes in motor loadings. Longitudinal **spoiler bars** can be placed inside the shell to increase the heat transfer rate.[13]

Though the pressures are low, steam heated rollers are nonetheless pressure vessels. Thus, they are subject to codes and required safety practices above and beyond those of other rollers. Steam heated rollers, particularly the large yankee dryers, require considerable more design analysis and maintenance.[14,15] There is a Yankee Dryer Safety Committee sponsored by TAPPI which has done much in this area.

Figure 15.1
Jacketed Steam Heat Roller

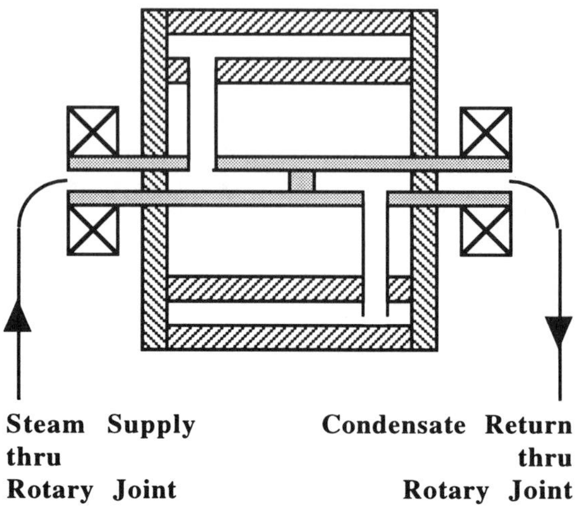

Figure 15.2
States of Condensate in a Roller

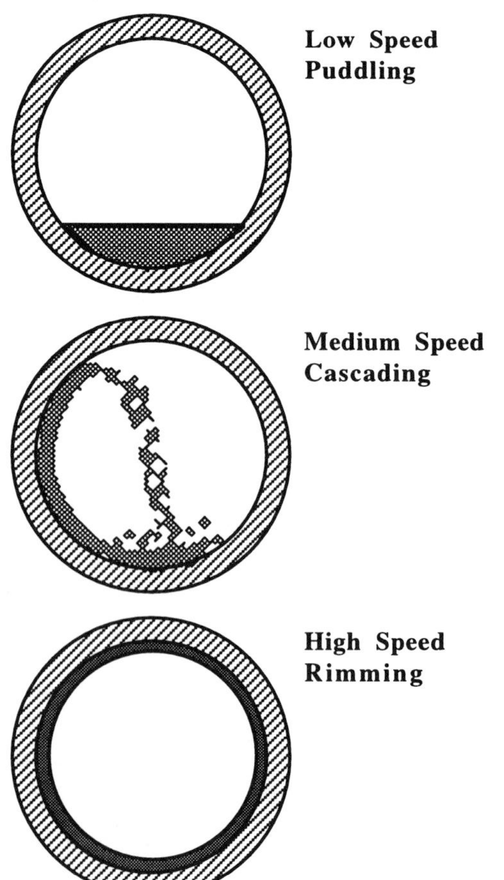

Circulating Liquid Heated Rollers

Liquids can be passed through a roller to provide cooling, heating or temperature stabilization. If temperatures are modest, such as below 200°F, a water based liquid may be the preferred medium. Intermediate temperatures, say up to 500°F, will require a specially formulated hot oil (or steam from last section). These temperature limitations are to keep water based fluids from freezing or boiling, and oil from thermally degrading.

There are other thermal properties of the fluid to consider as well. For example, **specific heat**, designated c_p, is the energy required to raise a unit mass a unit degree of temperature. Specific heat should be high so that temperature will be kept uniform across a roller without extremely high flow velocities for a given heat load. In this regard, water has a specific heat which is about double that of oil. This is also why a noncondensing gas is not a very effective heat transfer medium. However, **viscosity** is desirably low to reduce pumping costs and to promote turbulent mixing without high flow velocities.

Thermal properties of the roller are also important. While metal **shells** can easily tolerate temperatures common in most converting operations, composites and covered roller are much more temperature limited. The **thermal conductivity** of the shell is desirably high to achieve good temperature uniformity. In this regard, metals are far better than nonmetals. Even a thin layer of scale or contaminant on the inside of a circulating fluid roller can noticeably decrease heat transfer.

Finally, the **coefficient of thermal expansion** is desirably low so that temperature nonuniformities do not cause significant diametral uniformities. However, the true diameter increase will be less than calculated by a simple formula because adjacent material will resist nonuniform expansion/contraction. Also, lower coefficients will reduce the chance of damage to a roller due to shock cooling, such as by washing down a hot roller.

The temperature uniformity of a roller surface depends in part upon the circulation patterns of the liquid as seen in Figure 15.3. In general, the more passes of the fluid, the more uniform the temperature for a given flow rate. Alternatively, the more circuitous routes will maintain a given temperature distribution with a lower flow rate.

The most direct approach of simply passing a fluid straight through a hollow roller is not very effective except for small rollers. First, the circulation contains dead spots near the inside corners. Second, the flow velocity is small and thus heat transfer is reduced. Third, the roll's weight and inertia is high. A better approach is to use a double wall or diffuser to reduce these concerns. In one arrangement, a decreasing pitch spiral through the interwall passage creates a near constant temperature distribution as opposed to a linearly decreasing temperature found with other monopass arrangements.[16] The dualpass and tripass designs also improve end-to-end temperature profile.[17,18] These circulate a liquid in alternating directions along axial passages in the shell.

Figure 15.3
Fluid Flow Through a Roller

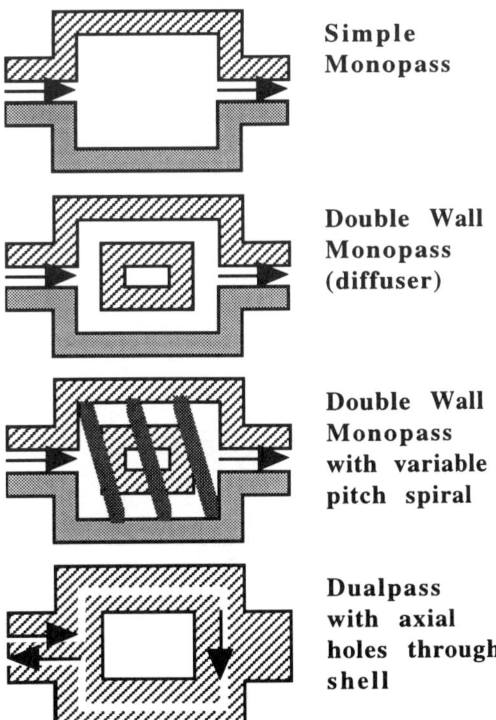

Heat Transfer

Estimating the power requirements for heating or cooling without phase change applications is relatively straight forward. It is the mass flow rate (basis weight x speed x width) times the specific heat of the web times the desired temperature rise of the web, plus losses. The majority of lost heat is due to convection off the roller shell, with lesser amounts lost on the ends of the roller (convection and conduction) and in the piping.

$$\text{(15.1)} \quad Power_{web} = \dot{m}\, c_{p(web)}\, \Delta T_{web} + \text{losses}$$

A similar equation could be written for fluids supplying the thermal power. Here, however, the temperature change is desirably small for maximum uniformity.

Calculating the temperature distributions of a thermal systems is done by finite element modeling.[19] While the modeling is straight forward enough, obtaining conductivities[20,21,22] for unusual materials, contact resistances[23] and convection coefficients for any application is difficult.

As seen in Figure 15.4, one surface of web gets heated almost immediately after touching the hot roller, while the other is still cool. With sufficient dwell time (wrap angle, radius, 1/speed), the web will heat through to the other side. However, the surface touching the roller will not quite reach roller temperature (due to contact resistance) and the outside of the web will be cooler than the inside (convection off of the web).

The temperature profile across the face is typically higher at the ends of the heated roller as seen in Figure 15.5. This is because heat is removed in the center by the webs. This can be minimized by reducing the heat input at the ends (insulation) or by forced air cooling.

Finally, roller temperature distributions across a section are shown in Figure 15.6 for peripheral hole heated rollers. Here, the roll heats at the passages forming an octagon shaped surface. The radial distribution from the hole outward is a function of distance and diffusivity.

Figure 15.4
Web Temperature Profile

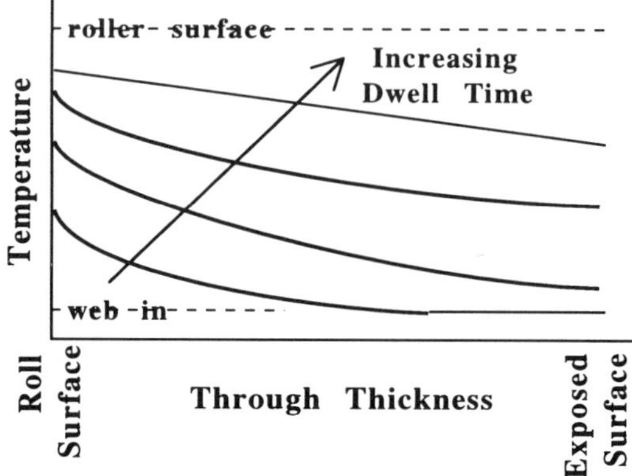

Figure 15.5
Across Roller Temperature Profile

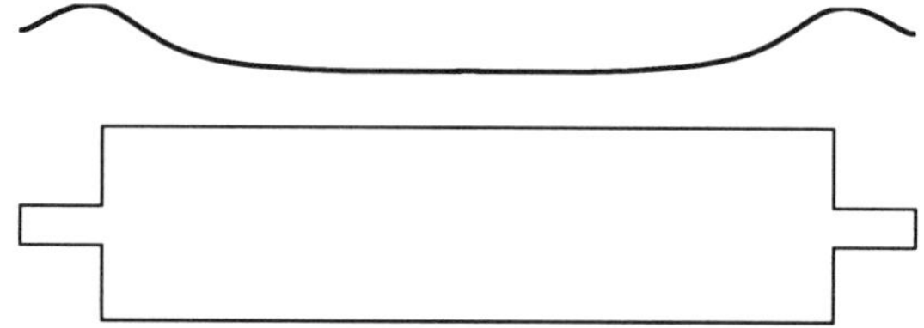

Figure 15.6
Temperature Vs. Circumference and Radius

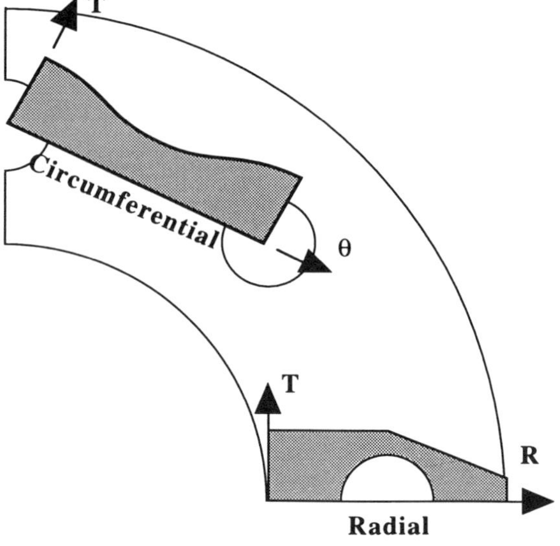

Warmup and Shutdown

Thermal dynamics might be considered when designing a machine for maximum productivity. This analysis determines how quickly cool rollers will heat once power is applied and how quickly rollers will cool when power is removed.

During a restart after a prolonged shutdown, for example, heated rollers and perhaps their supplies will be cool. If the web is applied at that time, the process and product will likely be out of specification. Thus we must determine how long it will take to heat the rollers sufficiently. The **warmup** curve for cool rollers is shown in Figure 15.7. The relative change in temperature versus time will roughly follow an exponential. The webbing temperature may be set below the lower temperature limit if the material is expected to be scrapped for a time for other reasons as well.

Alternatively, we may also want to know how long it will take for rollers to **cool down** sufficiently so that operators and maintenance personnel can gain safe access. This access temperature can be raised if exposed skin is covered with clothing such as gloves. The temperature versus time curve is shown in Figure 15.8 and also roughly follows the exponential (decay) law.

How rapidly temperatures can be changed depends in part on the thermal inertias of the roller and heating system. Thus, small diameter, thin walled rollers will warm up quicker. Also, the greater the excess power capacity of the supply over that required for running conditions, the faster the roller can heat. Finally, warmup may be quicker if the heat is supplied nearer the surface. Thus, ceramic rollers which are heated just under the surface will be faster than liquid flowing through peripheral drill holes, which will be faster than fluid through a central hole.

However, there may be limitations on how quickly temperatures should be allowed to rise or fall on a roller as determined by the builder. In particular, temperature/time gradients must be modest enough to avoid thermal shock which can crack metal. Thus, power might not be able to be turned off abruptly and hot rollers might not be washed down with cool water spray.

Figure 15.7
Temperatures During Warmup

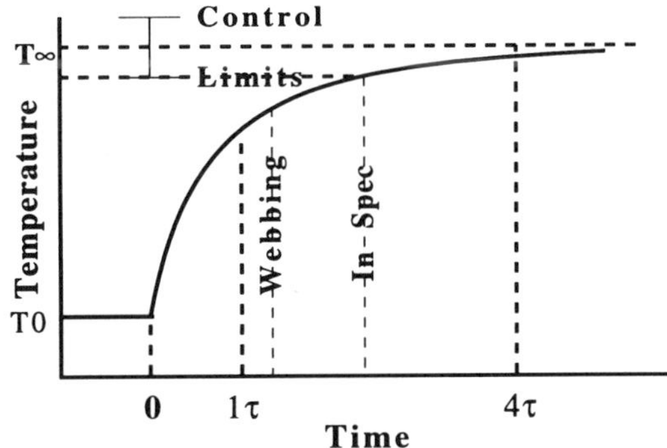

Figure 15.8
Temperatures During Cool Down

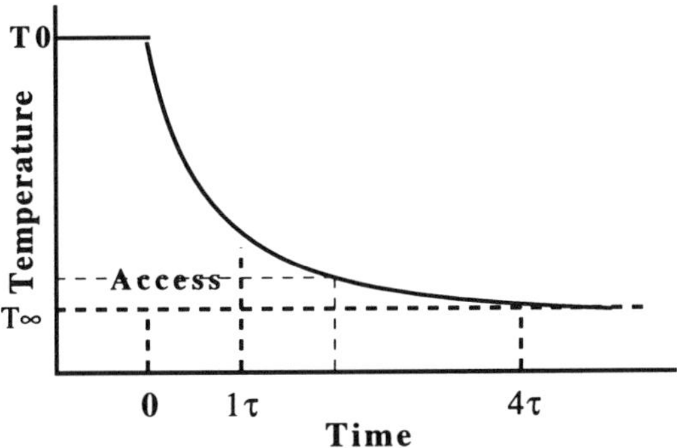

The warmup or cool down temperature response can be estimated for existing installations by fitting an exponential to a few measurements of temperature at various times. The temperature can then be calculated as:

$$\frac{T - T_\infty}{T_0 - T_\infty} = e^{-\frac{t}{\tau}} \qquad (15.2)$$

At one time constant, the temperature of the roller has made it 63% of the way to its final value. For all practical purposes, the system has reached steady state (within 2% of final temperature) after 4 time constants.

Bibliography

1. Doebelin, Ernest O. *Measurement Systems.* McGraw-Hill Book Co., 1975.

2. Dally, James W. and Riley, William F. and McConnell, Kenneth G. *Instrumentation for Engineering Measurements.* John Wiley & Sons, 1984.

3. Stenstrom, Stig G. et. al. *Equipment for Measurement of Cylinder Surface Temperatures.* Tappi J., vol 78, no 7, pp 245-246, July 1995.

4. Weiss, Herbert L. *IR Thermometers Invaluable for Accurate Temperatures.* Paper, Film & Foil Converter, vol 68, no 5, pp 166-167, May 1994.

5. Okamoto, K. *New Aspect of Temperature Gradient Calendering.* TAPPI Finishing and Converting Conf. Proc., Kansas City MO, pp 159-164, October 1-5, 1989.

6. Daane, Robert A. *An Experimental Study of Some Drier Drainage Siphons.* Tappi J., vol 42, no 3, pp 208-218, March 1959.

7. Wahlstrom, P.B. and Larsson, K.O. *Factors Determining Condensate Removal.* Pulp and Paper Magazine of Canada, vol 65, no 5, pp T203-T210, May 1964.

8. Calkins, D.L. *The Effects of Siphon Location on Paper Drying.* Pulp and Paper Magazine of Canada, vol 67, no 4, pp T225-T240, April 1966.

9. Wedel, Gregory L. *Performance of Dryer Syphons.* TAPPI Engineering Conf. Proc., 1982.

10. Appel, D.W. and Hong, S.H. *Condensate Distribution and its Effects on Heat Transfer.* Pulp and Paper Magazine of Canada, Tech Paper T51, pp 66-77, February 1969.

11. White, Robert E. and Higgins, Thomas W. *Effect of Fluid Properties on Condensate Behavior.* Tappi J., vol 41, no 2, pp 71-76, February 1958.

12. Wilhelmsson, B.I et al. *Condensate Flow Inside Paper Dryer Cylinders.* J. of Pulp and Paper Science, vol 21, no 1, pp J1-J9, January 1995.

13. Pulkowski, J.H and Wedel, G.L. *The Effect of Spoiler Bars on Dryer Heat Transfer.* Pulp and Paper Magazine of Canada, vol 89, no 8, pp 61-66, 1988.

14. Allevato, C. and Williams, J.D. *Acoustic Emission Evaluation of Yankee Dryer Shell Materials.* Tappi J., vol 75, no 7, July 1992.

15. Fernside, Richard L. *Yankee Dryer Joint Corrosion: cause, prevention and repair.* Tappi J., vol 78, no 7, pp 144-151, July 1995.

16. Roberts, Jack. *Computer Program Aids Heat Transfer Adjustments.* Paper, Film & Foil Converter, December 1990.

17. Zaoralek, M. *The Application of Different Designs of Heated Calender Rolls.* TAPPI Finishing and Converting Conf. Proc., Kansas City MO, pp 153-158, October 1-5, 1989.

18. Zaoralek, M. *Hot Rolls for Soft Calendering: Meeting the Operator's Needs.* TAPPI Finishing and Converting Conf. Proc., Lake Buena Vista FL, pp 41-48, October 7-10, 1990.

19. Keller, Sam F. *Heat Transfer in a Calender Nip.* J. of Pulp and Paper Science, vol 20, no 1, pp J33-J37, January 1994.

20. Hes, L. et al. *Theoretical and Experimental Analysis of Heat Conductivity for Nonwoven Fabrics.* INDA J. of Nonwovens Research, vol 3, no 3, 1991.

21. Schneider, A.M. et al. *Heat Transfer through Moist Fabrics.* Textile Research J., February 1992.

22. Tsutsumi, N. *Measurement of Thermal Diffusivity of Filler-Polimide Composites by Flash Radiometry.* J. of Polymer Science, August 1991.

23. Sanders, D.J. and Forsyth, R.C. *Measurement of Thermal Conductivity and Contact Resistance of Paper and Thin-Film Materials.* Rev. Sci. Instrum. vol 54, no 2, pp 238-244, 1983.

Chapter 16

Cores, Chucks and Shafts

In this chapter, we show that cores and their supporting chucks or shafts, used for winding and unwinding, share some of the same design concerns as rollers. However, they also have their own unique application considerations.

Applications

As seen in Figure 16.1, rolls are most commonly wound on mandrels or cores. Only rarely are the rolls wound with no interior support, such as with fiberglass insulation bats, or solid, such as capacitors and a few consumer rolls. As we will see, the core and mandrel both have similar design issues as do rollers. While the core and, to some extent the mandrel, have far poorer geometrical tolerances than most rollers, this difference is one of degree rather than principle.

In nearly all winding and unwinding operations, the inside of the roll must be held and/or guided during the operation in order to keep the roll edges square. As seen in Figure 16.2, the roll may be supported on core chucks (shaftless) or mandrels (shafted). The core chuck is similar in many ways to a cantilevered roller, while the mandrel is similar to a simply supported roller. In some cases the mandrel may also be cantilevered and would be more appropriately termed a chuck. While the core chuck and mandrel carry loads much greater than most rollers, this difference is also one of degree rather than principle.

However, this difference can be quite challenging when designing chucks or shafts. In the case of a dead shaft core chuck, the bearings normally fit within the inside of the core. Thus, it can be difficult to package enough bearing in that small confine that is needed to carry the necessary radial and possibly thrust loads. A similar situation arises in the case of the mandrel where the OD of the bearing housing might desirably clear the ID of the core so that the roll can be easily removed without taking off a bearing.

Figure 16.1
Rolls are Wound on:

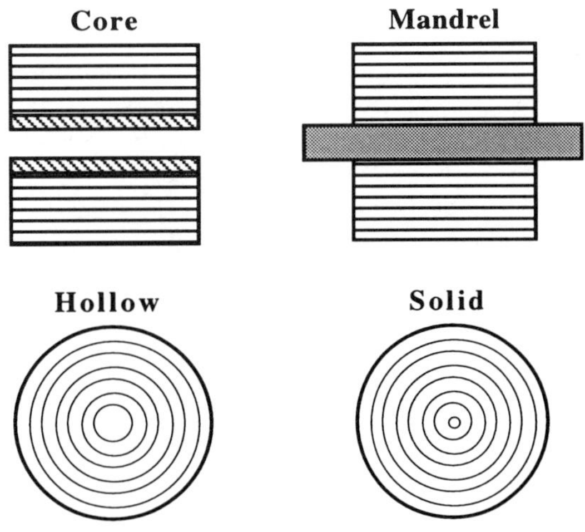

Figure 16.2
Wound Rolls are Supported by:

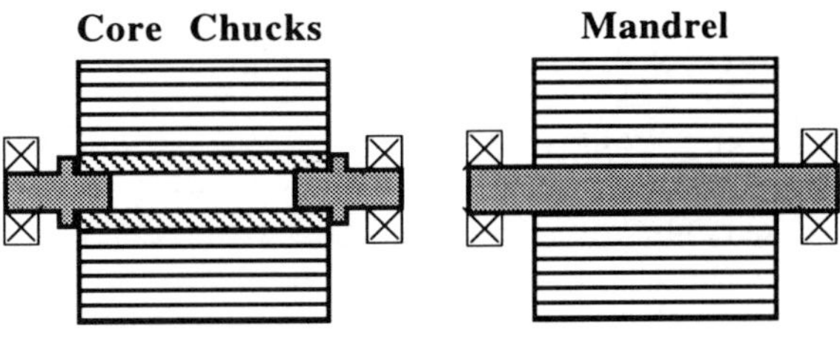

In addition to roll weight, significant nip loads may have to also be carried in the case of surface winder/unwinders. Also remember that any nip may generate thrust loads equal to the product of the web/drum friction coefficient and nip load. Since this friction coefficient may approach one, this means a conservative design will allow for thrust loads equal to nip loads. Finally, some chucks are axially preloaded to take up clearance on variable width cores. This load must also be carried by the bearings.

Deflection and Bending Stresses

To a web, a core or mandrel acts much like any roller during winding and unwinding. However, because small core sizes are often specified by the customer, a core/mandrel deflection criteria is dismissed in the rare event it is even considered. The worst deflections will be seen on chucked slender fiber cores, which have a modulus (≈500,000psi) only a few percent as stiff as metals (10-30 x E6 psi).

The deflection of cores is calculated by the procedures given in Chapter 3. A few example cases will show that the deflection of cores/mandrels are typically 1-2 orders of magnitude worse than a typical roller. Common problems caused by excessive core/mandrel deflection are shown in Figure 16.3. Also, nip loads on surface winders vary due to core deflection as shown in Figure 16.4. One way to reduce the core deflection due to tension is to orient the web so that it is over the winder and thus web tension will counteract roll weight. A full or partial width rider (lay-on) roller can reduce the effects of core deflection due to nips.

As the winding roll builds up over a core, the web material will add some stiffness to the system. It would be tempting to assume the effective core/roll beam stiffness were the sum of the stiffnesses (EI, or modulus x area moment of inertia). While this is a common composite beam assumption, it does not hold well for wound rolls. As seen in Figure 16.5, the core or coreshaft will deflect slightly more than the wound roll. This causes high pressure points over the ends of the core or spool and is the cause of many paper roll defects there.[1,2,3]

One reason that the wound roll behaves so strangely is because it is not a solid body. Rather, it is a stack of layers which are friction limited (interlayer pressure x coefficient of friction) in their ability to carry the shear stress which is maximum at the centroid of a beam. Thus, it is conservative to use an effective roll modulus that is somewhat smaller than the CD modulus of the web. This will mean that the stresses and deflection of cores and coreshafts will need to be checked at the core as well as with some amount of material as the roll builds and thus the roll weight increases.

Figure 16.3
Wrinkles and Core Deflection

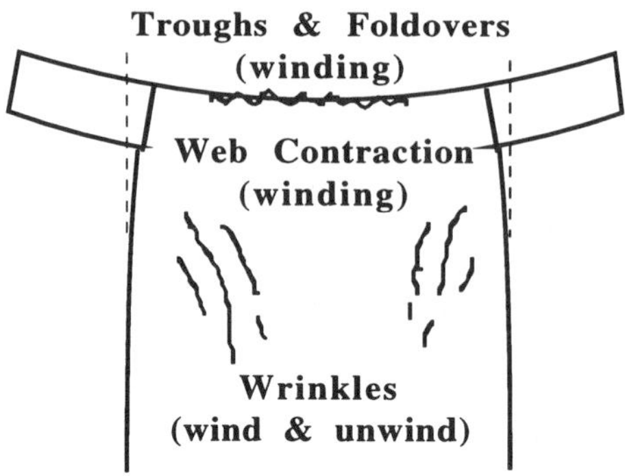

Figure 16.4
Nip Uniformity and Core Deflection

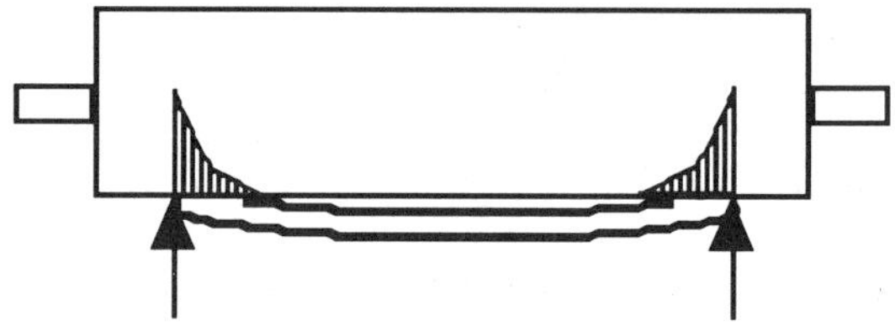

Figure 16.5
Deflection of a Core/Spool Supported Roll

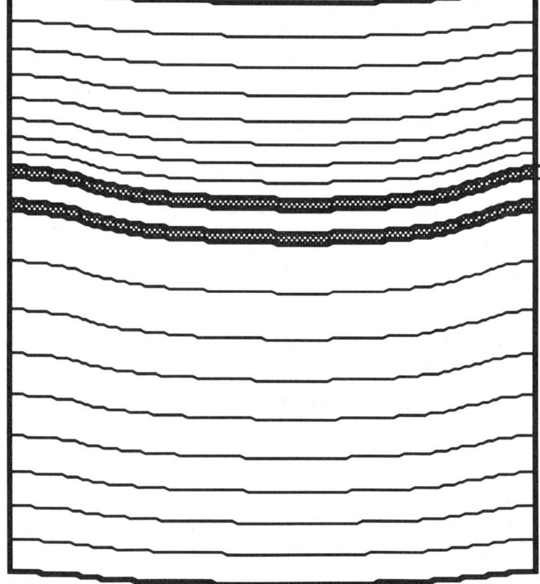

Deflection and Vibration

Rolls are much more prone to vibration during winding and unwinding than a roller. First, their low stiffness and thus increased deflection will reduce the resonant speeds. Also, there is often excessive chuck/core, shaft/holder, pivot or slide clearances causing increased amplitudes of vibration. In some applications, such as high speed newsprint printing, the unwinding roll could pass through a critical speed[4,5,6,7,8] and fail catastrophically. A typical frequency plot of unwinding and centerwinding is given in Figure 16.6.

Second, the cylindricity of rolls is much poorer than rollers,[9,10] especially for those that have sat on the floor or have been handled with clamp trucks.[11,12] If the roll is nipped, radial and rocking modes of vibration, as shown in Figure 16.7, are common. If the roll is unnipped, both roll runout and roll stand vibration are a common cause of web tension surges.[13,14]

Third, cores, coreshafts and certainly wound rolls are not explicitly balanced. The resulting imbalance from nonhomogeneous density and poor cylindricity can induce a strong 1x vibrating force. The only wound roll component that is commonly balanced is the spools or mandrels used on high speed paper machinery.

Perhaps the single most important vibration reduction technique is concentric chucking or shafting to reduce imbalance and runout. On chucks and shafts, this means reducing the clearance with the core. The clearance with the ID of the core should be such that the lower acceptance diameter will be a sliding fit with the chuck or shaft (approximately RC9, or 10-25 mils on a tightly toleranced 3" diameter core). This will require tight core manufacturing tolerances, (humidity) conditioning of fiber cores, and reconditioning of used metal cores to bring the ID into tolerance.

Larger clearances can be used without incurring an imbalance or runout penalty if concentrically expanding chucks or shafts are used. With this somewhat more intricate and costly style, the leaves or buttons are tapered and are driven outward by an axial movement of a cone shaped actuator. Also, tapered chucks can be used, but are hard on cores as seen in Figure 16.8.

Figure 16.6
Resonances of Unwinds and Centerwinders

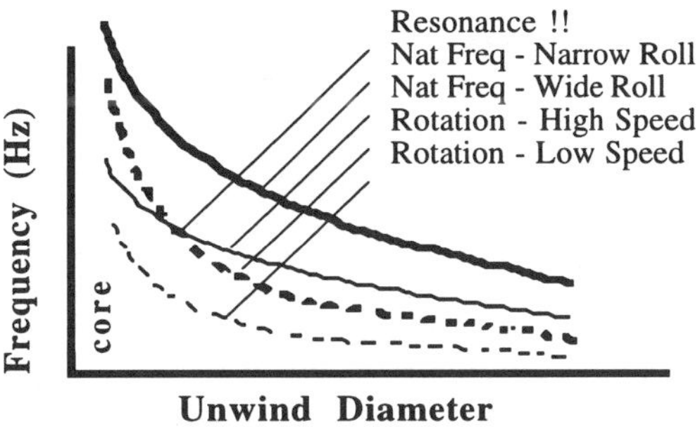

Figure 16.7
Winder Vibration Modes

Single Drum Radial

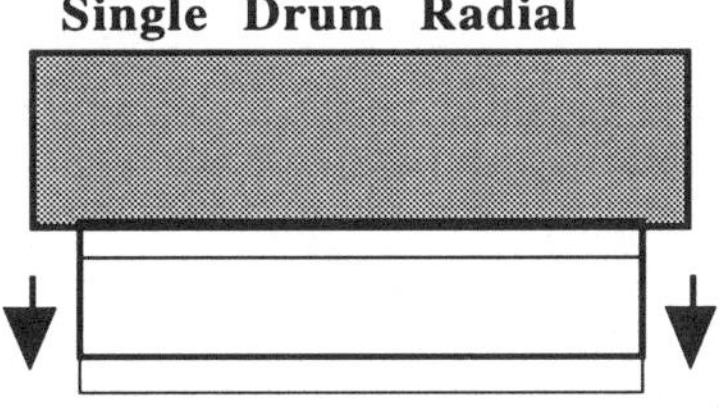

Single Drum Rocking

Figure 16.8
Concentric and Eccentric Chucking

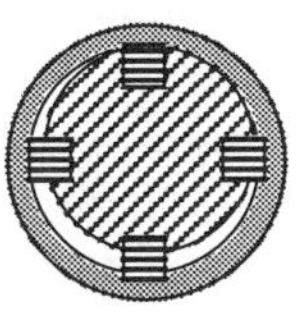

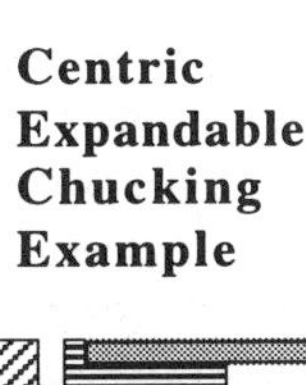

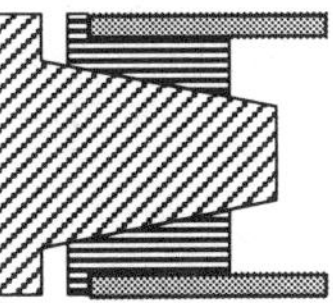

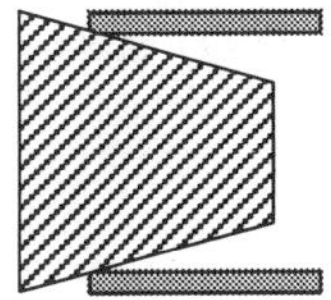

Fiber Cores

Fiber cores are wound continuously over a mandrel from a series of paper strips.[15,16,17] The strips are brought in on a helix by flat belts. The paper is often a recycled kraft which is dipped in an adhesive, such as sodium silicate, prior to winding. The cores are then automatically cut to a nominal or finished length as they exit the simply supported mandrel.

The core ID is nominally determined by the mandrel diameter, but is reduced by drying and hygroscopic shrinkage.[18] In fact, uncontrolled humidity and shrinkage is a major cause of wrong length cores[19] and loose cores after winding.[20,21] Core ID's are often as small as 1" for many consumer products such as tape, toilet tissue, kitchen towels, food wrap and so on. However, the more common size for shipping raw materials to converters is the 3" (which is now thankfully moving to 4") diameter. The core wall is determined mostly by the number and caliper of the plies of paper. Core tolerances are often specified for ID, length and bow (deviation from straightness).[22]

The industrial core, usually 3-4" ID, have a number of options for the ends. First, they can be notched to improve the torque carrying capabilities of a lugged chuck system. Second, they can be capped with a thin metal fixture that is swaged onto the ends to improve torque and bearing load capabilities. Third, the core may be plugged at each end to greatly reduce the risk of core collapse.

Of almost obsessive concern with core manufacturers is strength. Core crush strengths are measured routinely in both the axial and radial (2 point ring load) directions. Sometimes, weight and torque carrying capacities[23] or radial pressure crush[24,25] are also tested.

Always keep in mind, however, that not all core failures are so obvious as those related to strength. We've already mentioned excessive deflection causing nip and wrinkle problems, while excessive bearing load (unit roll weight / core diameter) damages web near the core.

Metal Cores

Fiber cores are inexpensive, generally single use, and thus disposable. Metal cores are very stiff and strong, and in nearly every way superior to fiber. Thus, it may be appropriate to use metal cores where the rolls are converted inhouse since they can be reused, with an occasional cleaning of the tape off the OD, until the ID wears out of tolerance. Sometimes metal cores are required because the application is demanding, such as heavy rolls for the printing (weights up to 5 tons[26]), coating and metals industries.

The metal core is usually heavy walled stock DOM tubing because it has much better cylindricity tolerances than conventional pipe. Very small cores can be handled by hand, while large cores require handling equipment. Intermediate weight cores are often aluminum to give some relief to the backs of the operators.

In some cases, the metal core may be used with no other treatment than being cut to length. In more demanding applications the core ends are counterbored to fit the chucks closely and the OD turned to increase nip uniformity. Also, the ends may be notched to increase the torque capacity when using lugged chucks.

While the life of the metal core is long, it is not indefinite. It needs to be checked and maintained[27] occasionally just as a roller would be. One common problem is that the ends tend to get belled after extensive use. While the belled ends could be cut off and the core used for a shorter application, they can also be salvaged by swaging both the ID and OD.

Mandrels and Spools

The design issues for wound roll mandrels or spools are identical in nearly all ways to those for rollers. The only significant difference is that mandrels must take the radial pressure at the interior of a roll which may be as little as 1 psi or can exceed 1,000 psi in some unique applications.

Core Chucks

If core chucks must transmit torque, such as on an unwind brake or centerwinder, they will be of the liveshaft style and either (externally) expandable[28] or torque actuated.[29,30] The challenge is to transmit sufficient torque (tension x width x diameter/2) without slippage. If the chuck slips, the inside of the core will be shredded at best.[31,32] The spinout failure can be insufficient traction between the chuck and core, or an interply shear failure of the core material. The torque capability of the chuck/core interface is primarily determined by diameter, the radial expansion force and the coeffient of friction between core and chuck. The torque capability of plies of the core are primarily determined by diameter, chuck length, and core material.

If the core chucks do not need to transmit torque, such as on many surface winders, they usually will be of the deadshaft style as given in Figure 16.9. The design challenge of deadshaft chucks is packaging enough bearing capacity and still fit inside the core. This is especially difficult because the chucks will have to carry both high radial loads and high axial loads in the case of surface winders. However, many chucks also have problems with chuck nose life, particularly when using metal or metal tipped cores. The metals used in chuck tips begin with flame hardened steels and proceed to the exotic.

Figure 16.9
Anatomy of a Simple Dead Shaft Chuck

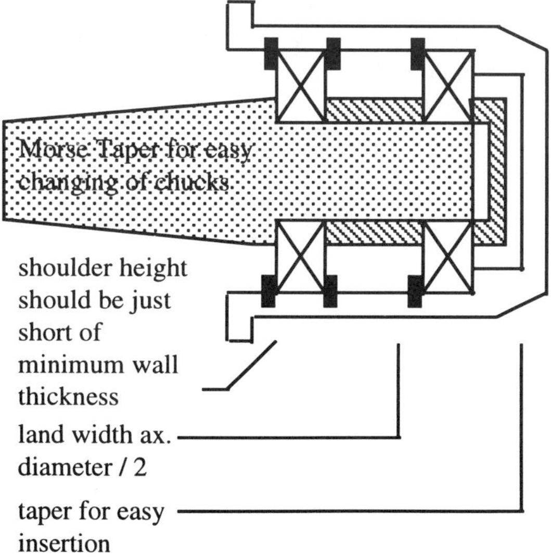

Core Shafts

In most cases, coreshafts will be of a liveshaft expandable style.[33,34,35] The major exceptions are surface driven unwinds (such as on tissue) or surface windups (such as reels or two-drums) which don't need to transmit torque (but do transmit large axial thrust). In these cases, the shaft will be aluminum or steel tubing with simple locking collars on the ends to provide a back side deckle location and a front side axial preload (often via an Acme thread) to remove the clearance between the core ends.

The primary sizing issues for expandable shafts are load capacity, torque capacity, and deflection for a given length and diameter of core. Often, the torque is transmitted by buttons, lugs or leaves riding in slots cut into the body of the shaft which expand radially when an internal bladder is pressurized with air. It is not uncommon to have a lot of maintenance associated with expanding mechanisms. Also, there must be a clutch between the brake/motor and the shaft to allow the shafts to be changed, which is another maintenance item.

With small cores, such as the ubiquitous 3", the bearings (or bushings) are often packaged on the shaft holder instead of on the shaft. The reason for this is to allow dressing of cores and roll removal without taking off the bearing housing. On larger cores, the bearing housing OD's may clear the ID of the core and thus could be permanently mounted on the shaft. Many facilities will have coreshaft pullers to remove the shaft from the finished roll set.

Aside from frequent maintenance, the other common issue with shafts is weight because many are lifted and handled manually by one or two operators.[36,37,38] For manual operation, the shafts should weigh no more than 50 lb per operator, better yet closer to 30 lb, and best is conforming to the NIOSH formulas. Reducing the risk of back injury is a good justification for using carbon-fiber shaft bodies which can cut shaft weight in half. Unfortunately, composites are not very durable to bumps and abrasion and thus may need to be sleeved with metal or coated with metal or ceramics.

Sizing Specifications

In this section, we will not detail the design and sizing of cores, chucks and shafts for two reasons. First, much of this is covered elsewhere in the book. Second, design of these components is primarily done by a handful of designers working for suppliers. Instead, this section will merely list some of the parameters that need to be supplied so that a builder can (hopefully) quote the appropriate hardware for the more common situations. Also, it indicates how these values will enter into the design so that appropriate tradeoffs and compromises can be made if needed. Finally, it is wise to include an elevation and gauge sketch with any quote request to eliminate any chance of confusion. Similarly, interchangeability with existing plant equipment will reduce spare parts inventories.

The minimum **core** diameter (for a nominal size) determines the chuck or shaft diameter with allowance for sufficient clearance for removal (RC9). The maximum core diameter (for a nominal size) determines the range of expansion for expandable or torque actuated chucks and shafts. In some cases, stepped chucks and shaft adapters can allow running multiple nominal sizes on the same unit, but this is not the norm.

The required core wall thickness is determined for the most part by resistance to core crush, although core grade will also have a minor effect. The core material should be specified as metal is quite strong, but is expensive and is very tough on chucks and shafts. Core ends may be specified with caps, notches, plugs and so on.

The maximum **roll** diameter, along with width, density and shaft weight will primarily determine the required load capacity of most of the elements. Also, with core diameter will determine the unit bearing load on the cores to avoid nip induced defects which are common on light paper grades.

(16.1)

$$\text{WgtRoll} = \text{WgtSpool} + \frac{\pi}{4}\rho W\left(\text{DiaRoll}^2 - \text{DiaCore}^2\right)$$

The maximum **web width** determines the working length, and in part, deflection and critical speed of a core or coreshaft. It also is a factor in the bearing capacity of any wound roll system. The bearing-to-bearing distance, of course, should be as short as possible to reduce deflection, stresses and vibration. The minimum slit width determines the maximum spacing of the buttons, lugs or leaves on an expandable shaft.

The maximum **tension**, along with roll and core diameters, will determine the required expandable or torque actuating traction of shafts, and with web width the torque capacity of mandrels, chucks, clutches, brakes or motors. Tension also plays a minor factor in deflection.

(16.2)

$$\text{TorqMax} = \text{TensMax} \times W \times \text{DiaRoll} / 2 \pm I\alpha$$

The maximum **nip load** affects load capacity and deflection. Also, there will be a friction limited axial load from nips (μxNxW) that must be carried by the bearings and chuck/shaft pivots or slides.

Speed, duty, and roll diameter will determine bearing life. Speed, as well as tension and other lessor factors will determine the horsepower requirements of the brake or motor. Maximum torques which must be transmitted by all elements may occur during accleration, deceleration or E-stop (emergency).

The **chuck** stroke determines the chuck land distance (with provision for clearance, tip taper and wound roll dishing allowance), but must be sufficient to make sure the rolls don't come off the chuck (> core ID/4) due to chuck and mounting slop or wound roll vibration.

Bibliography

1. Roisum, David R. *Nip Induced Defects of Wound Rolls.* TAPPI Finishing and Converting Conf. Proc., Chicago, pp 89-98, October 2-5, 1994.

2. Smith, Phil and Bagnato, Luigi. *The Relationship of the Paper Machine Reel to the Winding Process.* TAPPI Finishing and Converting Conf. Proc., New Orleans, pp 123-138, October 24-27, 1993.

3. Lindstrand, Bruce L. *Reel Spool Sizing and its Affects on Converting Performance.* TAPPI Finishing and Converting Conf. Proc., Chicago, pp 125-132, October 2-5, 1994.

4. Bank, L. C. and Gerhardt, T. D. and Gordis, J. H. *Dynamic Mechanical Properties of Spirally Wound Paper Tubes.* ASME J. of Vibration, Acoustics, Stress, and Reliability in Design, 111: 489, 1989.

5. Caccase, Vincent. *Dynamic Behavior of Steel Cores in the Unwinding Process.* TAPPI Finishing and Converting Conf. Proc., Nashville, pp 283-296, October 18-21, 1992.

6. Caccase, Vincent. *Dynamic Behavior of Steel Core in the Unwinding Process.* Tappi J., vol 76, no 10, pp 51-61, October 1993.

7. Gerhardt, Terry D. and Staples, John F. and Lucas, Robert G. *Vibrational Characteristics of Wound Paper Rolls: Experiment and Theory.* TAPPI Finishing and Converting Conf. Proc., Nashville, pp 263-282, October 18-21, 1992.

8. Gerhardt, Terry D. and Staples, John F. and Lucas, Robert G. *Vibrational Characteristics of Wound Paper Rolls: Experiment and Theory.* Tappi J., vol 76, no 6, pp 121-128, June 1993.

9. Eriksson, Leif G. *Deformations in Paper Rolls.* Winding Technology Conf. Proc., Swedish Newsprint Research Center (TFL), Stockholm, March 16-17, 1986.

10. Eriksson, Leif G. *What Happens to a Paper Roll in the Printing Plant.* Winding Technology Conf. Proc., Swedish Newsprint Research Center (TFL), Stockholm, March 16-17, 1986.

11. Merin, P. *How Mechatronic Engineering Led to Intelligent Paper Roll Clamps.* TAPPI Finishing and Converting Conf. Proc., New Orleans, pp 63-70, October 24-27, 1993.

12. Skinner, J.R. *Newly Designed Automatic Slip Controlled Paper Roll Clamp.* TAPPI Finishing and Converting Conf. Proc., Chicago, pp 37-42, October 2-5, 1994.

13. Roisum, David R. *Winder Vibration.* TAPPI Finishing and Converting Conf. Proc., October 1987.

14. Roisum, David R. *Winder Vibration Can Reduce Operating Efficiency and Increase Maintenance.* Tappi J., vol 71, no 1, pp 87-96, January 1988.

15. Johnson, D. *Taking Core Manufacture Out of the Dark Ages.* Calendering and Winding Technology, TAPPI PRESS, 1985.

16. Timell, Hans H. *State of the Art in Core Manufacture Handling and Preparations.* TAPPI Finishing and Converting Conf. Proc., 1985.

17. Harvey, D. Michael. *Operating Variables in Spiral Winding: Theoretical Interrelation.* Tappi J., vol 53, no 8, pp 1521-1524, August 1970.

18. Corbitt, Maurice R. *The Fiber Paper Mill Core.* TAPPI Finishing and Converting Conf. Proc., pp 125-126, 1982.

19. Anon. *Drying of Wound Tubes.* Algemeine Papier-Rundschau, no 24/25, 758-760, June 1980.

20. Hussain, S.M., Farrell, W.R. and Gunning, J.R. *Roll Winding - Causes, Effects, and Cures of Loose Cores in Newsprint Rolls.* Tappi J., vol 60, no 5, pp 112-114, March 1977.

21. LeBel, Roland G. *Core Conditioning: Elimination of Pressroom Core Slippage in Newsprint.* Tappi J., vol 61, no 10, pp 23-26, October 1978.

22. Kearns, Robert W. and Munroe, James H. *New*

Dimensions to Consider in Film-Core Specification. Paper, Film & Foil Converter, vol 67, no 9, pp 70, 72, September 1993.

23. Gerhardt, Terry D. *External Pressure Loading of Spiral Paper Tubes: Theory and Experiment.* Transactions of the ASME, vol 112, pp 144-150, April 1990.

24. Staples, John F. *Roll Weight Capacity of Paper Mill Cores.* Tappi J., vol 77, no 5, pp 190-194, May 1994.

25. Gerhardt, Terry D. *Radial Crush Takes the Measure of a Core.* Converting Magazine, pp 50-54, February 1994.

26. Angland, D.J. *Steel on Today's High Speed Winders.* TAPPI Finishing and Converting Conf. Proc., Kansas City MO, pp 141-144, October 1-5, 1989.

27. Nygaard, D.C. *Iron Core Processing.* TAPPI Finishing and Converting Conf. Proc., 1978.

28. Albee, Stephen H. *Pneumatically Operated Shafts and Chucks for Rewinding and Unwinding.* Southern Pulp & Paper Mfr, vol 40, no 6, pp 39-41, June 1977.

29. Batstone, David. *James River mill increases Core Reuse with Switch to Torque-Activated Chucks.* Pulp & Paper, July 1992.

30. Crooks, J. and Flagg, Edward. *Rolling Friction Lets Core Chucks Do Better Job.* Paperboard Packaging, vol 72, no 4, pp 58-60, 62, April 1987.

31. Diltz, Jack L. *Straight Talk on Spin-Out Solutions.* Paper, Film & Foil Converter, vol 58, no 4, pp 74 76, April 1984.

32. Diltz, Jack L. *How to Avoid Core Spin-Out.* PIMA, vol 67, no 2, pp 44-45, February 1985.

33. Albee, Stephen H. *The abcs of Expandable Air Shafts.* Graphic Arts Monthly, vol 49 no 4, pp 52 54 56, April 1977.

34. Albee, Stephen H. *Follow This Guide in Your Selection of Air Shafts.* Canadian Pulp & Paper Ind., vol 31, no 6, pp 75 77-78, May 1978.

35. Cartwright, D. *Air Shafts Give Converters a Grip on Roll Mounting.* Converter, vol 23, no 7, pp 8-12, July 1986.

36. Cox, Jacqueline. *Carbon Fiber Shafts.* American Papermaker, pp 25, October 1994.

37. Fortin, M.B. *Carbon Fiber Composite Tubes Provide Solutions to Heavy Shaft Handling Problems.* TAPPI Finishing and Converting Conf. Proc., Houston, pp 245-257, October 6-10, 1991.

38. Franklin, James. *Composite Materials Lighten the Load.* Converting Magazine, pp 64, 66-67, June 1993.

Chapter 17

Spreaders

In this chapter we define spreading and the mechanics that govern the behavior of spreaders. Then we describe the application and operation of the many types of spreading systems. Finally, some troubleshooting techniques are included.

Introduction

In a perfect world, there would be little need for spreading. In this world, the web would be manufactured uniformly flat and would remain so through subsequent handling.[1] The realities are, however, quite different. Spreaders are required on most web processes to prevent or remove wrinkles, to widen a web, or to separate slits during winding.[2,3,4,5,6] They are used in the converting of a variety of materials including film, foil, nonwovens, paper and textiles. However, they are also found in manufacturing as well. Forming wires and drying felts are just a few examples where spreaders are required on endless belts to keep them flat.

There are many types of spreading systems in common use including the bowed roller, dual bowed rollers, concave rollers, expander rollers, Pos-Z's and D-bars as given in Table 17.1. Some have a single element, while others have two or more elements. Some spreaders can be adjusted for local effects, while others can only be adjusted for overall spreading, or may be completely fixed in shape.

Despite their varying appearance and their many diverse applications, all spreaders follow a universal set of web handling principles that govern their operation. Understanding these principles allows us to make the most appropriate choice of spreader for a particular application, and then to operate that spreader most effectively. In doing so, we will see improved manufacturing and converting efficiency, while at the same time reducing waste due to causes such as wrinkles, foldovers, and stuck rolls.

Indeed, immediate gains can be made by removing so called spreaders such as wormed rollers and cigar shaped rollers which actually contract the web as we shall see. Other common pitfalls include undersized spreading systems that may not remove wrinkles or separate slits, and oversized spreaders that can defeat spreading or cause instabilities in the web's edge position. Finally, even well designed systems may not do the job if the bow is pointed in the wrong direction, for example.

However, before we begin to describe spreaders and their principles, it would be good to define what is meant by spreading.

Spreading is a converting process that widens the web on or near the spreading element.

Using this definition, much of the mystique of the benefits of spreading is removed. If the web is widened at a spreader, it will widen or open up the gaps between individual strips so that rolls don't intertwine during winding. If the web is widened, <u>some</u> types of wrinkles can be alleviated.[7,8] However, spreaders do not effectively correct baggy webs as is commonly believed.[9,10] In fact, spreading rarely leaves any permanent effects on the web. Rather, it tends to be very local and temporary, often lasting only for a span or so after the spreading unit. Finally, any roller or element can act like a spreader or contractor, if it exhibits a character described in this chapter.

Table 17.1
Types of Spreading

Type	Elements	State
Bowed Roller	1	traction
Concave Roller	1	traction
D-bar or Bent Pipe	1	sliding
Expander Rollers	1	traction
Dual Bowed Roller	2	traction
Pos-Z™	2	floating
Large Diameter Bar/Roller		float/slide
Tension increase or moisture/temp decrease		

Spreading Principles

An important web handling principle that governs many spreading systems is the **Normal Entry Law.**[11,12] As seen in Figure 17.1, a web will seek to enter a downstream roller at a right angle to the roller's axis at the point of first contact. If for a variety of reasons the web is not currently entering at a right angle, it will move sideways toward a normal entry at an ever decreasing rate. This universal principle not only governs most spreading, but also describes web behavior at misaligned rollers, guides, and winders. Similarly, the web will also tend to exit a roller in a normal direction. This normal entry/exit principle will be used to help determine the path of a web through a machine.

An important exception to the normal entry law however, is when there is slippage between a web and its roller. Thus, while it would not describe behavior at a sliding spreader such as the D-Bar or Pos-Z™, the web will enter and exit at right angles to the non-slipping upstream and downstream rollers. In any case, traction is such an important parameter that anything that improves traction will also improve the spreading potential. In fact, spreaders such as the concave and bowed roller can turn into contracters when traction is lost. Traction is described in more detail in Chapter 4 and summarized in Table 17.2.

As stated earlier, spreading occurs when the web gets wider at a spreader. A corollary to this is that the edges of the web must move sideways. Some of the more important forces that can **move a web sideways**, and thus help spread, are summarized in Figure 17.2. They will be discussed in more detail in later sections. However, at this time we will mention that the web will resist both lateral and in-plane bending as the beam stiffness of webs in that plane is quite high. Another limitation of lateral and in-plane bending is that there is a minimum amount of line tension required in order to keep the web under tension at the edges. Obviously, if the web is slack there will be no traction and thus no incentive to spread there. Folding is the basis for most dual element spreaders Folding/twisting is also the most powerful because the web isn't as stiff with this mechanism than it is with lateral forces and in-plane bending.

Figure 17.1
The Normal Entry Law
(for rollers in traction)

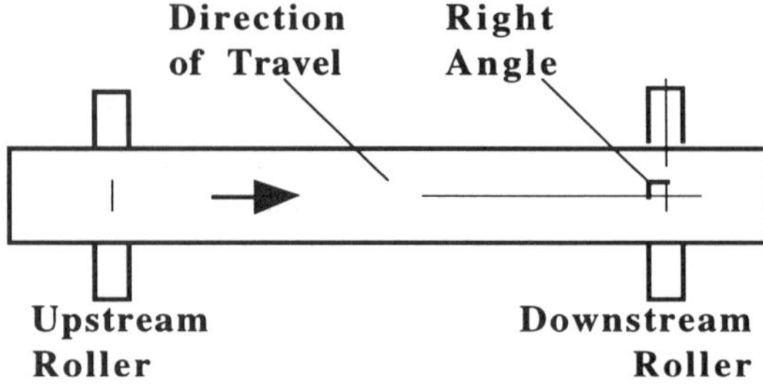

Figure 17.2
Spreading Principles

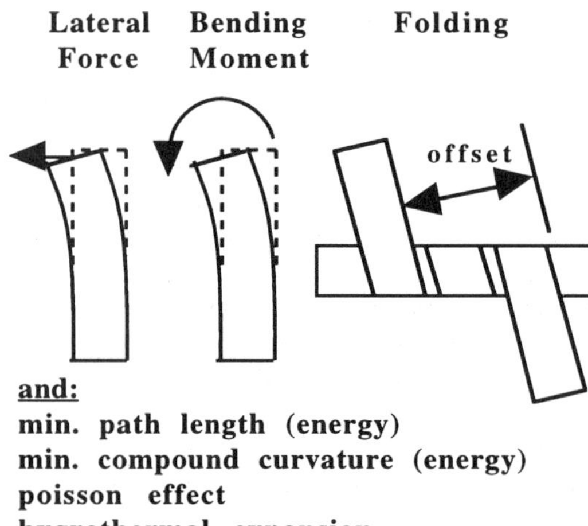

and:
min. path length (energy)
min. compound curvature (energy)
poisson effect
hygrothermal expansion

Because the web really does behave much like a beam, albeit a very tall and skinny one, we know that the path of a web through a machine will be a smooth and gentle curve. With this background, we can now explain the operation of several spreading devices.

Table 17.2
Traction and Spreading Can Increase With:

Direction	Variable	Speed
Increased	Wrap Angle	Any
Increased	Friction Coef.	Any
Increased	Tension	Any
Decreased	Speed	High
Decreased	Roller Diameter	High
Increased	Porosity/Groove	High

Concave Spreader Roller

One of the simplest and least expensive spreaders is the concave roller (sometimes called a **bowtie** roller). As seen in Figure 17.3, this spreader is a conventional roller whose diameter at the ends is slightly greater than at the center. In its ideal configuration, the diameter profile of the roller is cut as an arc of a circle. However, a simpler version can be made for unslit webs by cutting a roller with conical ends.

A crude but effective way to make a concave spreader is to add a couple of wraps of **tape** to the ends of a straight roller, which is sometimes referred to as **bumpering**. This can be a useful temporary field fix for flattening unslit webs, that is all too often misapplied. I've seen rollers so full of tape you couldn't see the shell, or tape well inside the edges of the web, or slippage (which turns it into a contractor) on the roller because the shell and/or tape was slippery.

A rough starting point for a diameter reduction at the center would be 10-25% of the MD strain induced by web line tension or draw control. These geometries would be quite easy to cut for stretchy materials, but end up to be quite small numbers for most webs.

While any solid roller will have a constant rotational speed (RPM) across its width, a concave roller will have a varying surface speed (FPM) that is proportional to the local diameter at every position. Thus, as seen in Figure 17.4, the surface speed of the concave roller is higher at the ends than in the center. The resulting tension distribution causes a small bending moment that pulls the web in (contracts) and a larger lateral force that spreads the web. Thus, the concave roller is not particularly powerful because these two forces are in opposition.

A concave roller will spread in traction and contract if it slips (acts like a backward D-bar). Conversely, a **cigar or barrel shaped roller** will contract the web in traction, but spread when slipping (acts like a D-bar). Importantly, diametral variations in any roller can cause localized spreading/contracting if the differences are sufficiently large. Finally, the mechanics of this roller has been modeled, but are exceedingly difficult.[13,14]

Figure 17.3
Concave Spreader Roller

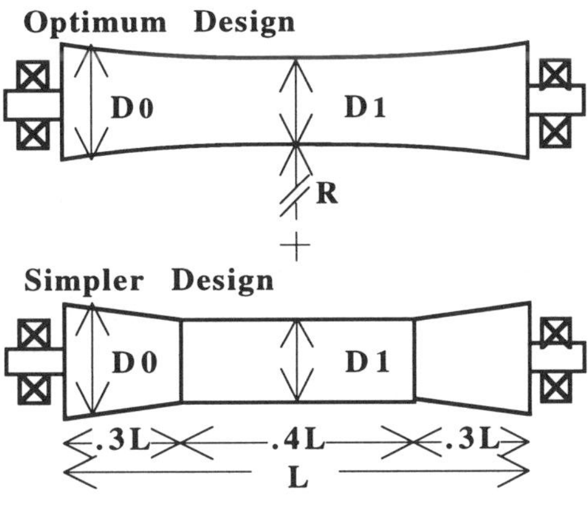

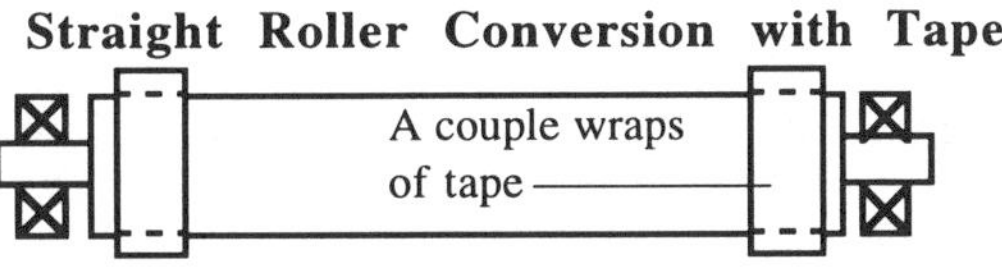

Figure 17.4
Principles of the Concave Spreader

Speed and Ingoing Web Tension Profile

Hi Tension
Hi Tension
Lo Tension

Induced Moment and Shape

F
M
℄

View of Left Half Web

Tension distribution is replaced by an equivalent tensile force at the centroid and a bending moment. However, the law of normal entry is violated by the angle of the web at the point of contact with the roller ...

Superposing the Lateral Force

V
F
M
℄
Spread
Normal Entry

View of Left Half Web

To get normal entry we must superpose an outward lateral force. Thus, the shape results from a nonuniform tension which causes an inward moment and outward lateral force.

Bowed Spreader Roller

The bowed spreader is formally known as a **curved axis roller**, but often referred to in slang as a **bananna roll**, and unfortunately and improperly by the **Mt. Hope** roll trade name. As seen in Figure 17.5, the bowed spreader has a curved stationary axle upon which a rotating sleeve(s) is mounted in numerous bearing sets. The axle may have a fixed bow, or can be variable though clever design arrangements such as split axles. The sleeve is typically a one–piece flexible tube of a soft synthetic composite. For high wear resistance and better cover life, however, the sleeve may consist of numerous narrow metal rings. The sleeve may be grooved to reduce air entrainment and thus improve traction, and metal sleeves may be tungsten carbide coated for abrasion resistance and traction. Finally, the bowed roller may be used in powerful dual element spreaders which will be described in a later section.

Since the bowed roller is more complex than other spreaders, there are more design considerations. For example, **bearing** design becomes very important on bowed rollers which may have dozens of bearing sets which are hard to replace, in contrast to conventional rollers which have only two bearings with relatively easy access. Also affecting bearing design is rpm limitations, lubrication, and the desire for minimal bearing drag. The application engineer must size the drives for motor driven spreader rollers, or size the tension upset on web driven spreader rollers. The torque to drive the roll is composed of bearing drag and sleeve hysteresis during constant speed, with an additional acceleration component from bearing and sleeve inertia during speed changes.

The synthetic **sleeves** are highly engineered composites. They must provide a reasonable tolerance to wear, bumps, cuts and other damage which is endemic to the industrial environment. The sleeve material must also have a minimal hysteresis because the continual flexing it takes as it rotates around the curved axis can cause increased roller drag, thermal degradation and fatigue. The sleeve must hold its shape to avoid disturbing the spreading uniformity due to diametral variations.

Figure 17.5
The Bowed Roll Spreader

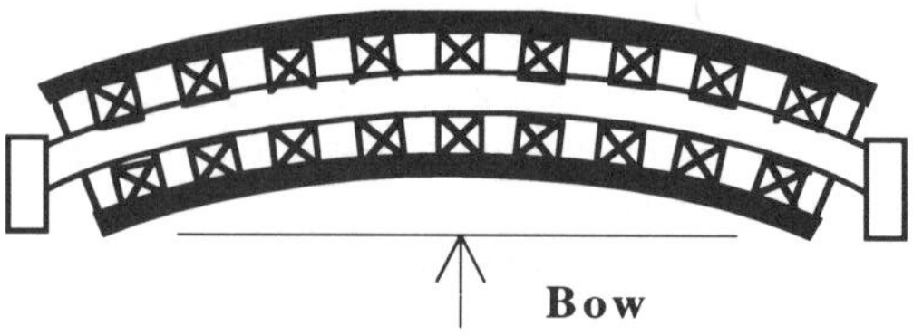

Also, loose sleeves are a common problem resulting from improper fits, centrifugally induced creep or thermal expansion. A loose sleeve will shift eccentrically and can cause severe spreader roller vibration. This tendency is exacerbated because spreader rollers are not dynamically balanced, and operate through resonance on high speed winders.

The **axle** serves two primary functions. First, it is the structural backbone of the spreader. Most of the effective bending stiffness of the spreader comes from the axle rather than the cover. The axle largely determines weight, deflection and critical speed. The second function is to determine the shape of the spreader. A critical aspect in manufacturing spreader roll axles is to maintain a very uniform in-plane (no twisting) arc of a circle (constant radius of curvature). The two most common methods to form the straight axle stock into an arc is to bend (three-point, four-point, or against a template) in a brake press, or to roll it in a roll bender. While bending may result in less axle twisting, rolling may result in more uniform curvature.

Sizing a spreader roller involves picking a length corresponding to machine width; diameter determined by deflection, critical speed, and bearing rpm considerations; and finally bow. While the bow shape is an arc of a circle with a 20-200 radius, the bow magnitude is usually specified as the distance between the chord and the arc of the roll at its center, instead of a radius of curvature. This is sometimes expressed as a %bow, which is the bow divided by face width.

The amount of **bow** is very important because too little will reduce the effectiveness of the spreader, while too much will destroy the effectiveness of the spread, and could turn it into a wrinkle making device. Unfortunately, the historical tendency has been to overbow.

The bowed roller in traction follows the Normal Entry Law. As seen in Figure 17.6, the center of the web travels exactly in the MD, while the edges travel perpendicular to the axis of the roller at the edges. While the surface speeds are uniform across the web (for a uniform diameter roller), the surface velocities are not. The velocity vector at any position has both an MD tension component and a lateral CD tension component. The MD velocity distribution is maximum at the machine centerline and steadily decreases toward the edges, which causes an outward rotation of the web due to in-plane bending. The CD velocity is zero at the machine centerline and increases as one progresses outward, which causes an outward lateral displacement of the web. Thus, while the bending moment fought the lateral force in the case of the concave roller, they work together on the bowed roller.

Additionally, there is also a carry out effect that increases bowed roller spreading with increased wrap angle because the web exits a point further out than where it entered. The carry out effect increases from zero at no wrap to a maximum at 180° wrap, and can be calculated from simple trigonometry. Thus, bowed roller spreading can be improved by increasing the wrap angle, which increases both traction and carry out. The only caveat is that this also increases the deflection, which is already much greater than most rollers.

Bowed roller spreading can be best visualized using the streamlines of Figure 17.7.[15] The web will leave the upstream roller at its original width, enter the bowed roller at a right angle (when in traction) and perhaps narrow slightly on the downstream roller. However, if the web is not in traction, it will take a path with a smaller spread or offset. The problem is, however, that this process is never uniform. The outside edges will be most likely to break loose first because of the larger required lateral force and the reduced tension force needed for traction. Thus, the slipping edges may run into or interfere with the still tracking inboard positions. The symptoms of this problem caused by excessive bow is: good spreading near the center, poor spreading (or contraction or wrinkle generation) near the edges.

Figure 17.6
Principles of the Bowed Roller

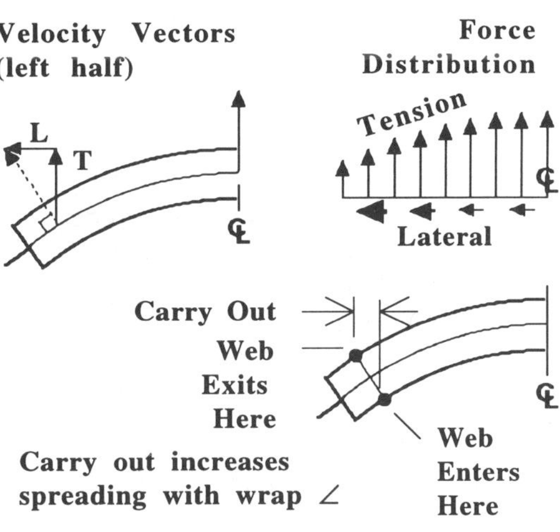

Figure 17.7
Bowed Roller Streamlines

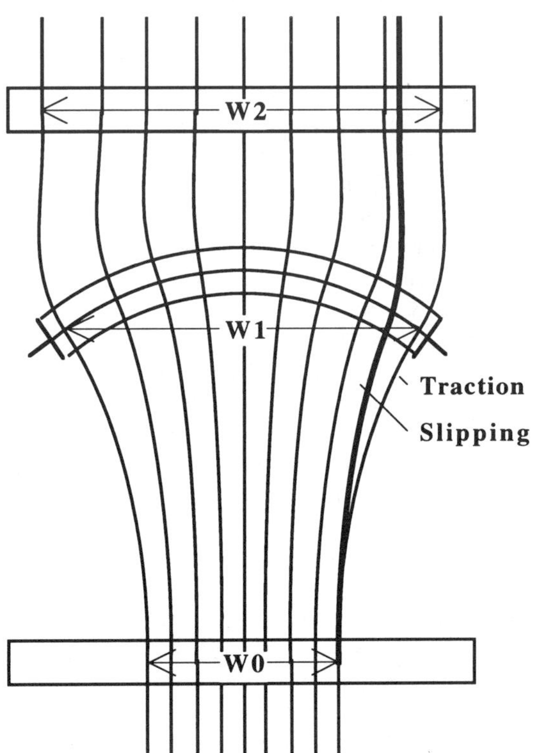

The effect of bow magnitude on spreading or contracting is best summarized in Figure 17.8.[16] As seen here, there is an optimum bow magnitude to achieve the greatest spreading which is largely a factor of material modulus and web/roller traction. Thus, it is better to err on the side of small bows rather than large.

As a rule of thumb, most bowed rollers should have a 1/8% net **bow**, unless the material is slit where the bow could increase to 1/2%, or if the material is extremely stretchy and flexible where we might specify 1%. Since bows can't be accurately made less than about 1/8", bowed rolls could perform nonuniformly or unpredictably for applications less than 100" wide unless the material is slit and/or stretchy. The net bow is the vector addition of pre-bow and tension deflection and weight deflection.

The setup of a bowed roller is straightforward. First, the **wrap angle** should be sufficient to maintain traction, and is typically 15-45 degrees. Additionally, the cover should be grippy and uniform in diameter. Second, the entering/exiting span length ratio is desirably about 2:1, but will work acceptably at considerably different ratios.

Finally, the **bow orientation** should be pointed downstream in a direction perpendicular to the bisector of the wrap. However, adjustments from this nominal orientation can be used to compensate for a **baggy** center or **slack** edges. As seen in Figure 17.9, a baggy center can be tightened by rotating the bow into the web. As the bow orientation is turned more into the sheet, the path that the center of the web must take is longer than the edges, and consequently tighter. Since the web tension/draw control sets the average tension across the width, the edges will loosen at the same time the center tightens in order to maintain that average tension. Conversely, slack edges can be tightened by rotating the bow out of the web, but care must be used in this direction to avoid edge wrinkling.

It is important to note that the web sees a different bow magnitude and a different bow direction than the specified bow and pointer direction that the spreader manufacturer provides. This is because web tension and gravity deflection both modify the bow's direction and magnitude.

Figure 17.8
Bow Magnitude and Affects on Spreading

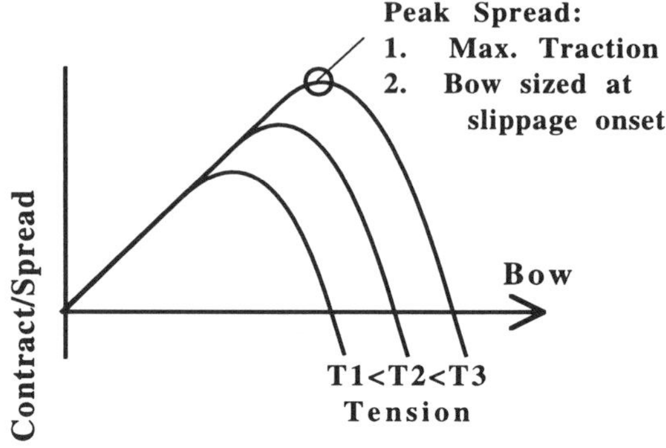

Figure 17.9
Bow Orientation

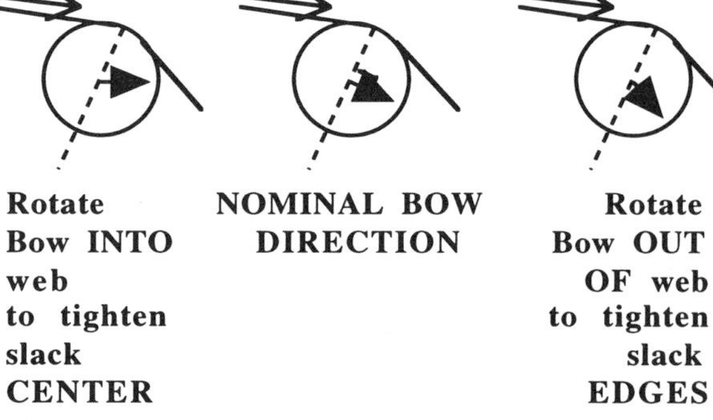

A bow turned into the sheet makes the center span length longer/tighter.

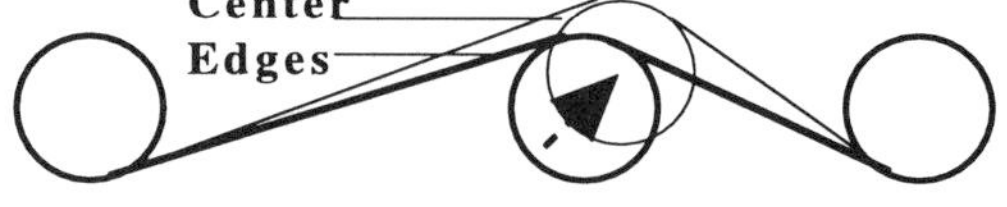

For example, a spreader pointing upward will see a net bow which is the pre-bow minus gravity deflection while the same spreader pointing down will see a net bow which is the pre-bow plus gravity deflection. To determine what pre-bow should be supplied to achieve a desired net bow, and to determine what direction to aim the pointer to give a desired bow direction requires simple vector math. This is a little more complicated for split axle designs which are less stiff in the bow direction than perpendicular to the bow.

Bent Pipe and D-Bar Spreaders

The popularity of the bent pipe spreader lies with its simple construction. Unfortunately, this simplicity often leads to crude manufacturing practices and dubious spreading qualities. A D-bar spreader takes its name from the characteristic D cross-section of the bar as seen in Figure 17.10. The only functional difference between a D-bar and bent pipe spreader is that the D-bar's shape can be adjusted by intermediate jacks. With this adjustment, the operator can spread a local baggy or wrinkled spot, or open up a particular slit position. The problem, however, is that the bar soon ends up looking like a snake after several adjustments, which adversely affects spreading uniformity.

The operation of the bent pipe or D-Bar spreader is based on two distinct principles. The first is to minimize the web's strain energy. This can be loosely interpreted as minimizing the path length through the spreading system, as shown in the example Figure 17.11. Here, a web at the quarter point may wish to take path A straight through, or B or C with increasing offset/spread. While it might appear that the straight path A is the shortest as seen in the plan view, it is the longest in the end view because that portion of the spreader protrudes into the web run farther than paths B & C. Not surprisingly, this means that path B, which has some spread offset, is the shortest and least energetic path. A good spreader setup will have a strain energy versus spread curve that has a minima located at a slight positive outward displacement. In simple terms, the web slides down the energy hill.

The second pipe/D-bar principle is bending and shifting toward the low friction side of a individual slit web. This may either increase or decrease spreading depending on several variables including bar shape, local tensions, and local coefficients of friction. Thus, if there are tight and loose areas of the web as manufactured, the spreading will be inconsistent from one position to the next. Furthermore, if these vary with time (MD) as well, the web will dither and wander. What saves us in these situations is the tremendous in-plane bending stiffness of the web that helps to keep spreading small and somewhat stable.

Figure 17.10
D-Bar Spreader

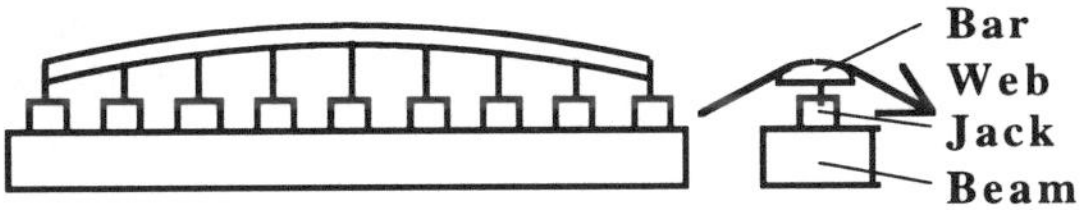

Figure 17.11
Principle of the D-bar Spreader

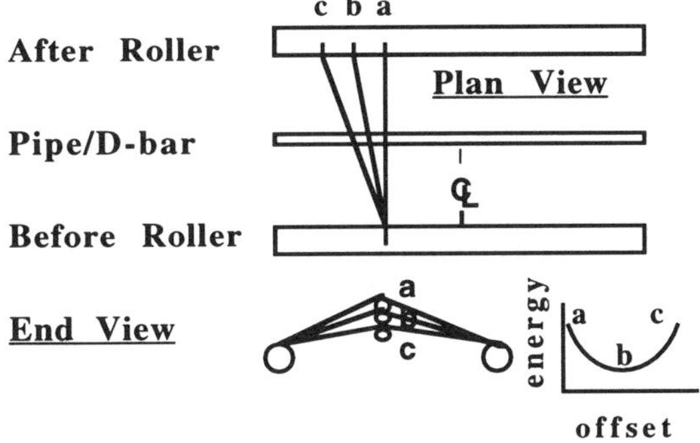

Like most spreader types, there are no analytical models and few application guidelines. One exception is for D-bars on wide paper machine winders which should have a 5 mm penetration into the sheet run and a 0.2% bow.[17]

One application limitation of pipe, D-bar and other sliding spreaders is that web scratching may be unacceptable for many grades, particularly those that are coated. Indeed, even the hardened steel bar may wear out prematurely on paper grades which can be so abrasive that they spark going over the bar.

Another concern is the shape uniformity of the pipe or bar, especially when spreading slits. The pipe can be bent using 3 point bending, but the result is a curvature and spreading that varies from a maximum at the center to zero at the ends. A more uniform curvature can be made either by rolling the pipe, or by bending it in 4 point bending and cutting off the ends. Uniform curvature on D-bars is best achieved by instructing the operators in the proper techniques for setup and adjustment. The local spreading power of a bent pipe and D-bar spreaders is proportional to the second derivative of shape.

Dual Bowed Roller Spreader

For more spreading power, two or more bowed rollers can be placed in series and oriented in the single bowed roller direction. However, the spreading power may be less than proportional to the number of elements. Sometimes, the bow magnitudes are increased progressively down through the series. An application that may use more than two tandem bowed rollers is to intentionally and permanently increase the width of flexible materials such as nonwovens.

As will be seen, however, the dual bowed roller operates on a different set of spreading principles than two tandem bowed rollers in the single roll orientation. The primary use of a dual bowed roller is after a slitter section which cuts many narrow webs. The dual bowed spreader is often mounted on a pivot so that the spreading magnitude can be adjusted for a varying number of slits by varying the wrap angle on the rollers. Unfortunately, it is a more costly spreader because there are two bowed rollers, which may require drive motors, and a pivoting stand. A variation of the dual spreader is a bowed roller in the first position and a D-bar in the second position which allows some adjustment of spread profile.

In its most common setup as seen in Figure 17.12, the pivot is located on the midpoint of a tangent connecting the upstream roller (typically an after slitter roller) and the downstream roller (typically the 1st winding drum). Both spreader roller bows are oriented perpendicular to the incoming and outgoing sheet runs, and thus are parallel to themselves. The first bow points out of the web, while the second points into the web. The bow orientations may need adjustment to maintain perpendicularity whenever a large pivot adjustment is made.

The primary spreading mechanism of the dual roller spreader is an outward geometrical folding and twisting of the web. As seen in the plan view of Figure 17.13, the web(s) are twisted in the entry span to the first roller. Then the web paths are redirected to diverge outward by folding over the first roller. The outward fanning paths then continue to the second roller. Finally, the second roller folds and then twists the web(s) back parallel to themselves.

Figure 17.12
Dual Bowed Roller Setup

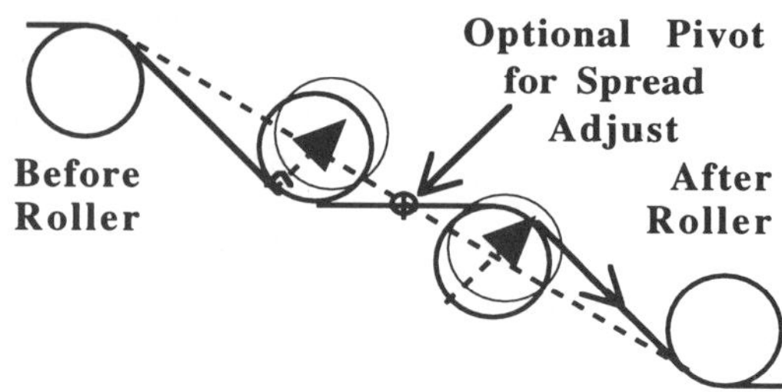

The reason why the dual roller spreader is so powerful, is that the web is being folded and twisted instead of bent sideways, which are much easier maneuvers for the web to follow than the in-plane bending of other spreaders. The astute reader may wonder why the webs don't appear to obey the parallel entry/exit laws in the figure. Rest assured that they actually do; however, this can only be seen from appropriate oblique views.

Of course, one can change from a tandem to a dual spreader and back again merely by changing the bow orientations. Thus, the two roller system has a considerable application flexiblity. However, it seems that most unslit applications are set up in the tandem mode, while many slit applications are set up in the dual mode.

Figure 17.13
Principle of the Dual Bow Roll Spreader

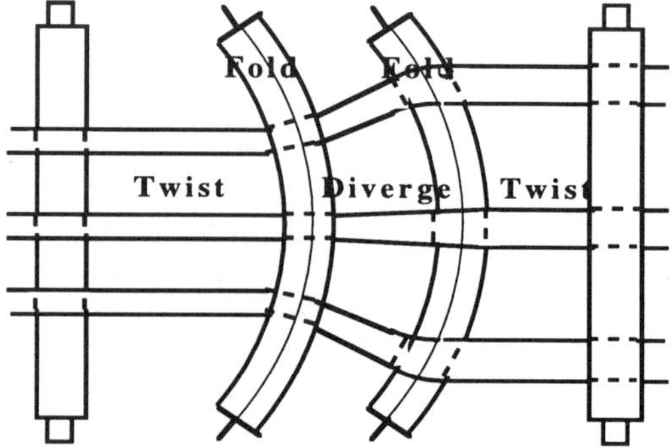

Pos-Z™ Spreader

The Pos-Z™ spreader shares some similarities with the dual bowed roll spreader as seen in Figure 17.14. First, it is a dual curved element spreader. However, this spreader has two stationary air floated bars instead of two rollers. Older designs were drilled fiberglass sleeves which fitted over axles, but now are hardened metal tubes to increase wear resistance. Though the webs are intended to float instead of slide, porous or uneven webs would occasionally contact the bars, even with very large blowers. Second, this spreader is also used primarily after a slitter section which cuts numerous small widths. Third, the dual element spreader is also an expensive spreading system.

However, the similarities end there. First, newer designs are adjusted by several jacks which change the bow magnitude. Second, the spreader bows point upstream and parallel to the incoming and outgoing web runs. Third, while the Pos-Z™ is also a twist/fold spreader, its operation is governed strictly by geometry instead of traction and the normal entry rule. The easiest way to demonstrate its spreading principle is to s-wrap two parallel but angled pencils with a long strip of paper as seen in Figure 17.15. The outward offset results from wrapping the bars on a very slight helix angle. The spreading magnitude of the powerful Pos-Z™ increases with bow magnitude, wrap angle, bar diameter, and bar separation.

The air float nature of this spreader has several disadvantages. First, the low pressure (5 psi) air blown through numerous holes in the bars can float only nonporous low tension (<5PLI) grades. Second, in some applications the air film is unstable, making it difficult to meet industrial noise standards. Third, spreading can be erratic if a tight portion of a web collapses the air film between the web and bar, causing the web to skid sideways slightly in response to the induced moment. Nonetheless, the Pos-Z™ is simultaneously a most powerful yet gentle-to-the-web spreading system.

Figure 17.14
Pos-Z Spreader Setup

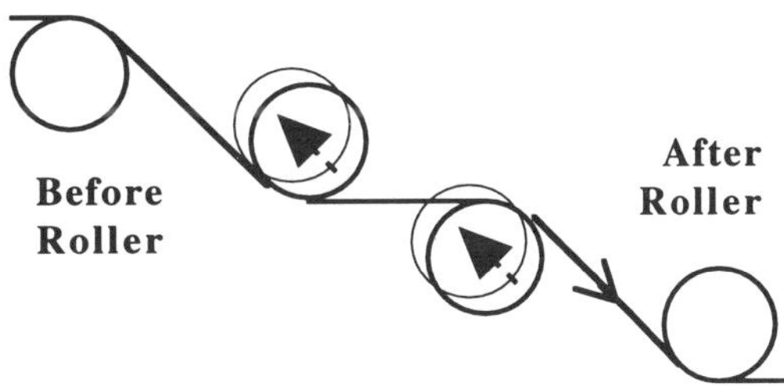

Figure 17.15
Principles of the Pos-Z Spreader

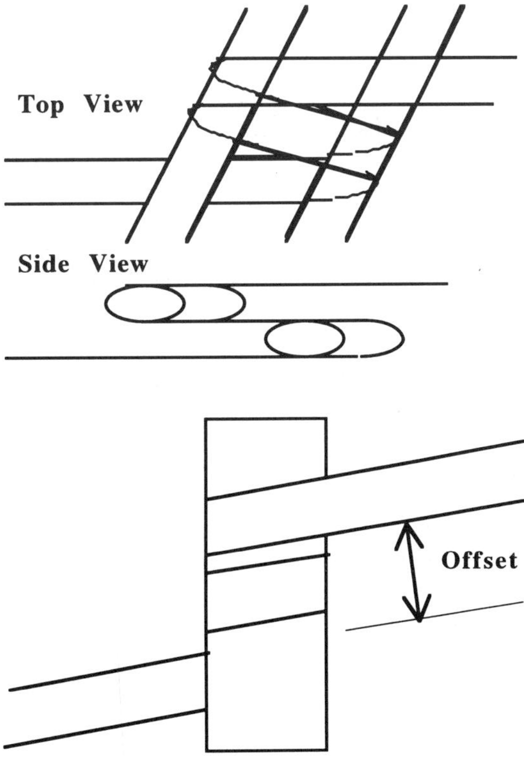

Expander Rollers

Expander rollers have shells or covers that move outward from the centerline as the web transits from the ingoing to outgoing tangents. The simplest version has cam operated half-width slats that slide at their junction in the middle. Another version has numerous elastomeric bands connecting across the machine to adjustable cams on each end. These bands, as seen in Figure 17.16, are arranged to approximate a cylindrical roller. Finally, another variant has a very flexible cover attached to the end cams and supported elsewhere by bristles mounted on a central shaft.

The principle of operation is really quite simple. The web's ingoing tangent is on the short band side, and the outgoing tangent is on the long band side of the spreader. (This is in contrast to the compliant cover roller whose ingoing and outgoing widths are similar.) Grade changes for many expander rollers can be made by either cam orientation or cam side angle adjustments.

While the expander roller is a very powerful spreader, it does have some severe limitations. First, speed is limited to a few hundred mpm before the centrifugal force pulls the elastomeric bands away from the roller. Second, the parts are quite tender compared with most covered rollers so that replacing worn or broken bands could be a maintenance nuisance. Thus, the expander roller is best suited to low speed converting operations on stretchy grades of nonwovens or textiles.

One very unusual characteristic of these spreaders is that they operate successfully in traction, sliding or even transitions between traction and sliding. The performance of other spreaders are destroyed by transitions.

Figure 17.16
Expander Spreader Roller

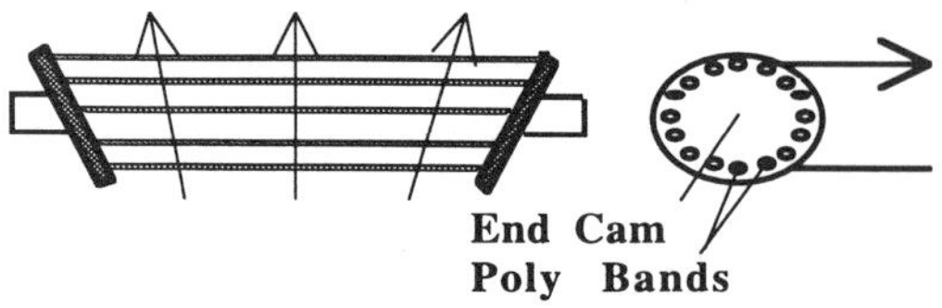

Edge Pull Web Stretchers

Edge pull stretchers are the most powerful spreaders, and are sometimes capable of increasing web width by hundreds of percent. Obviously, strains this large are limited to very stretchy materials such as films, nonwovens and textiles. Indeed, the intent may be to draw the web permanently wider.

Web stretcher rollers are a pair of narrow, soft covered nipping rollers on each edge of the web. They are canted outward, as seen in Figure 17.17, and accomplish spreading by the CD component of the rollers' velocity vector. Thus, they act like the the ends of a bowed roller spreader except that there is enormous gripping or traction. Because the MD component is reduced with misalignment angle, the rollers may need to be driven at speeds greater than web speed. Unfortunately, stretcher rollers impose very violent stresses to the web. This makes it tricky to set up drive speed, cant angle, and nip load to avoid wrinkling or tearing.

A similar edge pull spreader is the tenter, which is an endless track which guides a chain with numerous clips. These clips engage the web at the narrow upstream side, are pulled outward following the shape of the tenter track, and finally released on the downstream side. A primary disadvantage of the tenter is local web distortions near the clips which need to be trimmed away as waste.

Figure 17.17
Edge Pull Stretchers

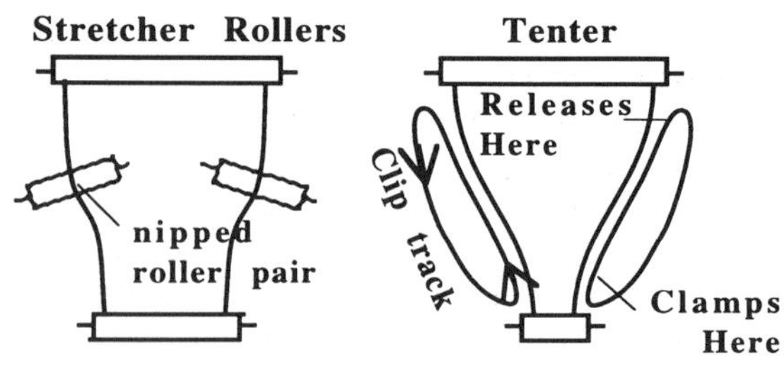

Compliant Cover Rollers

The compliant cover roller, as seen in Figure 17.18, is a straight roller which has a special grooving cut into a very soft outer cover. The grooving is undercut at an angle so that the inward radial pressure created by web tension deflects the lands outward, carrying the web with it to supposedly accomplish the spreading action. It is a very popular spreader on slow and narrow web converting processes because of its economy and simplicity. However, there is no published non-anecdotal evidence that this device is capable of spreading or wrinkle removal. This has led to significant doubt as to its benefits.

The web tension causes a pressure which deflects the lands outward as the web enters the wrap. This outward expansion may continue until it reaches a maximum at the center of the contact area as seen in Figure 17.19. However, guess what happens on the downstream side? Yes, it would appear that, as the pressure is released on the downstream side of the contact area, the land springs back inward carrying the web with it to exactly the same CD position.

This is because the compliant cover roller is symmetrical about the upstream and downstream sides, which is not true of any other spreader. One way the spreader could be unsymmetrical is if the web resists outward motion by slipping on the ingoing half of the contact area, which could easily happen on stiff webs. However, as seen in Figure 17.20, this would tend to cause the web to contract or buckle on the outgoing half of the contact area, which would be highly undesirable. There is also a slight asymmetry to the nip in the presence of rolling friction or other torques, but its effects are not clear.

The other reason the compliant cover roller may not operate as intended is due to the uniformity of CD movement. That is, if each land moves the web outward the same amount, spread would only occur at the center between the innermost grooves. True, there are evolutions of the grooving patterns which increase deflection from the centerline outward, but they are not a common style.

Figure 17.18
Compliant Cover Spreader

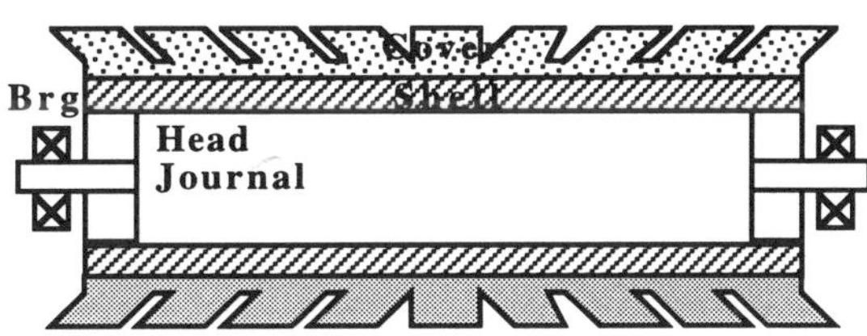

Figure 17.19
Principle of the Compliant Cover Spreader?

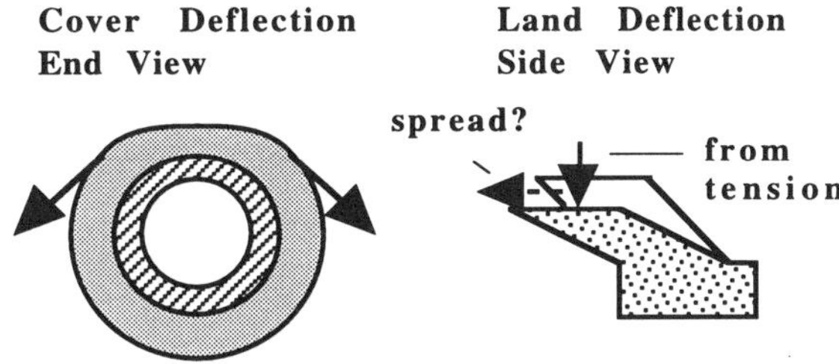

Figure 17.20
Contraction of the Compliant Spreader

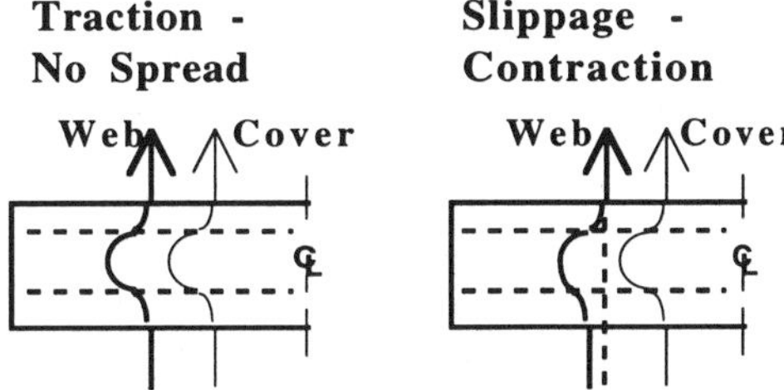

Thus, while the compliant cover may be beneficial (many people in the paper, film and foil converting industries swear by them), we do not see how it functions as a spreader. This is yet another case where both vendors and customers may have failed to do their homework to establish converting equipment performance.

In defense of this device, however, it seems that wrinkles do not tend to form on this roller as they might on a more conventional roller. Thus, it may serve in some web flattening fashion. The distinction between spreading and flattening is spreading <u>forces</u> a web to be <u>wider</u> than it would normally be while flattening <u>allows</u> it to be <u>as wide</u> as it would like to be.

Grooving

There is a near universal fallacy that has led many into the mistaken belief that a spreading function has been provided for. This fallacy is the so-called spreading from spiral grooved rollers. Unfortunately, this perception results from nothing more than a barber pole optical illusion. The spiral grooving, just as the barber pole, has no axial movement of the surface. Thus a web in traction with the spiral grooved roller locks in on the ingoing tangent, is carried around, and finally deposited on the outgoing tangent with absolutely no CD offset or movement. Also, the higher velocity air pumped through the grooves doesn't aid spreading because the aerodynamic forces are truly miniscule.

If there is (undesirable) slippage between web and roller, the spiral grooving may spread or contract depending on the direction of the grooves and whether the roller is undersped or oversped. The most common situation is an undersped roller, which can result from too little drive, wrap and tension, or too much machine speed and bearing drag. In the case of an undersped roller, the center arrow formed by the junction of the two different hand leads should point upstream to provide an outward plow. Unfortunately, this is not the way spiral grooving is usually set up because it gives a contracting illusion. Conversely, the oversped roller should be set up so the arrow points downstream.

So what is the best way to set up spiral grooved rollers? First, leave existing rollers just the way they are because it just doesn't make a significant difference (unless the grooves are too wide). In other words, annular, spiral front, spiral back, spiral center and spiral out will all behave the same in the desirable case of a fully tractive roller. Second, specify new rollers with the least expensive grooving, which is either annular or spiraled in one direction only. Finally, a true spreader may be needed for wrinkle removal, web flattening or slit separation since grooving just doesn't do it.

Figure 17.21
Contraction Caused by Grooving & Raised Threads

Wide Grooved Rollers

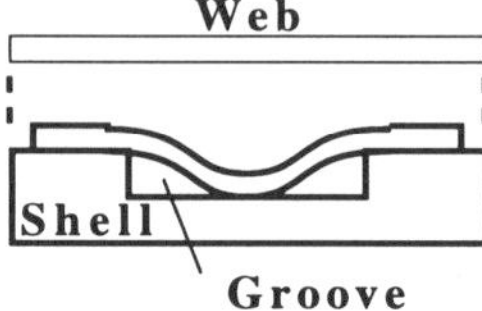

Raised Thread Rollers (aka worm rolls)

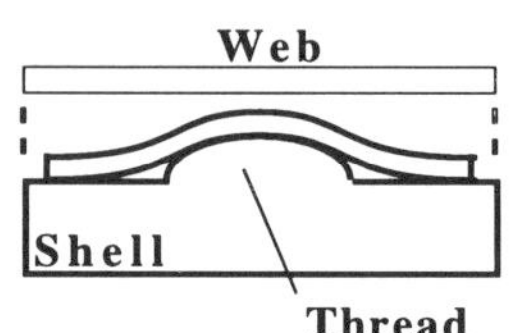

An even more serious issue is the contractive tendencies of all grooved and raised thread rollers. As a web of a given width is wrapped on a roller with grooving or raised threads, the ends pull inward as shown in Figure 17.21. This happens because the web tends to conform around the longer surface contour instead of forming a straight section across the width. The raised thread rollers, also known as worm rollers, are the worst because the web <u>will</u> conform to both the top of the thread and the bottom of the land, whereas a web may not conform significantly to grooves unless they are wide and the web is of light gauge. If the grooving between sectional rollers is too wide however, the web may be sucked into the roller gap until it bottoms out.

A simple estimate of the maximum contraction of a web due to grooves is to calculate the arc length of a half sine wave, which is the traditionally assumed cross sectional shape of a model wrinkle,[18] and then converting that value to a strain as,[19]

$$(17.1) \quad \varepsilon_{trough} \le \left(\frac{\pi \text{ x depth (or height)}}{2 \text{ x width}}\right)^2$$

Sample calculations will show that many of the grooving patterns commonly used could cause more contraction than most spreaders can spread. Furthermore, grooving can leave permanent marks in thin films as the web plastically deforms into the grooves.

To avoid web marking and web contraction:

Grooving should be no more than 10-20 times wider than the thinnest caliper web.

Spreading Is Temporary

For several reasons, the effects of spreading are ephemeral because they dissipate as the web progresses downstream from the spreading system. The first effect is from the necking of the web as it tries to return to its natural width as seen in Figure 17.22. This natural width is determined by

(17.2)

$$w_{natural} = w_{unstressed} \times \nu_{12} \times \left(1-\varepsilon_1\right)$$

$$w_{natural} = w_{unstressed} \times \nu_{12} \times \left(1-\text{tens} \times \text{mod/cal}\right)$$

This web necking is exactly the same phenomenon as specimen necking in a tensile testing machine.[20] Without resorting to measurement or model, St. Venant's Principle indicates all end effects dissipate by ten characteristic dimensions away from a discontinuity. However, the practical reality is that much of the spread is lost in the first web width downstream from the spreader. This is as far as a baggy web can be temporarily flattened, if it can be flattened at all.

A second cause of lost spreading is the random paths taken by different CD positions as the web moves downstream. These random fluctuations, more commonly known as **weave**, are caused by variations in factors such as traction, alignment, and web uniformity. In simple terms this means that without spreading intervention, wrinkling and overlaps will become more frequent as the web moves downstream through a converting process. Also, foldovers must be prevented at the source because they can't usually be cured downstream of their formation.

A common cause of weave is camber, which is a variation in the natural length of a web across its width caused by residual strains from the manufacturing process. A narrow web with a camber problem will have an arced in-plane shape when tension is removed, while a wide web will exhibit baggy lanes and tight bands. In any case, a cambered web will be difficult to keep flat and straight as it moves downstream.

Figure 17.22
Why Spreading is Temporary

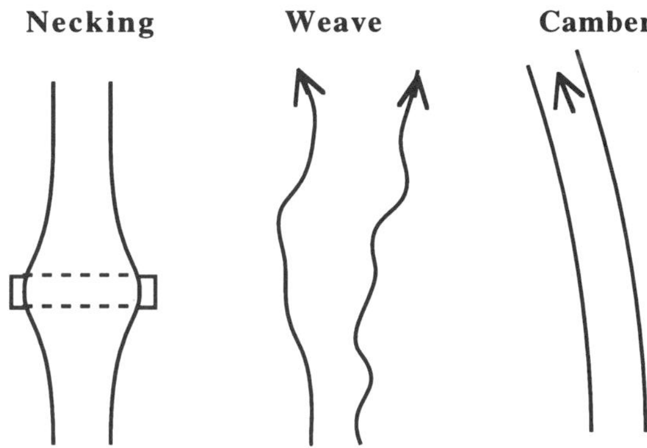

The temporary nature of spreading has three practical implications. First, spreaders should be located quite close to critical processes so that the maximum benefit is retained. Critical processes which demand a flat web include embossing, calendering, laminating, printing, slitting and winding. Note that many of these processes involve nips, because all **nips** are very intolerant of webs that are not flat.

Second, spreaders will often be required immediately after **expansive** processes such as heating, moisturizing, coating, treating or adding solvents.

Third, spreaders may be required elsewhere just to keep the web from getting too **troughed** as it proceeds down the machine. These locations can't usually be predicted beforehand, but can be easily seen once a process is running. Starting at the upstream end, follow the web until troughing gets to the point where it might cross a roller (as a crease etc.), and add a spreader there. Then continue downstream in the same fashion.

Thus, several spreaders may be needed on long machines with several critical processes. Be careful however, too much spreading could cause problems due to the bistable nature of spreading. Because spreaders pull both edges outward, the web will tend to toggle to one side or another of the machine centerline. While toggling is seldom noticed on wide webs, long narrow webs will tend to run off the edges of rollers if there is too much spread.

Bibliography

1. Roisum, David R. *Why is My Web Wrinkled?* Converting Magazine, Web Works column, pp 16, February 1994.

2. Roisum, David R. *The Mechanics of Web Spreading.* 2nd Int'l Conf. on Web Handling, Oklahoma State Univ., June 6-9, 1993.

3. Roisum, David R. *The Mechanics of Web Spreading.* TAPPI Finishing and Converting Conf. Proc., New Orleans, pp 87-108, October 24-27, 1993.

4. Roisum, David R. *The Mechanics of Web Spreading - Part 1.* Tappi J., vol 76, no 10, pp 63-70, October 1993.

5. Roisum, David R. *The Mechanics of Web Spreading - Part 2.* Tappi J., vol 76, no 10, pp 75-86, December 1993.

6. Roisum, David R. *The Well-Proven Principles of Web Spreading.* Converting Magazine, pp 60-64, November 1995.

7. Roisum, David R. *What Causes Machine Direction Trough Wrinkles?* Converting Magazine, Web Works column, pp 22, March 1994.

8. Roisum, David R. *What Causes Diagonal Shear Wrinkles?* Converting Magazine, Web Works column, pp 22, June 1994.

9. Roisum, David R. *What Causes a Baggy Web?* Converting Magazine, Web Works column, pp 22, April 1994.

10. Roisum, David R. *How Can I Fix a Crooked Web?* Converting Magazine, Web Works column, April 1995.

11. Pfeiffer, J. David. *Web Guidance Concepts and Applications.* TAPPI Finishing and Converting Conference Proc., October 1977.

12. Pfeiffer, J. David. *Web Guidance Concepts and Applications.* Tappi Journal, vol 60, no 12, pp. 53-58, December 1977.

13. Delahoussaye, Ronald D. *Analysis of Deformations, Stresses and Forces in Webs Encountering Spreading Rollers.* Ph.D Thesis, Web Handling Research Center at Oklahoma State Univ., December 1989.

14. Delahoussaye, Ronald D. and Good, J. Keith. *Analysis of Web Spreading Induced by the Concave Roller.* 2nd Int'l Conf. on Web Handling, Oklahoma State Univ., June 6-9, 1993.

15. Delahoussaye, Ronald D. and Good, J. Keith. *Analysis of Web Spreading Induced by the Curved Axis Roller.* 2nd Int'l Conf. on Web Handling, Oklahoma State Univ., June 6-9, 1993.

16. Lucas, Robert G. *Better Spreading of the Web in the Winder - and How to Achieve It.* Pulp & Paper, vol 30, no 3, April 1977.

17. Anon. *D-Bar Spreader Instructions.* Beloit Corporation Manual #J 045-10006, October 1973.

18. Friedrich, Craig R. *Stability Sensitivity of Web Wrinkles on Rollers.* Tappi Journal, vol 72, no 2, pp 161-165, February 1989.

19. Shelton, John. J. *Buckling of Webs from Lateral Compressive Forces.* 2nd Int'l Conf. on Web Handling, Oklahoma State Univ., June 6-9, 1993.

20. Seo, Y.B. et. al. *Tension Buckling Behavior of Paper.* Journal of Pulp and Paper Science, vol 18, no 2, pp J55-J59, March 1992.

Chapter 18

Specialty Rollers and Elements

In this chapter we briefly discuss a few specialty rollers and elements not covered elsewhere in the book, yet are commonly found on many types of web manufacturing and converting lines. The chapter begins with transport rollers and follows with process rollers and other important elements such as slitting and static elimination.

Additional Roller Resources

By some estimates, the web industry is the largest industry, surpassing other giants such as chemical and automotive. The reason its size has not been readily apparent is that there is such a diversity of components and products that are made from web materials such as paper, film, foil, nonwovens, textiles and so on. Since rollers are the foundation of all web manufacturing and converting processes, one might expect a considerable amount of information to be available. Unfortunately, the information is scattered, buried and often written in the vernacular of the particular industry it came from. Nonetheless, I offer a few roller resources of which I am most familiar and use regularly.

One source of roller expertise is component and machine **builders**. They are certainly the first place to turn for sizing and equipment selection recommendations, and possibly troubleshooting process problems. What you should be able to count on from a builder is to be able to bring a machine to a like new condition and performance. However, if the problem has been present from the beginning, chances are the builder has yet to find the solution.

There are several **organizations** worth following. Perhaps the most important is the CMM (Converting Machinery and Materials) which has a trade fair every two years. I am like a kid at a candy shop there because of the enormous scope of products and technology.

On a more technical level, the WHRC (Web Handling Research Center) at Oklahoma State University provides research funded by consortium companies, seminars about twice a year and an international web handling conference every two years. Occasionally web handling and converting research is also performed at other universities in support of the paper or printing industries. Finally, CEMA (Converting Equipment Mfg. Assoc.) and TAPPI (Technical Association of the Pulp and Paper Industry) also provide conferences and seminars.

The most widely accessible resources are **magazines** which everyone should follow regularly. My favorite (because I am one of its consulting editors), is the *Converting Magazine.* In all fairness, however, the *Paper, Film & Foil Converter* also has similar fare. A slightly more technical though still readable resource is the *Tappi Journal.* While there are a great many other trade magazines, most do not cover rollers frequently or in any depth.

There are only a few **books** which contain significant roller material. First, there is a series of four books written by the late Herbert Weiss and published by his company.[1,2,3,4] Second, there is an excellent overview of many types of converting machines by Donatas Satas.[5] Finally, there are several good books from the printing industry that speak about rollers.[6]

Finding information about rollers from a computer **database** search can be difficult because of all the different aliases or keywords which are used. The most common, of course, are roll and roller. However, "roll" can also mean a wound roll or coil. Other common aliases, such as drum, spool and mandrel, do not even carry the roll syllable.

Nonetheless, the reader is encouraged to look much further than the mere contents of this chapter or even of this book because there is much more out there.

Air Bars

One purpose of air bars is to reduce sliding friction over a stationary element. Specific application examples include folding bars, right angle edge guides and Pos-Z spreaders, all of which require a non-rotating element as opposed to a roller.

There are three requirements to float a web completely above a surface. First, the pressure under the web must be at least T/r where T is the lineal web tension and r is the radius of curvature. In general, this will be a few psi, and positive displacement blowers will be more cost effective than compressors for larger devices. Second, the air volumetric flow rate must be sufficient to make up the leakage on all four sides of the web (upstream, downstream and two ends) and through the web. Areas not under the web are usually blocked off so that no air is lost there. This requires estimating the required flotation height (perhaps 4X or more of surface roughness for flat webs, more for baggy webs). Finally, the hole spacing must be sufficiently close together. An ideal may be sintered metal tubes or porous plastic of the proper permeability.

Another use of air bars is to float a scratch sensitive web. The most common example is the bars in an air flotation oven. As shown in Figure 18.1, the bars are roughly rectangular in cross-section with air slots on either side. The bars in an oven are configured so that the web path is a mild sinusoid to improve effective in-plane bending stiffness and thus tracking. There is some science[7,8,9,10,11] accompanying considerable art[12,13,14,15] in the design of air bars, shoes and pans. Additionally, there is a lot of work in the design of web processes to avoid **flutter**.[16,17,18,19]

Figure 18.1
Bars in an Air Flotation Oven

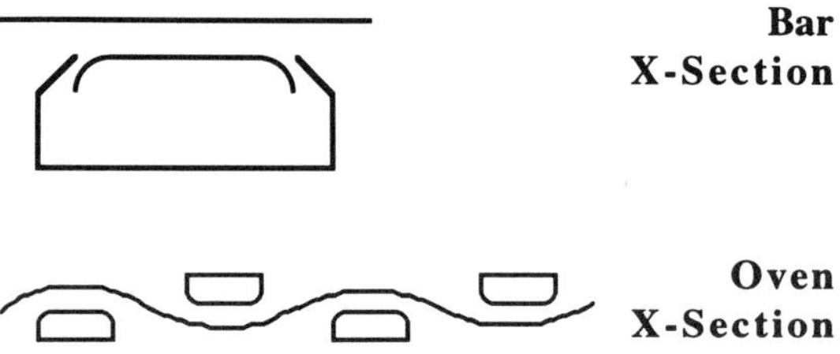

Vacuum Bars

Vacuum bars, rollers, pans and shoes may be used for a variety of purposes. A common application for vacuum rollers is to increase traction on a driven roller without adding a nipping roller. The sizing for this was covered in Chapter 4. However, vacuum rollers must be scrutinized closely because they are very expensive. First, is a drive point really needed there? Second, can traction be made sufficient by other means such as wrap angle or coefficient of web-roller friction? Finally, is the location appropriate? A common design mistake is to place a vacuum roller at the end of a long air flotation oven, and just upstream of an edge guide. A vacuum roller there will often compromise the guiding function, and it would have been better to pivot the vacuum roller as a guide instead.

Vacuum rollers may also be used to draw air through a web (drying) or to pull a liquid out (paper making). These applications require fluids and thermodynamic considerations in addition to the web handling ones.

Finally, vacuums may be used to pin or hold a web or tail against a bar, roller or pan. For example, a web tail can be temporarily secured by vacuum during a splice. The holding power is the vacuum times the acting area times the coefficient of web-element friction.

Vacuum systems tend to be quite noisy. First, the blower produces so much noise that it may need to be located outside of the building or enclosed in a sound-proof doghouse. Second, thin metal rectangular ducting can be quite noisy unless flat panels have shape stiffeners to prevent oil canning and drumming. Finally, the vacuum holes or slots can produce high pitched siren and whistle noises. The shape of the hole may need to be rounded instead of sharp-edged to avoid whistling. Indeed, a drilled shell can be noisy when rotating even if the vacuum system is off.

Concave Rollers

Concave rollers, as seen in Figure 18.2 have a slightly smaller diameter at the center than at the ends. The concave roller is also known as a bowtie roller or a reverse crown roller.

The primary application of the concave roller is as a mild spreader[20] as described in Chapter 17. In this mode, the center should be cut down only about 10-50% of the strain of the web to be run. A relatively inextensible material such as paper may have a breaking strain of only about 1%, and may commonly be run at a web strain of perhaps 0.2%. Thus, the diameter should be cut down 0.02% to 0.10%, or 2-10 mils on a 10 inch diameter roller. Roller diameter tolerances are important because inadvertent variations as small as a couple of mils can cause large, erratic and undesirable effects (gathering at the high diameter locations) on inextensible webs.

The concave roller will act as a spreader if the surface is in good traction, but causes a tendency for bistability of web path. Opposite effects occur if the concave roller is slipping, so that the roller will tend to cause a CD contraction of the web and the web path to seek the center of the roller.

Figure 18.2
Concave Spreader Roller

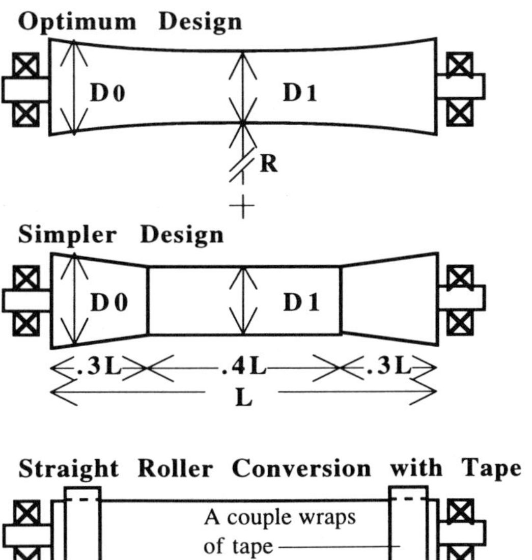

Crowned Rollers

Crowned rollers, as seen in Figure 18.3, have a slightly larger diameter at the center than at the ends. A roller may be crowned to even out a deflected nip or to passively center guide a thick belt.

In a nipped application, the crown is calculated as the beam deflection of the rollers as given in Chapter 3. Crowns are quite small for properly sized rollers. Thus, even a 100" wide roller with a Class B deflection should deflect at most 15 mils under load, thus putting a maximum crown of only 15 mils on this long slender roller. However, it is extremely difficult to accurately cut crowns less than about 2 mils. Note from the figure how the crown will fit for only a specific value of nip loading.

Another application for crowning rollers is to passively center guide a web. However, the web must be very thick and/or stiff to withstand the resulting compressive CD stresses without buckling (MD wrinkles). Thus, crowned guides are used for applications such as endless belts and in the metals industry.

Figure 18.3
Crowning for Nipped Roller Deflection Compensation

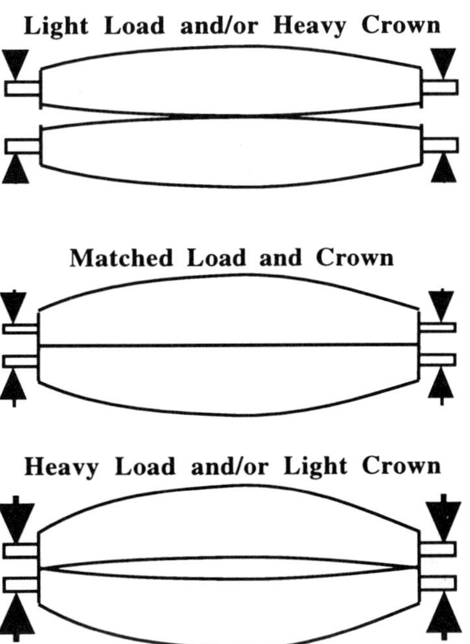

Edge Guides

Guides are used to bring the edge or center of a web to a specific CD position.[21,22,23,24,25] Sometimes the guide may only need to correct the web path to keep it on the rollers after it has traveled some distance. In other cases, CD registration may need to be as fine as a few mils to register plies of a laminate.

Passive guides do not require sensors, controllers and actuators.[26] As shown in Figure 18.4, passive guides will bring the web toward center merely by virtue of their shape. However, these guides are not always suitable for thin, weak or flexible webs.

Active guides systems are composed of a sensor, an actuator, and a controller. The sensor can be any detector which can reliably pick up the edges of a web. The most common are pneumatic (nonporous webs), photoelectric (opaque webs) or paddles (thick webs). If two sensors are used, the web could be guided to the front edge, back edge or center. The output of the sensor goes to a controller which outputs a movement of the actuator. If the gain of the controller is too low, the response of the guide will be sluggish and slow to correct. If the gain of the controller is too high, the guide will be hot but overshoot or may even be unstable. The actuator which moves the guide mechanism may be a stepper motor and ballscrew for smaller assemblies or a hydraulic cylinder for larger assemblies.

The **steering guide**, as shown in Figure 18.5, is a roller which has a controlled misalignment in the plane of the incoming web. Any web edge position error picked up by the sensor located very near the guide roller will cause the controller to pivot the roller. As described in Chapter 12, the web will follow a path conforming to normal entry into the cocked guide roller. This misalignment will move the web back toward center. Unfortunately, this will also cause web bending stress to be superposed onto web tension after the upstream roller.

Steering guides are common whenever there is a long entry span, such as following an oven. Indeed, this type of guide requires the entry span to be at least 3 (if not much more) web widths long to minimize excessive bending stresses.

Figure 18.4
Passive Guides

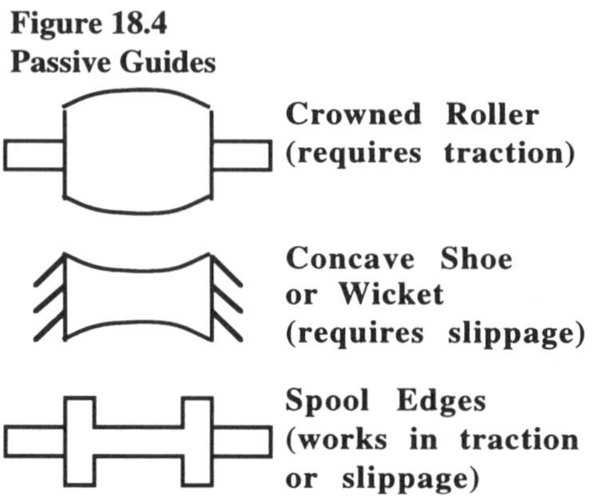

Figure 18.5
Steering Guide

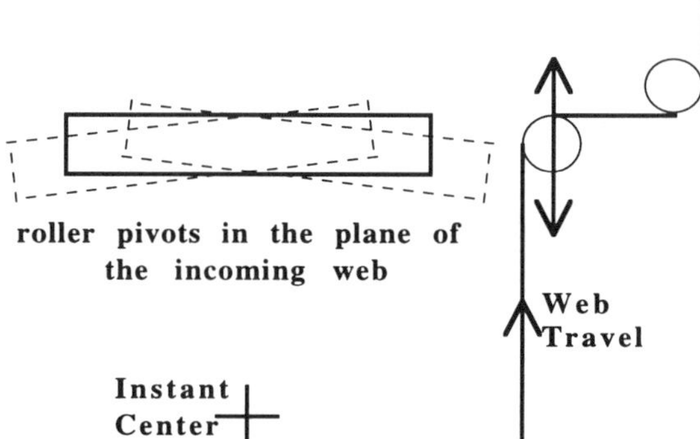

Other requirements for a healthy steering guide setup is to allow at least 1/2 web widths downstream of the guide to accommodate the out-of-plane web twisting. Finally, the pre-entry span must be shorter than the entry span.

There are two types of guide movement. The simplest is to simply pivot the roller from one end, which may be adequate for some endless forming belts.[27,28,29] However, a better response can be achieved by locating the instant center about 1/2-2/3 of the entry span upstream. This arrangement causes the roller to also move sideways, thus carrying the web with it. An upstream instant center is achieved by mounting the ends of the rollers on linear bearings canted at a slight angle. The static and dynamic behavior of these web guide systems is truly a science due to pioneering work by Dr. John Shelton[30,31,32] and those who have followed.[33,34,35,36]

When a sufficiently long entry span is not available, a **displacement guide** may be a better choice. The displacement guide, as seen in Figure 18.6, is a system of four rollers. The first and last are fixed, while the center two rollers pivot on a common frame about the center of the upstream roller. The web sensor must be located very close to the exit of the 3rd roller.

Figure 18.6
Displacement Guide

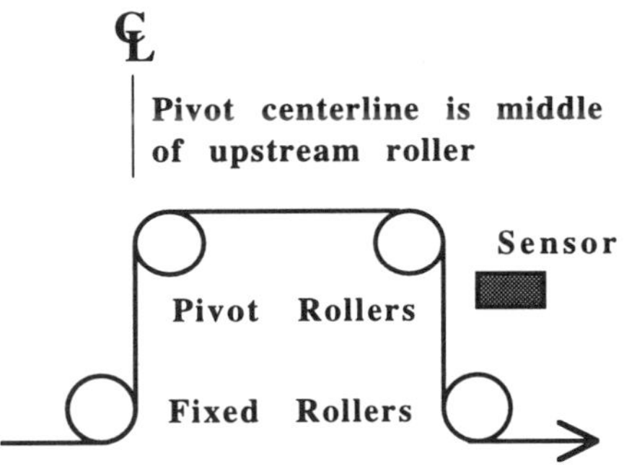

While this is the most common depiction found in advertisements, the system can be arranged in at least 16 distinct ways, each of which have four orientations. These numerous equally effective possibilities allow a great flexibility in designing a machine. The only real constraints are that the spans entering and leaving the pivoting table must be at least 1/2 web widths, but the three legs can be of different length. Note that the length between the pivoting rollers determines the effective sidelay range for a given twist of the web.

Figure 18.7
Unwind Guide

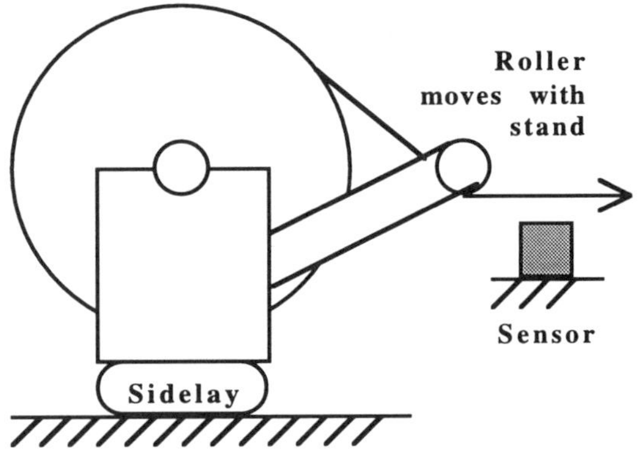

There is an important difference between the behavior of the displacement guide and the steering guide. While the steering guide will move the edge of the web upstream and consequently also downstream of the guide, the displacement guide will only move the downstream web path. Thus, the upstream process must be tolerant of uncontrolled web CD position.

Sometimes it is important on roll-to-roll web lines to start the web down an appropriate path despite poor edge quality of the unwinding roll.[37,38] In this case, an **unwind guide** would be used as seen in Figure 18.7. The unwind guide consists of an unwind which can be moved sideways on linear bearings, a following roller which (desirably) moves with assembly, and of course, sensor, controller and actuator.

Figure 18.8
Winder Guide

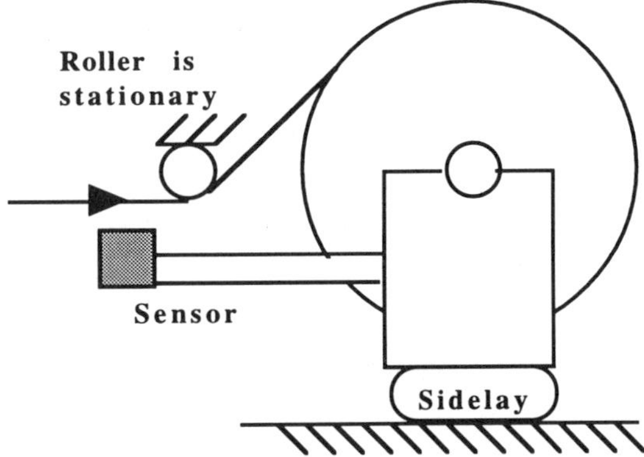

Sometimes, straight edged rewound rolls are desired despite edge problems with the unwind rolls or the web path between. In this case, often a salvage operation, the rewind may be guided. As shown in Figure 18.8, winder guiding system closely resembles unwind guiding. In fact, the only substantial difference is that the edge sensor is fixed to ground on the unwind guide and fixed to the winder frame on the **winder guide**.

Footage (Length) Counters

Length is a commonly advertised or specified feature of a web product. While length is easily measured, it is more difficult to define.[39,40] This has led to numerous disagreements between suppliers and their customers.

Length is almost invariably measured by counting revolutions of a wheel or roller traveling at web speed. When in traction, every revolution of the roller means that (Pi x Diameter) of web has passed. All that needs to be done is to reset (zero) the counter prior to run, and read it after the run is completed.

The wheel mounted footage counter is perhaps the least expensive option and is ubiquitous on older machines. However, it suffers from many accuracy problems. First, the rubber surface of the wheel will wear to smaller diameters, making the resulting length measurements erroneously larger. Second, the rubber compresses so that the footage may read lower with higher nip loadings. Third, the edge of the wheel needs to be ground to a special arc to reduce the effect of camber alignment errors. Finally, there will be a reduction of footage with toe-in alignment errors. More pragmatically, however, the wheel mounted footage counter can be a nuisance. It tends to get in the way of operator access and threading, and doesn't always hold up well in plant environments.

A better method of length measurement is to count revolutions of a roller in traction. Revolutions can be counted by (direct) drive motors or an incremental encoder if the live shaft roller is so equipped. The encoder has the advantage of higher resolution, but is slightly more complex or expensive. Alternatively, any roller can be fitted with a metallic (for inductive proximity switches) or optical (for photoeyes) target. These small sensors are best mounted on the ends rather than the OD of a roller to minimize interference. The only issue with roller footage counters is to make sure that the web is always in traction on that roll, and that roller diameter is measured to a few mils or whatever resolution is required. The effective diameter is the roller's (or wheel's) physical diameter plus web caliper.

Defining length is a bit trickier however. With most products, dimensions are specified under no stress. A common example is a 3" rubber band which is the advertised length when laying on the table under no load, but is capable of expanding to at least twice that during use. However, suppliers of roll products will usually measure length under web line tension, while their customers may measure it under a different tension, or under no tension at all. This difference in length definition, as well as errors in the measurements themselves, contribute to discrepancies between suppliers and their customers.

However, there are two other factors that change length elastically. The first is temperature changes. A length measurement coming out of an oven will be longer than the on the same product when cooled. The second is moisture/solvent differences between shipped and received products. Paper and some polymers will swell significantly with increasing moisture/solvents.

All of these factors can be calculated by the following equation which accounts for elastic strain (length) changes due to tension, temperature and moisture. These correction terms, in some cases, can be as much as several percent. In order to use this equation, however, we need to know the material's modulus and coefficients of hygrothermal expansion.

Thus, accurate measurements of length necessarily requires accurate sensors. However, this may not be sufficient for tight tolerance specifications. We may have to accurately control and factor out changes in tension, temperature and moisture.

(18.1)

$$\frac{L_B}{L_A} = 1 - \left(\frac{\Delta \mathrm{Tension}_{A-B}}{E_{11}\ \mathrm{Caliper}} + \alpha_1\ \Delta \mathrm{Temp}_{A-B} + \beta_1\ \Delta \mathrm{Moist}_{A-B} \right)$$

Spring Mounted Rollers

Sometimes you will find idler rollers mounted with some type of independent spring mechanism at each end as seen schematically in Figure 18.9. In the paper industry, the spring may be an automobile tire where the roller is mounted in the hub and the tire's outside is restrained by a circular clamp.

The rationale behind using spring mounts is to allow the roller to deflect more on the high tension side of a web and vice versa. This might then even out tension profiles caused by baggy webs, misalignment or other causes. While it may accomplish this, the result is crude, local and quite temporary. Furthermore, the spring mounted roller can cause several problems.

First, the roller is difficult to align unless the springs are exceedingly stiff (which is somewhat self defeating). Second, the roller will induce tension profile disturbances if the web is run even slightly off center or if the springs are slightly different. This is because the roller becomes effectively misaligned. Third, the spring mounted roller is destabilizing to web path. If, for example, a strip runs slightly to one side, the forces will be higher there causing the strip to want to run even further to that side.

A roller need not be mounted on springs to behave similarly. If framework is flimsy, or cross ties on pivots or slides inadequate, the same undesirable results may be achieved.

Figure 18.9
Spring Mounted Roller

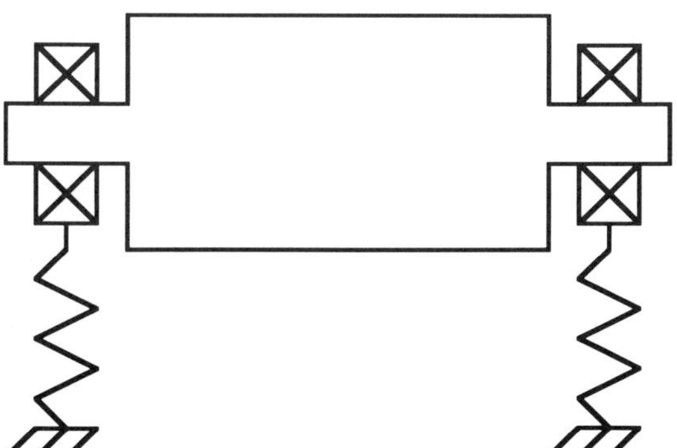

Skewing or Squaring Rollers

Often you will find idler rollers which have one end that is moveable via handcrank, screw or similar mechanism. The direction of adjustment may be oriented in a number of ways with respect to both the incoming and outgoing web paths as shown in Figure 18.10, although the first is most effective for tension balancing.

The usual justification of these types of roller adjustment is to balance web tension of the front with respect to back in order to even out tension profiles caused by baggy webs, misalignment or other causes. In most cases, however, these devices cause more tension upset than they cure. Also, the effects are crude, local and temporary. End adjustment is also used in niche applications such as to move the edge of a narrow web sideways (much like a guide). The reader should refer to Chapter 12 on alignment to find out more how the web behaves under these adjustments.

Several factors must be in place in order for an end pivot adjustment to be effective in balancing tensions. First, the device must be located just upstream of a critical location because the effects (much like spreading) are quite local. In most web lines, this would probably be but a single location. Second, there must be a load cell in proximity which can read and output the difference of front and back tensions independently (and the web must be centered or center position must be accounted for). Even the best operators can't read tension profile well by sight or feel, and thus the need for a tension difference indicator. Third, the roller must have an aligned home indicator accurate to a few mils (not a sheet metal pointer). The gain must be such that a fine adjustment can be made over the small required working range. Obviously, the control movement is via a no-backlash displacement rather than force.

Figure 18.10
End Adjustable Skewing or Squaring Rollers

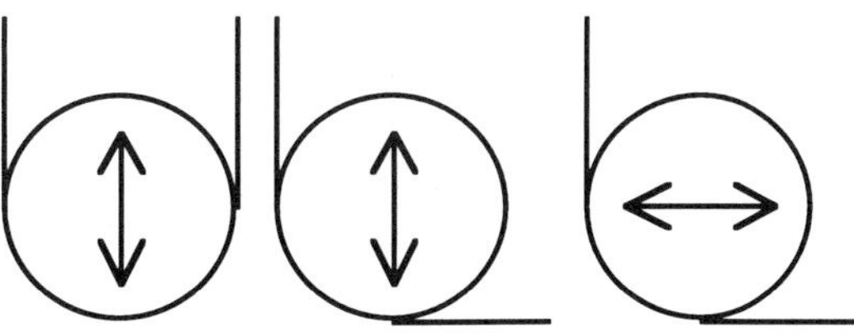

Calendering

Calendering is the process of running a web between two nipped rollers and which is used for a variety of purposes in all of the web industries. As described in Chapter 5, the rollers are often loaded together with a cylinder as shown schematically in Figure 18.11. Precision control of nip loads and roller cylindricity are vital.

The most common effect of calendering is a decrease in web thickness or caliper. Other effects, given in Table 18.1, include an increase of ZD bond strength and surface smoothness of fibrous materials. Also shown here are parameters that commonly affect the strength of the effects. These parameters are often interchangeable over a range of operation. For example, a decrease in nip load can be compensated for by an increase in temperature, nip width or a reduction in speed.

Multiple nips can increase the total calendering effect, though at ever diminishing returns. Multiple nips can be achieved by a tandem arrangement of two-roll calenders or as a calender **stack** as shown in Figure 18.12. The calender stack has the advantages of a reduced roller count and economics of heating/cooling. However, the stack has a more complex framework and loading mechanism.

Figure 18.11
Calender Schematic

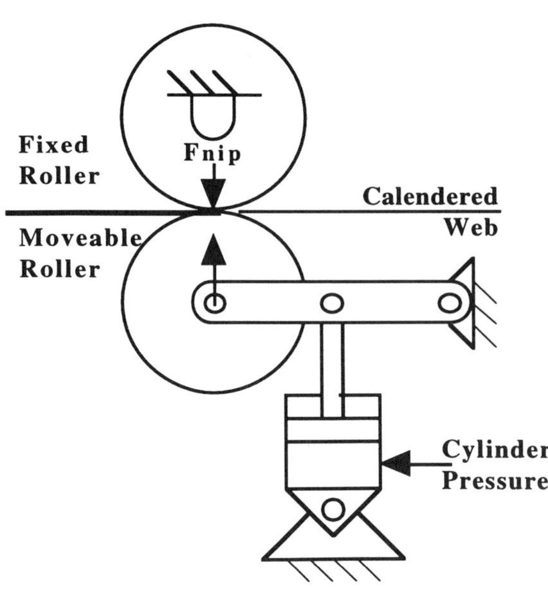

Table 18.1
Calendering Summary

Calendering Effects:
- bulk and caliper decrease (most webs)
- bonding (fibrous webs and lamination)
- gloss increase (fibrous webs)
- smoothness increase (fibrous webs)

Effects Increase With:
- nip load increase, gap decrease
- cover hardness increase
- multiple nips
- roller diameter decrease (small effect)
- speed decrease
- temperature/moisture/solvent increase

Modes:
- metal-metal (intolerant on thin webs)
- metal-soft or soft-soft (covers, filled)

Control:
- fixed gap (intolerant on thin webs)
- fixed load (most common)
- crown
- temperature

Figure 18.12
Multiple Nips Via Tandem or Stack Calenders

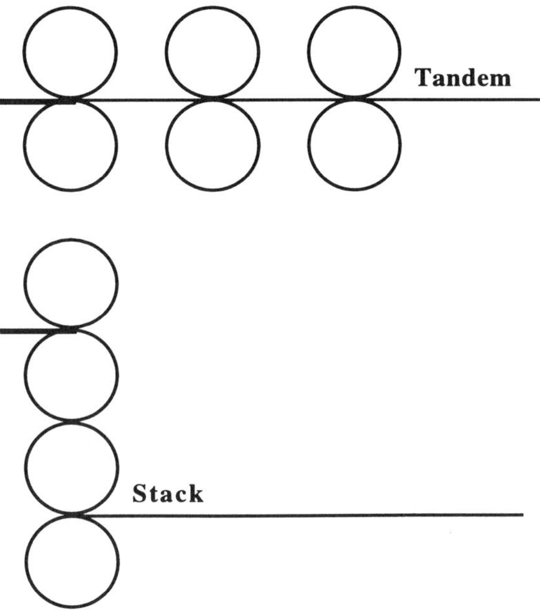

In the **converting** industry, calenders are most commonly used to laminate or bond similar or dissimilar materials together. One example would be labels which are laminates of a printable paper on a release paper. Other examples include countertop and flooring which are stacks of various dissimilar materials. The laminate bond may be achieved by a combination of nip pressure, temperature/moisture and/or adhesives. One of the biggest concerns with lamination is avoiding product curl which is covered in the next section.

In the **metals** industry, calenders are called rolling mills. Their purpose is to take a thick material and roll it to a thinner finished dimension. The metal may be heated for metallurgical purposes or to reduce the required nip forces. Hot or cold, however, metal rolling nip loads are huge. This may necessitate elaborate means of minimizing caliper profile errors due to roller deflection. For example, profiling wheels might be loaded against backup rollers which load the work rollers. Alternatively, the backup or work rollers might be controlled crown rolls.

In the **nonwovens** industry, calenders are commonly used to consolidate and bond polymer fibers together. As manufactured in air laid or wet laid or other processes, the nonwoven web tends to be very fuzzy, something like the texture of a cotton ball only much thinner. While this may be desirable for some filter and personal care products, consolidation and bonding may be needed for other products such as disposable garments and coverings, textile substitutes, upholstering and packaging.

While calendering will decrease the bulk and caliper of the web, it will also tend to make thickness more resistant and stable to creep under subsequent loading such as seen during winding, packaging and handling. It will also tend to increase tensile strength and related properties.

In the **paper** industry, calendering is used on most magazine grades (LWC or light weight coated) paper to impart a gloss or smoothness to the surface in order to improve printability. Calendering may be used on other grades for similar purposes and on multi-ply tissue for bonding. There are almost uncountable empirical articles on paper calendering by researchers from paper companies and machine builders, pulp and paper universities and independent research organizations. Gloss is measured optically with a light source and detector configured in diffuse, reflected or specular modes. Smoothness is something of a misnomer because it may be measured as the rate of escape of air around the edges of a pressurized open ended cylinder nipped against the web.

Calenders may be located inline with the paper machine, offline on separate calendering assemblies, or inline of rewinders. Calenders in the paper industry tend to be very full featured systems which may include basis weight, caliper and moisture web scanners; separate end load controls and crown compensation for gross profile control; and temperature controls for fine profile control.

Calender rollers are metal, covered or filled, but in paper are almost always in load rather than gap control. Paper calender nip loads are usually in the 200-2000 PLI range. **Filled** rollers are made with either textile or paper disks which are slipped over a mandrel and compressed axially to enormous loads, making the surface almost bone hard, which is then finished on a lathe/grinder.

Supercalenders are extremely large offline calender stacks, often supplied in pairs following a single paper machine, and usually follow a rereeler which edits out paper defects which might damage the super. **Soft nip calenders** are covered rollers nipped against a metal roller and are largely supplanting supercalenders because of reduced equipment and operating costs.

Laminator Curl

With few exceptions, the product coming out of a laminator is desirably flat and free of curl. Controlling curl requires that the ingoing webs are matched in both MD and CD strains. This laminator design principle of strain matching is based on the observation that if the webs are evenly stretched in both directions, they will contract evenly when process tension is removed.[41]

However, if the MD strains are not balanced, then the product will curl into a trough that is oriented in the CD as seen in Figure 18.13. In the absence of gravity or support stresses, the trough will be an arc of a circle. The requirement of MD strain matching can be converted to process tension settings for the unixial tension case (of no CD stresses) as:

$$\frac{T_A}{T_B} = \frac{t_A E_{1A}}{t_B E_{1B}} \tag{18.2}$$

where T is the unit tension, t is the web thickness, E is the MD modulus and the subscripts A and B refer to the two constituents. A practical difficulty results when laminating constituents that have widely different moduli such as film and foil.

A similar curl problem can result if the CD strains are not matched as seen in Figure 18.14, where the trough is oriented in the MD. However, in this case the CD strains are the result of Poisson contraction. The requirement of CD strain matching can be converted to process tension settings for the unixial tension case (of no CD stresses) as:

$$\frac{T_A}{T_B} = \frac{\upsilon_{21B}}{\upsilon_{21A}} \frac{t_A}{t_B} \frac{E_{1A}}{E_{1B}} \tag{18.3}$$

where ν_{21} is the anisotropic Poisson contraction in the CD resulting from a MD stress. Comparing the two strain matching equations shows that an exact MD and CD strain match can only occur when the two material's Poisson ratios are identical.

Figure 18.13
MD Strain Mismatch During Lamination

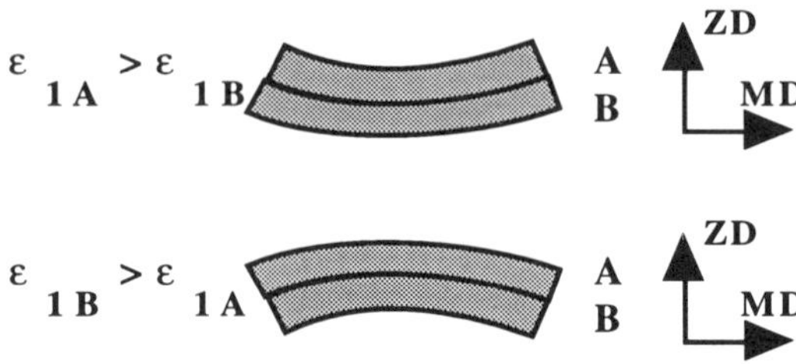

Figure 18.14
CD Strain Mismatch During Lamination

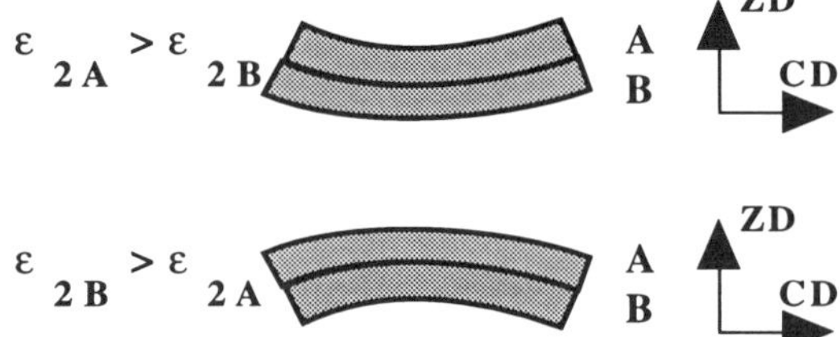

Two other laminating constraints result from the requirements of web tensions being tight enough to avoid floppy problems and loose enough to avoid damage as was discussed in Chapter 2.

$$10\% < T_A / T_{Ayield} < 25\% \tag{18.4}$$

$$10\% < T_B / T_{Byield} < 25\% \tag{18.5}$$

1. Draws do not provide strain information and thus can't be used for design calculation.
2. There may be small CD contractions resulting from constrained expansion, CD expansions resulting from spreading in the vicinity and ZD stresses in the nip that can complicate this simplified design procedure.
3. Product curl resulting from strain mismatch can be calculated from elementary composite beam theory (the classic reinforced concrete beam).
4. Curl that is not uniform (such as just near the edges) has other causes.
5. Once curl has formed, the only way to flatten the product is to bring it to sufficiently high temperatures (film), moistures (paper) and/or stresses (bending over a sharp radius) to yield and reset the material.[42]

Embossing

Embossing, much like calendering, is the process of running a web between two nipped rollers. The only significant distinction is that one or both of the rollers is intentionally patterned. Indeed, the machinery, controls and effects are quite similar to calendering except that embossing tends to increase bulk and caliper rather than decrease. However, there is very little published information on embossing so that the process is very much of an art form.[43,44,45,46]

There are several reasons why an additional embossing step might be included in a converting operation. Perhaps the most common is to impart a visual pattern to a product to improve the aesthetics. Examples here could include synthetic upholstery and furniture laminates which might simulate the grain of leather or wood respectively. Flooring and other architectural products may be embossed to simulate a natural material, or to reduce reflected gloss. Certain tissue grades or other materials may be embossed to simulate a textile texture, and thus increase its perceived value. Other patterns might be used to create product identity via trademarked designs, lettering or logos.

Other reasons to emboss a web, particularly for tissue products such as paper towels, include increasing the bulk and associated properties such as softness and absorbent capacity. However, tensile strength may be reduced. Other functional reasons may be to bond laminates together at selected locations instead of overall to allow easy separation or decrease bending stiffness.

Patterns can be imparted to embossing roller(s) via any of the methods given in Chapter 10 such as conventional tooling, acid etching and laser engraving. In addition, blank filled rollers can be patterned by running them against a male patterned metal roller. **Filled** rollers are made with either textile or paper disks which are slipped over a mandrel and compressed axially to enormous loads, making the surface almost bone hard, which is then finished on a lathe/grinder. The two rollers on male-female systems such as this may need to be externally geared together to preserve MD registration of the matching rollers.

The three major embossing modes are shown in Figure 18.15. Soft covered rollers may need to be cooled to increase life. The **nested** or matched mode may be capable of the deepest patterns, but the rollers must be closely registered (MD through gears, CD, and ZD engagement) so that the patterns do not touch and cut the web. The **pin-to-pin** method is primarily used for selective bonding.

As seen in Figure 18.16, a product may look less streaky and roller life might be extended by minimizing the CD alignment of elements. However, there is no need to be concerned about using whole number ratios of roller diameters. Minute differences in diameter will cause the rollers in nongeared systems to go in and out of phase so that distress on one roller is not transferred to the same place on the other.

Finally, embossing (and calendering) are kinematically equivalent to stamping as seen in Figure 18.17 because rolling action is negligible. Knowing this allows one to quickly prototype products on a stamp or press rather than using web equipment, and effects go as log time. More advanced studies can be done with nonlinear finite element modeling similar to what has been done with metal forming.

Figure 18.15
Embossing Modes

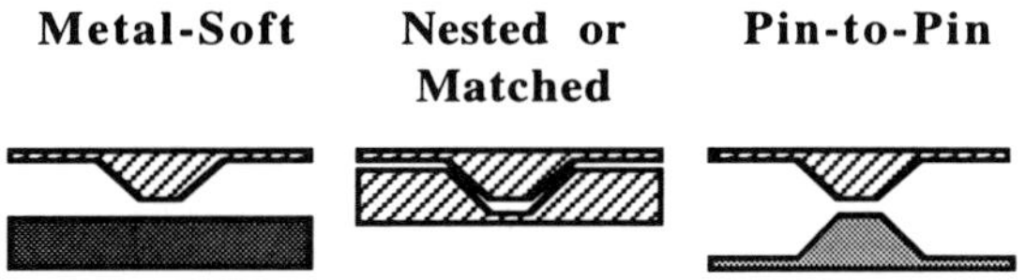

Figure 18.16
Emboss Pattern and Roller Design

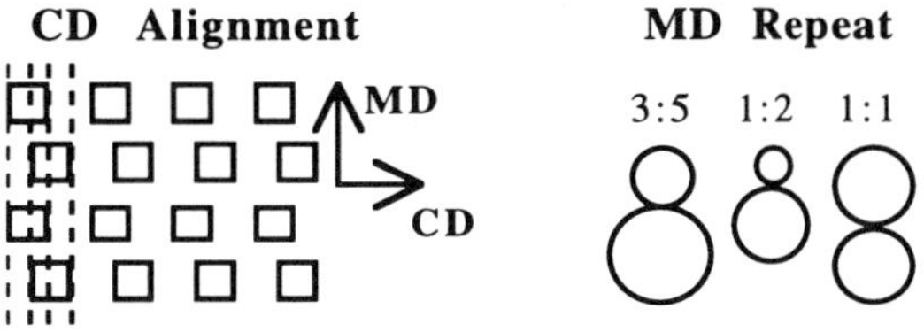

Figure 18.17
Kinematics of Embossing

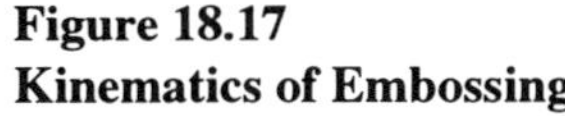

Slitting

Most webs are slit during the course of manufacturing, if for no other reason than to remove the edges which aren't perfect. However, the usual motivation for slitting is to produce the widths required by the customer, which are in general much narrower than can be manufactured economically.[47]

Slitting should sever the web completely and cleanly. The quality of the cut is defined as the absence of roughness, hairiness, fuzz, overhung fibers or dust.[48,49,50,51] Slit quality can be measured with a microfiche reader calibrated with a centerline and rejection limits.

The quality of the slit depends on a number of material factors, the method of slitting, as well as setup and maintenance of the slitters. Material properties which can make slitting more challenging include abrasiveness, bulk (thickness), bulk (inverse of density), brittleness (web breaks) or plasticity (extrude rather than cut). Also, laminates or webs with adhesives can be difficult. Finally, slitting at high speeds or slitting very narrow webs demands precision.

Blade life is another continual slitting issue. Often, cuttability comes at the cost of life. In other words, the harder you slit the sooner the blades will need to be replaced. Thus, time for slitter setups, blade changes, and maintenance must be considered in the economics of slitting. Surprisingly, the initial cost of slitting equipment is similar to most of the methods, and tends to be low in comparison with ongoing costs. The major slitting options are given in Figure 18.18.

One very important design consideration for slitting machines is where to place the slitting units. The answer to this is always the same: as close as possible to the windup or other critical process it feeds. Slitter sections more than a few feet from a windup will never produce sharp looking roll edges because the lanes will have ample opportunity to weave and wander.

Finally, slitting safety must be carefully observed. Obviously, a slitter which can cut a web can also cut operators. At a minimum, slitters are guarded and labeled. Also, gloves are used whenever handling blades.

Figure 18.18
Slitting Methods

Log or Roll Saw **(not shown)**
for small diameter rolls

Razor Blade
for thin (plastic) films

Razor Blade

Web

Blade Holder

Score (Crush) Slitter
adhesive tape & tissue.

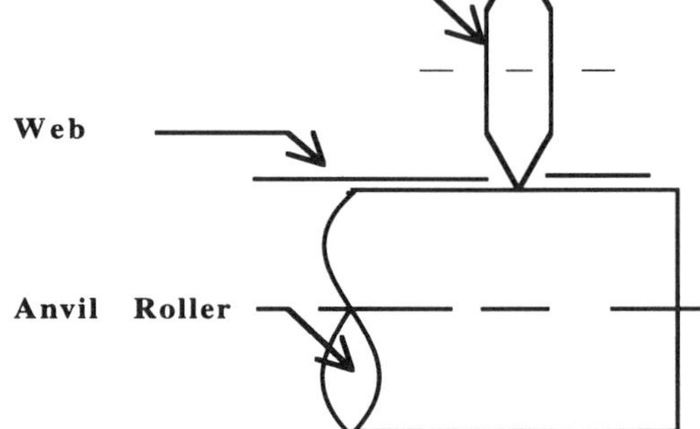

Shear Slitter
most common slitter.

Top Blade

Web

Bottom Blade
or Band

Water Jet Slitter ***Laser Slitter***
difficult materials **special applications**

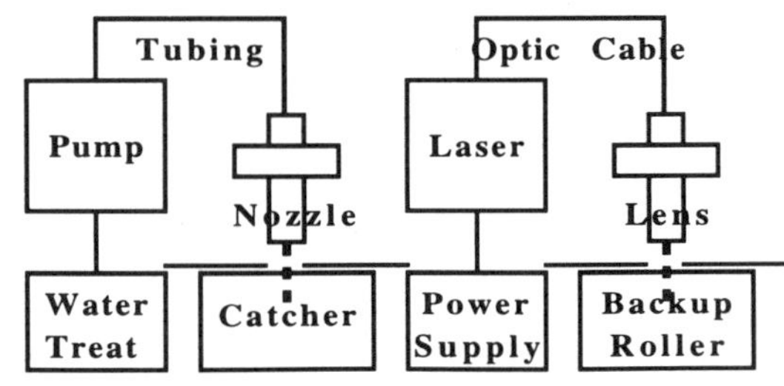

Log or roll saws (aka baloney slicers) may be used on small diameter rolls. The log saw is similar to a radial arm saw in that the rotating blade translates with respect to a stationary wound roll supported in a trough. The roll saw is similar to a grinder on a lathe in that both the blade and wound roll supported on a mandrel are rotating. Saw blades for rolls usually do not cut a kerf, and thus soft rolls may be cut on a scallop instead of having straight edges because the roll deforms under the blade. Log saws are capable of production rates up to 1200 roll per minute.

Score or crush cut slitters are pointed rotating blades loaded radially against a driven anvil roll. It is important to only use the minimum load necessary to cut the material because higher loads only wear out the equipment faster. Also, the radius of the tip should be very small but not sharp. Unfortunately, the tip is dressed by a hand held stone and thus is sensitive to the craftsmanship of the grinderman. Score cut slitters are capable of operating at speeds up to 5,000 FPM if the anvil and blade radial runout is kept to a couple of mils or less.

Shear slitters are used on a wide variety of materials and at speeds in excess of 10,000 FPM.[52,53,54] Shear slitters come in a variety of styles including individual mount (wide and/or high productivity machines) or twin shaft (narrow cuts or heavy materials), tangent or wrap. Also, shear slitters have many more setup parameters. First, the bottom band must protrude slightly into the web run to ensure sheet stability. Second, the overlap between the top and bottom blades must be high enough to avoid slitter jump and skippy cuts on heavy materials and low enough to reduce blade wear. Third, the cant (toe-in) angle ensures that the leading as opposed to trailing edge of the top blade is what touches the bottom band, but should not be more than a couple of degrees else blades will wear quickly. Fourth, the rake (camber) angle or back bevel of the top blade also ensure proper cutting geometry. Finally, the side loading between the blades must be minimum and controlled.

Trim removal options from fall on floor to pneumatic conveying are given in Table 18.2.

The web must be supported and rock stable on all sides of any slitter to maximize cut quality.

Blade life depends on many factors including the cutting difficulty of the material, the intensity of the slitting parameter adjustments, machine speed, quality of the slitting elements and the required quality of the slit edges. Typically, blade life is at least one shift, but may be more than a year between regrinds. Blade life is also extended by using more exotic blade materials such as tungsten carbide or ceramic[55,56] instead of high speed tool steels.

Waterjet slitters are used on a very wide variety of difficult applications for materials as soft as food products to those as hard as granite.[57,58,59,60] In fact, waterjets can cut any web material. While heavy materials can only be cut at slow speeds, light paper can be cut at over 10,000 FPM. Waterjets are also compact and well suited for edge trim applications. However, cut quality must be considered. In some cases, such as Kevlar or circuit boards, the waterjet will cut cleaner than mechanical methods. In other cases, such as most paper grades, blades will give a cleaner edge. One should also consider that waterjets are very noisy and are not familiar to most maintenance departments.

Laser slitting is limited to a few niche applications because of the greater equipment complexity and costs.[61] Also, lasers are not as suitable for heavy materials at high speeds. Lasers should be tuned to a material much like a microwave oven is to water so that cutting efficiency is maximized. Common applications for lasers are to perforate rather that slit products such as laser perfed computer paper and the tiny holes on cigarette plug wrap.

Table 18.2
Trim Removal Systems

Any Speed
- butt roll (on axis of wound rolls)
- weed roll (off axis of wound rolls)
- log saw

Low Speeds
- fall on floor
- fall into basket
- fall through floor into dumpster

High Speed Pneumatic
- all negative (vacuum)
- in-line fan trim
- eductor
- ejector

Static Electricity

Static generation and discharge is common in many processes because most web materials are non-conductors and some rubbing of the web is inevitable.[62,63,64,65,66,67,68,69,70] The severity of static concerns vary from a benign nuisance to a life threatening hazard. At the nuisance level, operators may not work as effectively because static discharges are unpleasant and even painful. If the potential is further increased, static can be a **safety** issue. Static discharges have caused injuries directly, such as an arc into the eye, or more often indirectly, such as startling an operator into a hazard. Static discharges can cause explosions of volatile solvents.[71,72]

Static may not be desirable for the process either. Though the the forces are weak, static can cause a web to cling to a surface. On light webs this may cause the web to leave a wound roll or roller erratically and with tension surges. Laminating of light materials such as tissue may become more difficult because the plies may cling to each other or separate during transport. Also, static may interfere with the proper stacking and destacking of sheeted material. Finally, a statically charged web will attract dust and lint. However, static electricity can be useful on occasion. Static can be used to pin a web tail to a core during an automated turnup on a winder.[73] Similarly, static can be used to pin a tail to a roller during sheeting. Finally, some web cleaners are statically assisted.

Static electricity begins with an electrostatic charge generation where there is sliding or other relative motion between materials. A common example of static charge generation is to walk (scuff) across carpet in a very dry room. This can lead to a static discharge when one then touches a lightswitch, register or other grounded object. Common demonstrations of static are to rub a balloon on one's hair, or rub a wool cloth against a glass rod. In all of these cases, electrons were peeled off one material and deposited on the other. This leaves a potential (voltage difference) between the objects. If the potential is high enough, the electrons will cross a small air gap as an arc. Though dependent on humidity and electrode shape, it takes about 50,000 volts to jump an inch of air. There is, however, very little current or power to the arc.

The amount of static **generation** is roughly proportional to the speed of relative motion between two objects, such as between a web and roller. Sliding is one type of relative motion which is always present, even on rollers in full traction. This occurs prior to the exit tangent where the web transitions by the band-brake equation from the upstream to downstream tensions. Thus, one way to reduce static generation might be to reduce the tension differential across rollers. However, static can still be generated at the departure of a roll(er) as the web is peeled away from it. Another large source of relative motion are process nips and wound roll nips. While static might dissipate on a web in a matter of minutes or at most hours, wound rolls can store a static charge for months or years.

Another factor for static generation is the separation on the **triboelectric series** of the two rubbing materials. As seen in Table 18.3 for example, paper will generate more static running across a polyurethane roller than across a steel roller. Also note how air flowing across any other material can generate a charge (which is why you can get a shock from a vacuum cleaner hose). To minimize static generation, we might choose our roller surfaces to be close to our web on the triboelectric series.

Table 18.3
Triboelectric Series

+ Positive +
Air
Hands
Glass
Nylon
Aluminum
Paper
Cotton
Steel
Hard Rubber
Copper
Polyester
Polyurethane
Teflon
Silicone Rubber
- Negative -

While a static charge can be generated by rubbing, this charge can quickly **dissipate** away through several mechanisms. If the web is even slightly conductive, the charge will quickly bleed off by traveling to the ground through rollers and framework. Thus, one way to reduce the dangers of static is to embed or coat the web with a conductor if it can be done economically and without compromise to product attributes. Similarly, rollers can dissipate their charge to ground through their bearings and supporting framework. In most cases, the contact pressures in a bearing are enough to yield enough metal-metal contact through the bearings and do not require graphite additive to increase lubricant conductivity. Only in rare cases is a grounding brush to the roller surface needed. However, the larger problem with charges on rollers are covers which are, for the most part, non-conductive.

Static can be bled off of the web and into the air, provided that the humidity is high enough. This is the reason why static is much less noticeable in our homes and plants in the humid summer months. Static is easily dissipated and rarely a problem when the relative humidity is 50% or more. However, there seems to be a great reluctance by most plants to humidify their air. The usual reason given is that the cost is prohibitive. However, humidification is much cheaper than the heating or cooling systems that most plants enjoy, and may well be cheaper than some static removal systems. A side benefit to maintaining a 50% RH is that it is more comfortable to plant personnel.

One way to help operators in a static prone line is to provide them with **grounded wands.** These are useful to discharge a web or wound roll that must be handled or worked around. The wand is a rod with a insulating handle that is connected to a grounded frame via a conducting chain or braided wire. An alternative is to use a hand held anti-static blower.

Of the three methods of continuous static elimination, the **passive** means is the least expensive. This is nothing more than conductive tinsel, garland or brushes which lies on or just above the web. The **tinsel** should have sharp needle tips, much like lightning rods, to be most effective. Finally, the tinsel must be well grounded.

There are three primary disadvantages to tinsel. First, it is only able to bring down potentials to about 5kV. While this is more than enough to eliminate nuisance static, it may not be sufficient to eliminate the threat of solvent explosions. Second, the tinsel can come loose if it is not securely mounted. Third, the tinsel might scratch some exceptionally soft coatings.

Air can be ionized by bars powered by a high voltage AC or DC supply. Ionizing the air makes it a better conductor so that static can be bled off the web and onto the bar's numerous tiny needles. Electrical **ionizing bars** are most suitable for reducing the potential of solvent ignition because they can reduce static potentials more than passive eliminators. Disadvantages of electrical ionizing bars are higher cost, bulkiness and the potential for electrical shock on some designs. Also, they can produce ozone and nitric acid vapors.

Low grade radioactive material generates alpha particles which ionizes the air close to the web, and thus operates similar to electrical bars. **Nuclear bars** may be used on photographic film where other methods may fog the film. Disadvantages of nuclear bars are that they need to be exchanged (renewed) about once a year, and the negative perceptions about nuclear materials.

Static can build up on a web when it passes any other element in a process line, and thus several eliminators may be needed. Their **location** is determined when the measured static potential exceeds a desired maximum. Static potentials may need to be as little as 1 kV in an explosive environment or less than 5-10 kV for operator comfort. Static potentials of 30 kV or more may be considered dangerous (the startle factor).

The level of static potential can be easily measured with inexpensive electrostatic field meters. These can be used to survey the line while running different grades, speeds and at different relative humidities. They can also be used to evaluate the efficacy of static eliminators and treatments. Static bars and tinsel should not be placed directly over idler or other rollers as it will impair their effectiveness. Rather, the bars must be much closer to the web than to any other metal or grounded object.

Bibliography

1. Weiss, Herbert L. *Coating and Laminating Machines.* Converting Technology Co, Milwaukee WI, 1977.

2. Weiss, Herbert L. *Technically Speaking.* Converting Technology Co, Milwaukee WI, 1978.

3. Weiss, Herbert L. *Control Systems for Web-Fed Machinery.* Converting Technology Co, Milwaukee WI, 1983.

4. Weiss, Herbert L. *Rotogravure and Flexographic Printing Presses.* Converting Technology Co, Milwaukee WI, 1985.

5. Satas, Donatas. *Web Processing and Converting Technology and Equipment.* Van Nostrand Reinhold Co. Inc., 1984.

6. Anon. *Gravure Process and Technology.* Gravure Association of America, Rochester NY, 1991.

7. Chang, Young Bae and Moretti, Peter M. *Ground-Effect Theory and its Application to Air Flotation Device.* 3rd Int'l Conf. on Web Handling, Oklahoma State Univ., June 18-21, 1995.

8. Moretti, Peter M. and Chang, Young Bae. *Coupling Between Out-of-Plane Displacement and Lateral Stability of Webs in Air-Support Ovens.* 3rd Int'l Conf. on Web Handling, Oklahoma State Univ., June 18-21, 1995.

9. Pinnamaraju, Ravi. *Measurements on Air Bar/Web Interaction for the Determination of Lateral Stability of a Web in Flotation Ovens.* Oklahoma State Univ., WHRC, M.S. Thesis, December 1992.

10. Swanson, Ronald P. *Air Support conveyance of Uniform and Non-Uniform Webs.* 2nd Int'l Conf. on Web Handling, Oklahoma State Univ., June 6-9 1993.

11. Young, Gary E., and Shelton, John J., and Fang, Bin. *A Lateral Web Limit Cycle Caused by an Air Bar.* ASME Winter Annual Meeting, Anaheim, CA, November 1992.

12. Anon. *Noncontact Drying and Web Handling for Sizing and Coating.* Equip. Mach. Mater., May-June 1992.

13. Fraser, W. A. R. *Air Flotation Systems: Theoretical and Practical Applications.* Paper, Film & Foil Converter, vol 57, nos 5-6, May-June 1983.

14. Sutton, Judy. *Going with the (air) Flow.* Package Printing & Converting, vol 41, no 4, pp 46-49, April 1994.

15. Wimberger, R. J. *New Flotation Drying Techniques.* International Review for the Pulp and Paper Industry, 1991.

16. Chang, Young Bae and Moretti, Peter M. *Interactions on Fluttering Webs with Surrounding Air.* TAPPI Engineering Conf. Proc, Book 1, pp 175-182, Seattle, September 1990.

17. Chang, Young Bae and Moretti, Peter M. *Interactions on Fluttering Webs with Surrounding Air.* Tappi J., vol 74, no 3, pp 231-236, March 1991.

18. Pramila, Antti. *Sheet Flutter and the Interaction Between Sheet and Air.* Tappi J., vol 69, no 7, pp 70-74, July 1986.

19. Uno, Minoru. *Fluttering of Flexible Bodies.* J. of the Textile Machinery Society of Japan, vol 14, no 4, pp 103-109, 1973.

20. Roisum, David R. *The Mechanics of Web Spreading - Parts I & II.* Tappi J., vol 76, no 10, pp 63-70, October 1993 and vol 76, no 12, pp 75-86, December 1993.

21. Feiertag, Bruce A. *Intermediate Guiding on Steel Strip Processing Lines.* Iron and Steel Engineer, September 1967.

22. Feiertag, Bruce A. *Selecting a Web Guiding System.* Converting Machinery/Materials. 2nd Conf. Proc., Philadelphia, October 17, 1979.

23. Larsen, S. *Web Guides - Friend or Foe.* Flexo, October 1989.

24. Markey, Frank. *Edge Position Control for the Steel Strip Industry.* Iron and Steel Engineer, February 1957.

25. Platt, D. *Automatically Guiding the Web.* Printing Trades J, no 1067, pp 23-24.

26. Wright, Gerard. *How to Make Belts Track.* Machine Design, May 11, 1967.

27. Edgar, Clement B. Jr. and Jefferis, Raymond P. III. *Computer Control of Tension and Lateral Position of Wires and Felts.* Tappi J., vol 57, no 6, pp 113-116, June 1974.

28. Gallahue, William M. *Web Guiding and Control in the Paper Industry.* Paper Trade J., pp 38-41, March 29, 1965.

29. Thompson, Ron and Cappell, Dan. *How to Guide Synthetic Forming Fabrics.* Tappi J., vol 70, no 1, pp 65-67, January 1987.

30. Shelton, John J. *Lateral Dynamics of a Moving Web.* Oklahoma State Univ., WHRC, Ph.D. Thesis, 1968.

31. Shelton, John J. and Reid, K.N. *Lateral Dynamics of a Real Moving Web.* Transactions of the ASME, 1971.

32. Shelton, John J. and Reid, Karl N. *Lateral Dynamics of an Idealized Moving Web.* J. of Dynamic Systems Measurement and Control,Transactions of the ASME, 1971.

33. Kardamilas, C.E. *Stochastic Modeling and Control of Lateral Web Dynamics.* Oklahoma State Univ., WHRC, Ph.D. Thesis, May 1990.

34. Kardamilas, C.E. and Young, Gary E. *Stochastic Modeling and Control of Lateral Web Dynamics.* American Control Conference Proc., San Diego, May 23-25, 1990.

35. Fang, Bin. *Lateral Web Behavior of Interactive Web Spans.* Oklahoma State Univ., WHRC, Ph.D. Thesis, December 1990.

36. Young, Gary E., and Shelton, John J., and Fang, Bin. *A Lateral Web Limit Cycle Caused by an Air Bar.* ASME Winter Annual Meeting, Anaheim, CA, November 1992.

37. Hopcus, Kenneth I. *Unwind and Rewind Guiding.* 2nd Int'l Conf. on Web Handling, Oklahoma State Univ., June 6-9, 1993.

38. McWilliams, L. *Selecting Web Guiding Control for Unwinding.* Paper, Film & Foil Converter, vol 64, no 4, April 1990.

39. Roisum, David R. *What is the Best Way to Measure Length?* Converting Magazine, Web Works column, pp 22, June 1995.

40. Roisum, David R. *How Can I Control Width Accurately?* Converting Magazine, Web Works column, pp 22, July 1995.

41. Roisum, David R. *How Can I Reduce Laminator Curl?* Converting Magazine, Web Works column, pp 22, March 1995.

42. Roisum, David R. *How Can I Fix a Baggy Web.* Converting Magazine, Web Works column, pp 20, April 1995.

43. Anon. *Steel-to-Steel Embossing Rollers.* Converter, June 1990.

44. Duffy, Alan. *Specialty Embossers.* British Paper Machine, July 1990.

45. Ranoia, Marie. *The Doctor is In.* Package Printing and Converting, pp 26-33, October 1994.

46. Vivier, F.P. *Embossing Fine Papers, the Evolution from Basement or Back Room to Main-Floor Finishing and Converting.* TAPPI Finishing and Converting Conf. Proc., Nashville, pp 229-236, October 18-21, 1992.

47. Largess, M. *Slitting Techniques.* Paper, Film & Foil Converter, vol 64, no 1, January 1990.

48. Bezigian, Thomas. *Why do we experience slitter dust, and how do we eliminate it.* Converting Magazine, Converters Consultant, pp 19, September 1994.

49. Brandt, Eckhard and Herrig, Friedheim, and Krolle, Arne and Ramcke, Bernd. *Controlling Dust on the Sheet.* PIMA Magazine, October 1992.

50. Frye, Kenneth G. *Finishing Defects: Trailing Edge Dust.* Tappi J., vol 69, no 8, pp 81-84, August 1986

51. Frye, Kenneth G. *Web Separation and Slitter Dust.* Tappi J., vol 69, no 12, pp 52-54, December 1986.

52. Laumer, Edward P. *How to Improve Slitter Performance on High-Speed Winders.* TAPPI Finishing and Converting Conf. Proc., 1987.

53. Schable, Reinhold. *Slitter Performance, Problems and Solutions.* TAPPI Finishing and Converting Conf. Proc., Lake Buena Vista FL, pp 65-82, October 7-10, 1990.

54. Schable, Reinhold. *Spotlight on Slitting.* Series of monthly articles in the Paper Film Foil Converter Magazine beginning in January 1992 and ending in December 1993.

55. Dohr, N. *Advanced Ceramic Material Offers Advantages for Slitting and Sheeting Operations.* TAPPI Finishing and Converting Conf. Proc., Nashville, pp 135-142, October 18-21, 1992

56. Mellor, A. and Winter, A.G. *Ceramic Knives for Winders.* Wochembl. Papierfabr., February 1991.

57. Amundsen, B.L. and Kleissler, B. *Ultra High Pressure Waterjets - Are They Right for You?* TAPPI Finishing and Converting Conf. Proc., Lake Buena Vista FL, pp 91-104, October 7-10, 1990.

58. Corcoran, M. *Technical Aspects of Waterjet Slitting.* TAPPI Finishing and Converting Conf. Proc., Lake Buena Vista FL, pp 111-116, October 7-10, 1990.

59. Olsen, John H. *Waterjet Slitting.* TAPPI Finishing and Converting Conf. Proc., 1978.

60. Olsen, John H. and Kapcsandy, Louis E. *Waterjet Slitting of Corrugated Board.* Tappi J., vol 60, no 4, pp 83-85, April 1977.

61. Hulsbusch, W. *Laser Cutting of Plastic and Fibrous Materials.* Pap. Kunstst. Verarb., December 1991.

62. Alonso, Marie Ranoia. *The Neutral Zone.* Packaging Printing & Converting, pp 20-24, January 1995.

63. Anon. *ESD Handbook.* Electrostatic Discharge Assoc, Rome, NY, 1995.

64. Arthur, C. *Controlling that Static.* Print World, October 1989.

65. Beasley, N.O. *Generation and Control of Static on Converting Machinery.* Tappi J., vol 55, no 3, pp 363-365, March 1972.

66. Horvath, T. and Berta, I. *Static Eliminations.* John Wiley and Sons, Letchworth, UK, 1982.

67. Peterson, H. *Static Electricity on Sheet and Rotary Printing Presses.* Dtsch. Drucker, April 1990.

68. Picciano, F. *Technical Forum: Static Control.* Converting Mag., March 1990.

69. Roisum, David R. *Static Electricity.* Converting Magazine, September 1995.

70. Schweriner, H. Allen Sr. *Static Electricity.* TAPPI Finishing and Converting Conf. Proc., 1982.

71. Wettermann, R.P. *Static Control Methods in Hazardous Areas in Converting Operations.* TAPPI Polymers, Laminations and Coatings Conf. Proc., August 28 - September 1, 1994.

72. Wettermann, Robert P. *Static Control Methods in Hazardous areas in Converting Operations.* Tappi J., vol 78, no. 2, pp 192-200, February 1995.

73. Wettermann, Robert P. *Electrostatically Assisted Starts on Automatic Winders.* Tappi J., vol 76, no. 11, pp 231-237, November 1993.

Chapter 19

Wound Rolls

In this chapter we show that while wound rolls share the same issues as rollers, the wound roll has unique concerns because it does not always behave as a solid body. Covered are cylindricity, deformability, vibration, hardness, air entrainment, interlayer slippage, and drive sizing.

Rolls Versus Rollers

The term "roller" uniquely refers to the subject of this book, though there are certainly many aliases. One of these aliases is "roll" which may refer to a roll<u>er</u> or, unfortunately, a wound roll. To eliminate ambiguity, we will use roller to designate a web machine component, roll to designate a wound roll, and roll(er) to designate either or both.

Most of the laws of web handling will apply equally well to rolls or rollers. For the most part, the web leaving a unwinding roll or entering a wound roll does not know about these distinctions. Rather, it leaves an unwinding roll or a roller with a stress state that is locked in by traction to the underlying surface (roller or web), but is modified by the span and next downstream roller. Conversely, the addition of a wrap to a winding roll is in many ways similar to a web entering a roller.

Thus, we would expect that the winding web also conform to web handling laws such as normal entry, and will respond to variations in cylindricity in a similar fashion as described in the chapter on spreading, and so on. What makes wound rolls different than rollers is often just a matter of degree rather than principle. It should not be surprising then to find web behaviors exaggerated when entering a wound roll whose cylindricity is one or two orders of magnitude worse than a roller. It should not be surprising to find that wound rolls tend to vibrate far more than their metallic roller cousins. Finally, it should not be surprising that winding is often more destructive to the web than rollers.

However, there are some aspects of wound rolls that are truly different than rollers. In most cases this is because rolls can be quite deformable. For example, merely setting a wound roll on the floor will almost inevitably result in a permanent flat spot. However, setting a covered roller and certainly a metal roller on the floor will not usually produce a permanent impression. The irony of this is that few people in their right mind would set a roller on the floor, whereas this is quite common for wound rolls.

Another aspect of the wound roll's deformability is variously called gearing, J-lining, interlayer slippage, or clockspringing. Here, a rotating nip force on either the inside or outside of the roll can cause the layers to slide one upon the other. This slippage is an ingredient of perhaps most of the wound roll defects. For example, the telescope (sideways movement of an unwinding roll) is a defect where the layers first slipped in the MD due to the nip or torque, and then are freed to more easily shift to the side.

Finally, the wound roll has drive and tension control issues far more challenging than any roller. For example, the changing radius requires a drive/brake torque range that will challenge the best of drives. The changing radius also means that the drive must change its gain in order to control tension properly, which is called inertia compensation. Finally, many winding and unwinding machines are often discontinuous as opposed to continuous operations. This means they must hold tensions tightly during stall tension, thread, acceleration and deceleration in addition to the steady state that is the mode of duty for most web machine elements. The result is that winder drives, particularly regenerative unwinds, are the most complicated drive applications in any industry.

This chapter merely highlights some of these issues. The interested reader should consult Roisum[1] for more information on winding and Smith[2] for more information on winding defects or other general sources on winding.[3,4,5,6]

Roll Cylindricity and Defects

The wound roll is composed of upward of 10,000 layers of web which will have a CD varying profile (but relatively stable in the MD). Thus, the thick lanes of the web will build up into high diameter positions on a roll as seen in Figure 19.1. However, the relative variations on diameter will be much smaller than the original caliper variations. This is because the high caliper areas are more highly stressed and thus wind tighter than they otherwise wound and vice versa.

These diametral variations can cause a number of problems on wound rolls. First, it is possible under severe situations that the web may be permanently stretched on the high diameter areas, thus causing a **baggy web** for unwinding. Second, narrow lanes of high diameter known as **ridges** can gather material there during the winding process, much like the opposite of a concave spreader roller. These ridges will build even larger and thus increase the gathering force further and so on in a potentially vicious and unstable cycle. This is a severe problem with the winding of many thin films.

Third, a nipped winding roll will have much greater pressures under the high caliper area and could cause blocking, pressure sensitive material damage, and web breaks. Fourth, nipping rolls which have high caliper gradients into a roller is a cause of **corrugations**.[7,8,9]

Wound rolls are nominally solid bodies rotating at a uniform rotational speed, which will result in high surface speeds at the high diameters and vice versa. The problem is when the winding roll is nipped against a roller whose surface speed is uniform across the width. As seen in Figure 19.2, the web surface speed will tend to be retarded at the high diameter area and advanced at the low diameter area. This generates inplane shear. If the caliper variations are closely spaced (1-5 cm), the web may shear under the nip into a classic rope or corrugation band. The worst caliper profile variation for ropes is a high-low-medium pattern. Unfortunately, the caliper variation which induces corrugations is often too small and too closely spaced to either measure or control.

Figure 19.1
Caliper and Diametral Profiles of a Wound Roll

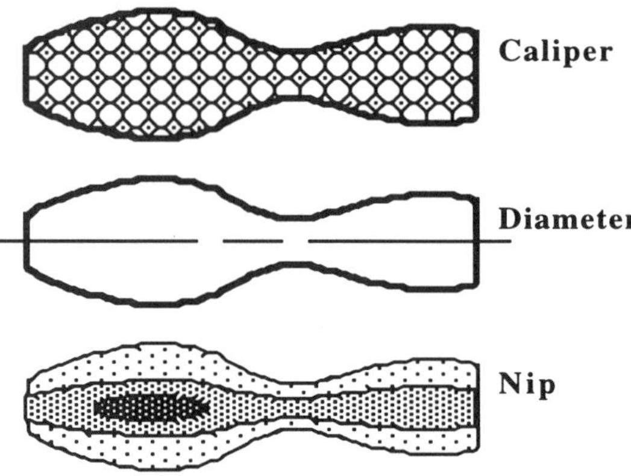

Other problems with diametral variation are more subtle. For example, a roll nipped against a roller will tend to thrust axially and twist, which imposes large loads on the machinery. For multiple slit rolls rotating on the same axis or nipped against a common drum, the large diameter rolls may wind so tightly as to break while the low diameter rolls may be so loose as to cause floppiness, wrinkling or interweaving.

The unyielding nature of winding is not often well appreciated. While the machinery can certainly influence defects, more often than not it is the product profile or product properties that underlie many wound roll defects.

Figure 19.2
Corrugation Mechanics

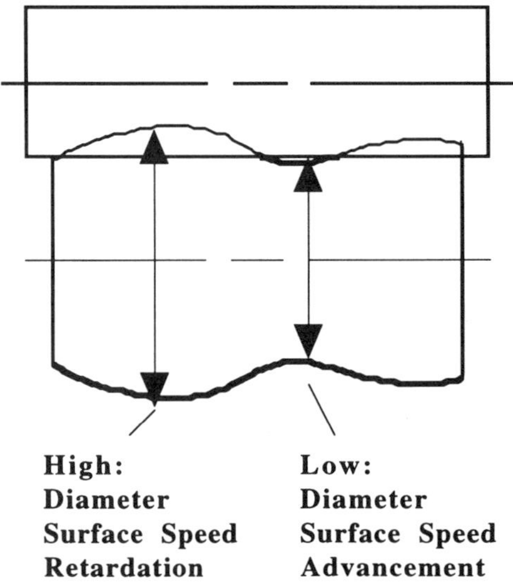

Roll Runout, Vibration and Handling

The wound roll can have radial runouts far worse than any roller. I've seen many cases where the core was eccentric from the center of a 40" roll by as much as an inch. Obviously, this would cause severe vibration when wound against a nipping roller and tremendous web flutter and tension surges when unwound. Even more moderate or typical runouts can cause vibration problems. Tiny bumps as small as 10 mils can induce a pounding resonance against a winder drum. As was covered in Chapter 11, there is often little that can be done to effectively reduce rewound roll vibration.[10] Thus, here we will focus on caring for the roll during storage and handling to reduce the additional insult that could compromise unwinding.

Merely setting a wound roll on a floor can leave a flat spot from 0.1" deep on a hard roll to almost an inch on a soft roll of nonwovens or tissue. Roll support options are shown in Figure 19.3. The choice of support is not particularly influenced by the length of time in storage because the damage goes roughly by log time. Thus, an hour on the floor is only slightly worse than a minute and so on. Rather, it is on the cost benefit tradeoff between higher value product and higher cost handling.

There are also many issues with transporting rolls as well. For example, as shown in Figure 19.4, a clamp truck will almost inevitably leave its mark on a roll.[11] This can be minimized to an extent by using a minimum clamping pressure or intelligent clamps.[12,13] Many automated roll handling stations accelerate rolls against pushers and decelerate against stops. The damage from these can be reduced by: spreading out the force, reducing forces, and reducing accelerations. It is easy to diagnose the sources of this type of damage as it retains a fingerprint of the component that produced it.

One final technique to keeping rolls round is to wind them as tight as possible (about 25% of the tensile strength). Increased winding tightness translates to an increased interlayer pressure which holds the layers together against shifting.[14,15,16,17,18]

Figure 19.3
Roll Storage Options

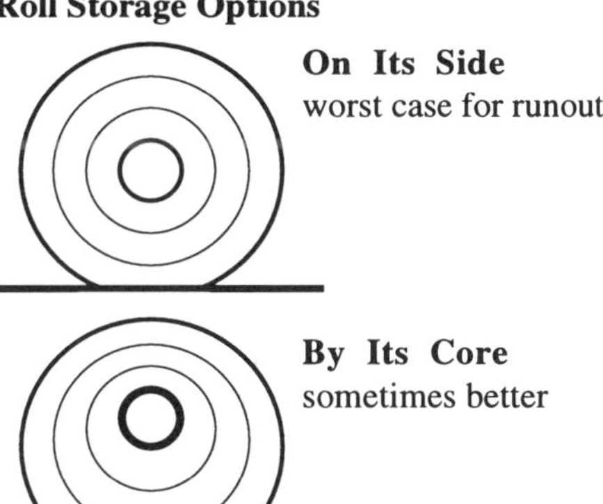

On Its Side
worst case for runout

By Its Core
sometimes better

Sunday Drive
usually best, but very expensive

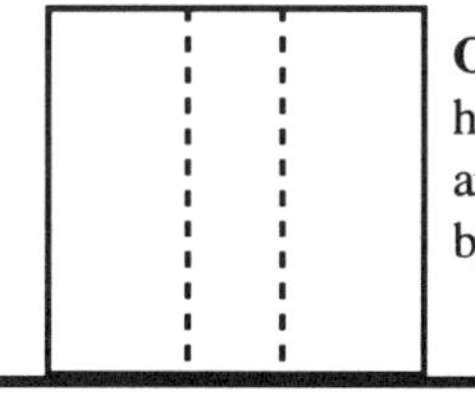

On Its End
hard on edges unless they are quite square and padded by headers

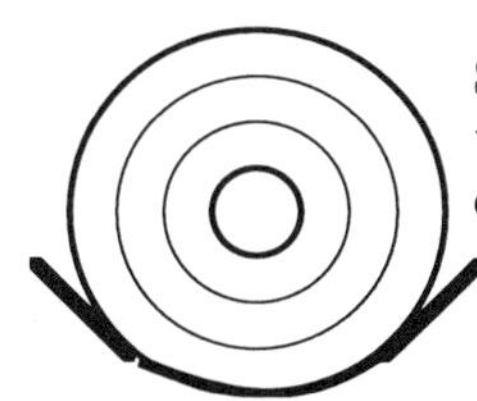

Sling
very good, but racks more expensive than floor

Figure 19.4
Roll Handling Damage

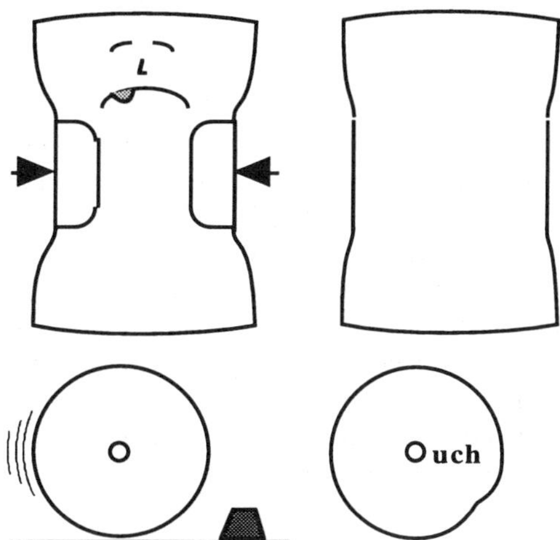

Roll Hardness

Roller covering hardness is often specified and measured because it is an important process parameter. Cover hardness is usually measured with an indentor. Wound roll hardness is also an important parameter for winding as it indicates how tightly the roll is wound. If the roll is too loose, it is prone to handling damage as well as telescoping during unwinding. If the roll is too tight, the web can be damaged.

Figure 19.5 shows the idealized structure of a wound roll as a function of radial position. First, the roll must be wound harder (tighter) at the core because it serves as the foundation upon which to build the rest of the roll. If the roll is too soft (loose) at the core, it may telescope upon unwinding. The second is that the roll must be wound softer at the finish diameter. If the finish is too hard or tight, then that material might be damaged or be prone to telescoping. Finally, the roll must be wound with a smooth and gentle transition of hardness from the tighter core to the looser finish diameter. If there are abrupt changes in between, such as lifting a nipping roller or other steep TNT (Tension, Nip and Torque) change, then the roll may be prone to starring.

Besides through roll structure, the roll must be uniform across the width of the roll(s). In almost all cases, nonuniformities across the roll are caused by base web manufacturing or processing (coating, embossing, laminating, etc.) problems, the most common of which are variations of basis weight or caliper across the deckle. However, base sheet profile problems are sometimes easier to diagnose by measuring the wound roll rather than the web itself!

However, while the term roll hardness has become almost synonymous with tightness, there are a number of other ways to measure how the roll has been built as given in Table 19.1. The impact measurements are instrumented versions of the billy club that is commonly used to sound a roll by striking it. The interlayer pressure methods may measure directly as with pressure transducers, or indirectly such as the acoustic method and in-roll caliper. The remaining infer pressure from force and coefficient of friction. The strain based methods look at the elongation

Figure 19.5
Roll Structuring Principles

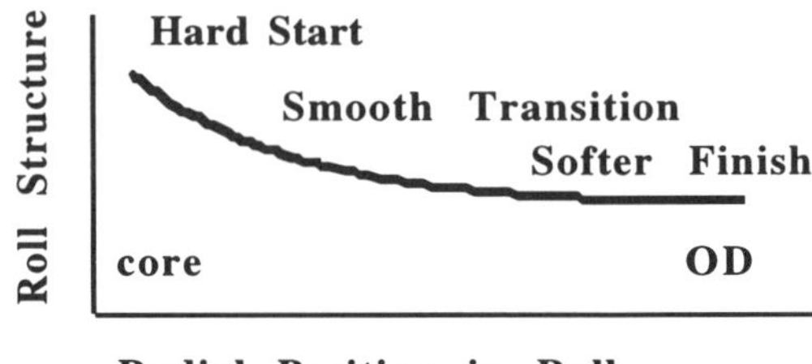

of the material inside a roll, from which tension could be calculated based on the material's MD modulus and Hooke's law. Tension can be directly measured on a WIT-WOT pilot lab rewinder. These methods can be evaluated in terms of accuracy, sensitivity, ability to profile across the width and/or through the diameter, ease of use, automatic data recording, etc.[19]

Table 19.1
Roll Structure Measurements and Principles

Impact Hardness
- Backtender's Friend
- Billy Club
- Paro Roll Tester
- Rhometer
- Schmidt Hammer

Interlayer Pressure
- Acoustic Time-of-Flight
- Axial Press
- Caliper in the Roll
- Core Torque
- Pressure Transducers
- Pull Tab
- Smith Needle

Strain
- Cameron Gap
- J-Line
- Radially Drilled Holes
- Slit Roll Face
- Strain Gages

Tension
- WIT-WOT

Other
- Density Analyzer
- X Ray Tomography

Interlayer Slippage

There can be no doubt that the layers of a roll can slide with respect to each other during the winding and unwinding of many web materials. The most convincing evidence of this is obtained from the J-line test.[20] As seen in Figure 19.6, an initially straight radial line struck on a winding roll can bend over into an arc or J shape. This is not an abrupt motion but rather the accumulated microslip from each passage of a nipping roll. Also, this slippage is confined to a narrow band near the surface of the roll, and interior layers remain locked together. However, a roll can shear anywhere there is a nip as seen in Figure 19.7. Finally, the slippage magnitude and direction varies with the grade of web as seen in Figure 19.8.

The reason we are concerned with interlayer slippage is that it is the cause of many wound roll defects such as bursts, corrugations, telescoping and wrinkles[21] to name a few. Since most of these defects are only formed in the presence of a (interior or exterior) nip, they have a common mechanics.[22] The solutions to these defects include: reducing the nip load, increasing web-web friction, increasing caliper or basis weight, increasing web uniformity, and or decreasing deflection.

Only the clockspringing form of telescoping does not require a nip, but can slip merely because of tension induced torque. The torque tension required to break the layers loose can be calculated from the radius (or just above) of the core, the interlayer pressure and the coefficient of web-web friction.

Figure 19.6
Interlayer Slippage and the J-Line

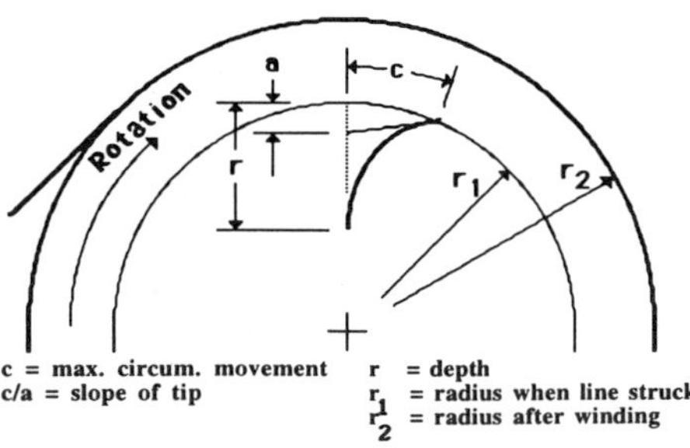

Figure 19.7
Slippage Cases for Bulky Materials

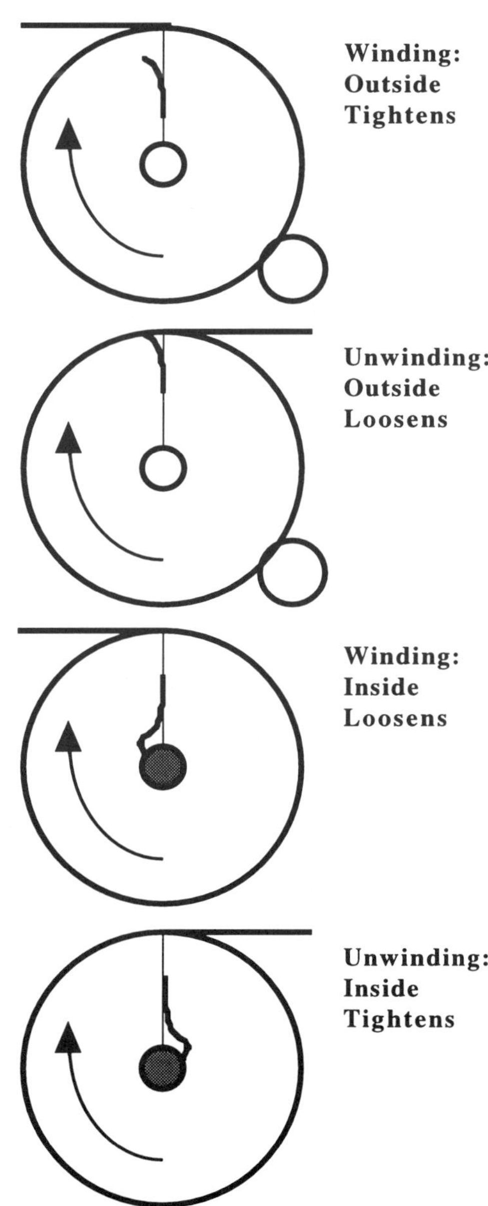

Figure 19.8
Slippage Direction versus Density

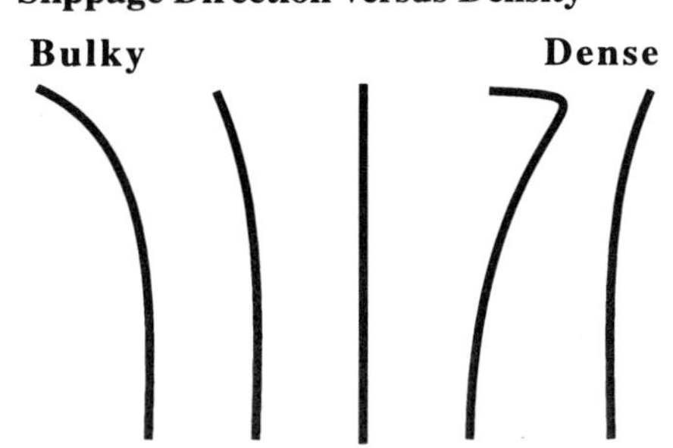

Air Entrainment

One problem with air entrainment is its layer-to-layer lubricating tendencies. This can cause the web to almost explode sideways as it loses traction and slips at higher speeds. Also, air entrainment reduces the tightness of the wind.[23,24,25] Curiously, however, some air might be needed on some webs to keep the layers from blocking (sticking) together.

Lowering the speed or raising the web tension is the only means to reduce air brought into a centerwound roll. A much more effective method is to use a nip. Light webs will usually require a wrapped nipping roller to go much over 300 m/min (1000 FPM). Nipping rollers after the ingoing tangent are less effective.

However, if the roll and roller do not tightly conform, much more air can be brought in than will be estimated by the methods of Chapter 4.[26,27,28,29] (The actual amount brought in can be measured as the ratio of roll density to web density for stiff nonporous webs.) As seen in Figure 19.9, air is brought into the roll between the drums. The problem occurs when this air gets trapped in front of the next roller as a bubble. Instead of metering through smoothly and evenly, it tends to go through in gulps. This is the cause of the **air shear burst** or wrinkle defect. Letting the air go through by lighter loading of the second roller and possibly adding a wide grooving pattern are two remedies.

Many have thought that grooving the first roller would help remove air. Unfortunately, it is on the wrong side of the sheet and actually makes the problem worse. As seen in Figure 19.10, the web pulls into the drum **grooves** slightly. This leaves an open tunnel between the first and second layer for air to easily get into the roll. A similar problem will occur if the roll has a profile problem. Thus, a highly loaded and soft covered nip roller may be indicated.

Figure 19.9
Air Shear Burst and Wrinkle

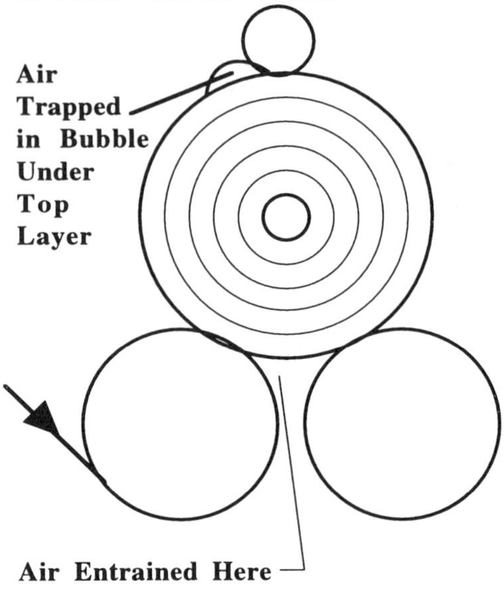

Figure 19.10
Grooving, Roll Profile and Air Entrainment

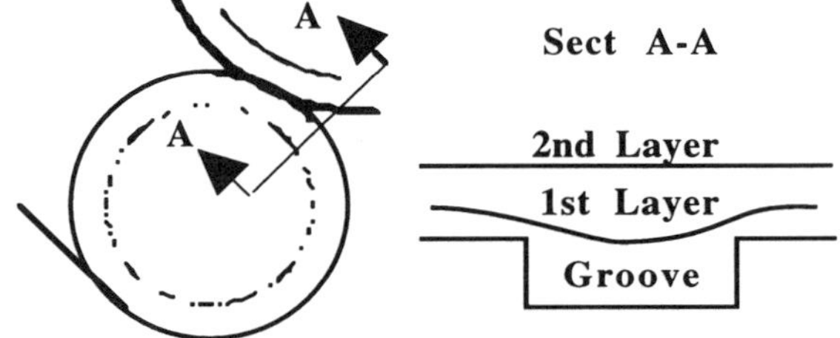

Figure 19.11
Starring Caused by Air Outgassing

A last problem we will mention is another type of **starring** caused by air entrainment, or rather air outgassing.[30] Interlayer pressure also increases the air pressure between the layers above the ambient atmosphere. With time, this will bleed off through the edges of the roll. As the air is bled off, it will cause the effective caliper to decrease which causes the roll to collapse inward. Unfortunately, in the process of collapsing the MD compression is increased, often to the point of a starring type buckle as seen in Figure 19.11. The time it takes to bleed off enough air to cause the roll to collapse is hours to days.

Unwind Drive Sizing

The procedure for sizing a drive/brake for an unwind or winder is in no way different than sizing for a roller. However, it is more involved by virtue of changing diameters and inertias. In the next two sections we will detail the sizing procedure with an example in both English and Metric units. The English example will include the built in conversion factors that will speed up the calculation while the Metric example will be built from first principles. To show multiple approaches, the English example will work primarily with power, while the Metric will use a perhaps more sensible torque approach.

The application example is chosen to illustrate most of the common drive issues encountered. For example, the range of widths, strengths and calipers will push the torque range of the motor/brake. Also, the multiple accel/decel rates will illustrate some additional application techniques.

The example is for a narrow web unwind running a medium weight material at moderate speeds but respectively high acceleration rates (for productivity). We will first determine torques, horsepowers, RPM's and so on. Then, we will be able to tell whether a brake, a motor or a brake and motor combination is most appropriate. (Note that most rollers and all winders will be motor instead of brake actuated). Most of the problem can be worked in a single width (maximum) until the absolute torque range of the device must be checked (#4 below).

The things that need to be checked are ultimately:

1. Maximum continuous tension horsepower
2. If there is sufficient torque at maximum diameter, width and tension to slow down a roll at a prescribed rate
3. If the minimum tension is able to accelerate a full roll without a motor
4. If the torque range is achievable by the device (Not checked on this example)
5. If RPM's are so low as to indicate a reduction unit on a motor or are so high as to exceed a drive train component specification.

Drive Sizing Example - English Units

material yield strength min - 3000 psi
material yield strength max - 7000 psi
caliper min - 4.5 mils
caliper max - 5.5 mils
roll diameter max - 30"
core - 6" ID x 1/2" wall fiber
width min - 8"
width max - 12"
roll weight max - 650 lb
acceleration rate - 125 FPM/sec
deceleration rate - 100 FPM/sec
emergency stop rate - 250 FPM/sec
speed max - 600 FPM
RPM limit of drive train - 1800 RPM
WR^2 of drive train - 2 lb-ft^2

Tension Range

Since this is a new material for which the customer had no previous experiences, we will need to estimate tension ourselves based on the 10-25% rule of Chapter 2. Also, it is assumed that the 10% value will also be suitable for stall tension.

$$\text{TensMin} = (0.10)(0.0045\text{in})\left(3000\frac{\text{lb}}{\text{in}^2}\right) = 1.35 \text{ lb/in}$$

$$\text{TensMax} = (0.25)(0.0055\text{in})\left(7000\frac{\text{lb}}{\text{in}^2}\right) = 9.63 \text{ lb/in}$$

Tension Horsepower

$$\text{(19.1)} \quad \text{HPtensMax} = \frac{\text{FPM x PLI x } (\text{Width } in)}{33{,}000}$$

At max speed and max width;

$$\text{HPtensMin} = (-)\frac{600\text{x}1.35\text{x}12}{33{,}000} = -0.295\text{HP}$$

$$\text{HPtensMax} = (-)\frac{600\text{x}9.63\text{x}12}{33{,}000} = -2.10\text{HP}$$

Note that the minus sign is used to indicate the braking instead of the motoring direction. Also note that this is not the total horsepower requirements because we have not considered acceleration, deceleration and mechanical losses.

RPM

$$(19.2) \quad RPM = \frac{3.82 \text{ x } (\text{Speed } ft/min)}{(\text{Diameter } in)}$$

for calculating accel/decel torq/hp

$$RPM(\text{max spd dia}) = \frac{3.82x600}{30} = 76.4 \text{ RPM}$$

for calculating max drive train RPM

$$RPM(\text{max}) = \frac{3.82x600}{7} = 327 \text{ RPM}$$

Note that 327 RPM is quite slow and should not exceed rotational speed limits of most equipment, including our example. If it did, a tach and master reference speed slowdown circuit could be provided to automatically keep unwind RPM's within limits as the unwinding roll expires if it were run off at speed. However, many unwinds would be slowing down to a stop before the core. Also note that if this unwind were motor driven, the low RPM's could quite likely indicate the need for a gear or belt reduction unit. If so, we would need to consider mechanical torque and horsepower losses.

WR^2 (rotational inertia in English units)

(19.3)

$$WR^2 = (0.000682)(\rho)(w)\left(Do^4 - Di^4\right) + wr^2_{\text{drive train}}$$

where

WR^2 = inertia $\left(lb \cdot ft^2\right)$

ρ = density $\left(lb/in^3\right)$

w = web width (in)

Do = diameter outside (in)

Di – diameter inside (in)

$$\rho = \frac{650lb}{12in\frac{\pi}{4}\left(30^2 - 6^2\right)in^2} = 0.0798lb/in^3$$

The density is twice that of typical films or dense paper (0.04), but slightly less than aluminum (0.100). Must be exotic stuff.

$$WR^2max = (0.000682)(0.0798)(12)\left(30^4 - 6^4\right) + 2$$

$$WR^2max = 531 \text{ lb} - ft^2$$

Time

times will be used for inertial calculations

$$t_{accel} = \frac{600FPM}{125 \text{ FPM/sec}} = 4.8sec$$

$$t_{decel} = \frac{600FPM}{100 \text{ FPM/sec}} = 6sec$$

$$t_{estop} = \frac{600FPM}{250 \text{ FPM/sec}} = 2.4sec$$

Acceleration HP

(19.4)

$$HP_{accel} = \frac{\left(WR^2\right)N^2}{16.15E5(t)}$$

N = speed (RPM)

t = time

$$HP_{accel} = \frac{(531)76.4^2}{16.15E5(4.8)} = 0.400 \text{ HP}$$

$$HP_{decel} = (-)\frac{(531)76.4^2}{16.15E5(6)} = -0.320 \text{ HP}$$

$$HP_{estop} = (-)\frac{(531)76.4^2}{16.15E5(2.4)} = -0.800 \text{ HP}$$

Again observe that the minus sign indicates a braking as opposed to a motoring.

Horsepower Run

The maximum horsepower requirements during the steady state run are quite easy. It is merely the maximum tension horsepower of -2.10 calculated above, where the minus sign indicates a braking direction. Obviously, lower horsepower will be consumed when the speeds, tensions or widths are reduced.

Horsepower Acceleration

It takes 0.400HP to accelerate a full diameter (and full width) roll. However, a minimum web tension will only be able to brake at -0.295HP. Thus, a motor must provide the difference of up to +0.105HP. The only way we could get by solely with a brake would be if our acceleration rate were reduced or we raised our minimum design tension.

Horsepower Deceleration

During deceleration, we must brake at as much as -2.10HP to provide maximum tension and another -0.320HP to slow down the roll for a total of -2.42HP. However, this does not mean that we must be able to continuously dissipate 2.42HP because this is only for a short period. In fact, many devices can take 50-100% more than rated power for periods up to a minute because of thermal inertia. However, we need to check to make sure that the torque range is covered.

Torque due to Tension

(19.5)

$$\text{Torq}(ft-lb) = (\text{Tens } lb/in)\text{x }(\text{Width } in)\text{x }(\text{Dia } in)/24$$

$$\text{TorqTensMin} = -(1.35)\text{x }(10)\text{x }(7)/24 = -3.94\text{ft}-\text{lb}$$

$$\text{TorqTensMax} = -(9.63)\text{x }(12)\text{x }(30)/24 = -144.5\text{ft}-\text{lb}$$

Torque Ratio

The ratio of the tension torque requirements in this example is 37:1. If this were strictly a braking operation, this range would be difficult to hold with finesse on a conventional pneumatic brakes unless multiple calipers were automatically valved in/out as needed to keep the required brake pressure above 10 psi. Torque ratios of up to 100:1 can be well controlled only on very good electric drive motors.

Torque due to Deceleration

(19.6)

$$\text{Torq} = \frac{(WR^2)N}{307.6\,(t)}$$

Torq = Torque (ft–lb)

N = speed (RPM)

t = time

$$\text{TorqDecel} = \frac{-(531)\,76.4}{307.6\,(6)} = -21.0\text{ft}-\text{lb}$$

$$\text{TorqEstop} = \frac{-(531)\,76.4}{307.6\,(2.4)} = -55.0\text{ft}-\text{lb}$$

Torque Requirements during Decel

The torque requirements during decel could be as much as 144.5ft-lb to pull maximum tension and another 21.0ft-lb to slow down a full sized roll for a total of 165.5ft-lb.

Torque Requirements during E-Stop

During an emergency stop, the drives must brake sufficiently, at 55.0ft-lb, to bring the roll down at least as fast as the given E-stop rate or time without attempting to pull or control tension. From the above calculation, however, a brake sized for decel is adequate for E-stop (not always the case) as long as tension is not being pulled (web break or downstream roller decelerating faster). In an E-stop situation, all drives brake at maximum available torque.

Horsepower From Torque and Speed

(19.7)

$$\text{HP} = \frac{(\text{Torque } ft-lb)(\text{N } RPM)}{5250}$$

$$\text{HP}_{totdecel} = \text{HP}_{ten} + \text{HP}_{decel}$$

$$\text{HP}_{totdecel} = \frac{-144.5\text{x}76.4}{5250} + \frac{-21.0\text{x}76.4}{5250} = -2.42\text{ HP}$$

or

$$\text{HP}_{totdecel} = \frac{-165.5\text{x}76.4}{5250} = -2.42\text{ HP}$$

Design Summary

Since a motor is required, it could be used to provide the entire range of horsepowers (-2.42 to -0.3) and torques (up to -165.5ft-lb). Since the RPM's are quite low, a gearbox or belt reduction might be used to trade motor RPM's for torque as given in the last section. The losses of these devices may cause an effective 5-10% additional brake capacity.

Alternatively, perhaps a 1/2 HP motor and a 2 HP brake in series may provide the required range at less cost. However, a mechanical brake is seldom used with a motor except for the limited case of providing additional E-stop torque because of the difficulty in coordinating the two devices' controls.

If this application did not require a motor, we might use a pneumatic or electric brake. We would need to check the maximum torque, horsepower and RPM's as well as whether the device can hold low tensions suitably on a narrow light web during acceleration (torque range).

Sizing a Regenerative DC Drive Motor

The above example shows how we calculate the torque and horsepower requirements of an unwind (or center-driven rewind) during conditions of acceleration, run and deceleration. Now, at the risk of really losing the reader, will show an example of sizing a DC motor.

The important thing to bear in mind is that the DC motor is only able to put out full rated horsepower at its base speed. At lower speeds, the motor will only be able to develop a proportionally smaller horsepower. This relationship between developed horsepower and actual speed (RPM) is given in the DC drives section of Chapter 14.

Thus, we will gear in for the maximum rotational speed of the drive train which occurs at the core diameter and design speed. From (19.2), the maximum RPM is 327. This RPM can be achieved with a 5.35 ratio gearbox and a 1750 RPM motor. Then, the required horsepower at the full roll during run must be

(19.8)

$$HP_{required} = HP_{tens} \frac{RPM_{core}}{RPM_{full\ roll}} \times \text{efficiency}$$

$$HP_{required} = 2.1 \times \frac{327}{76.4} \times 0.95 = 8.55\ HP$$

Since there are no 8.55 HP motors, we will have to use a 10 HP motor.

Now we need to check for decel power. We will assume that the motor is capable of a 150% overload factor for the short period of a normal stop.

$$HP_{decel} = -2.4 \times \frac{327}{76.4} \times 0.95 = 9.67\ HP$$

This is OK since the 9.67 HP during decel is less than the 15 HP (1.5x10) the 10 HP drive can put out for a short time.

Similarly, we check for E-stop power. We will assume that the motor is capable of a 200% overload factor for the very short period of a normal stop. From equation (19.4), we calculate an E-stop power of -0.8 HP, which is far less than the 20 HP the 10 HP drive can put out for a very short time.

From this example, we will note several important observations about drives used in center-driven unwinds or rewinds. First and foremost is that the required frame size or horsepower rating of a drive will be much larger than the physical power required. This is because the motor will only put out rated power at its base speed. In our example, a 10 HP drive was required when a maximum of only 2.4 physical horsepower were required. Second, the required size of the motor is very sensitive to the maximum roll diameter to minimum core diameter ratio. Thus, it is helpful to use large cores to keep drive sizes down. Third, this example only highlights a few of the complications of sizing and selecting drives, particularly electric drives.

Thus, as the reader has probably begun to suspect, drive sizing should be left to drive engineers. Fortunately, this is easy to do since any reputable drive supplier would be happy to size the drive as part of a quotation or order.

Drive Sizing Example - Metric Units

material yield strength min - 20.7 MPa
material yield strength max - 48.3 MPa
caliper min - 0.115 mm
caliper max - 0.140 mm
roll diameter max - 0.750 m
core - 152 mm ID x 15 mm wall fiber
width min - 200 mm
width max - 300 mm
roll weight max - 300 kg
acceleration rate - 40 m/min/sec
deceleration rate - 30 m/min/sec
emergency stop rate - 75 m/min/sec
speed max - 180 m/min
RPM limit of drive train - 1800 RPM
WR^2 of drive train - 0.08 kg-m^2

Tension Range

Since this is a new material for which the customer had no previous experiences, we will need to estimate tension ourselves based on the 10-25% rule of chapter 2. Also, it is assumed that the 10% value will also be suitable for stall tension.

(19.9)

$$Tens_{Run} = (\text{Fraction of Yield})(\text{Caliper})(\text{StressYield})$$

$$Tens_{Min} = (0.10)\left(0.115\text{x}10^{-3}\text{m}\right)\left(20.7\text{x}10^{6}\frac{N}{m^2}\right) = 238\ N/m$$

$$Tens_{Max} = (0.25)\left(0.140\text{x}10^{-3}\text{m}\right)\left(48.3\text{x}10^{6}\frac{N}{m^2}\right) = 1690\ N/m$$

Tension Power
(19.10)

$$Power_{Tens} = (\text{Speed})(\text{Tension})(\text{Width}) =$$

$$\left(180\frac{m}{min}\right)\left(1690\frac{N}{m}\right)(0.300m)\left[\frac{min}{60sec}\right]\left[\frac{kW}{1000\ \frac{N\cdot m}{sec}}\right]$$

$$= -1.52\ kW$$

where the minus sign is used to indicate the braking instead of motoring direction. This is the maximum thermal power that the brake/motor needs to achieve continuously.

Density

$$\rho = \frac{mass}{volume}$$

$$\rho = \frac{300\ kg}{\frac{\pi}{4}\ 0.300m\left(0.750^2 - (0.152 + 2*0.015)^2\right)m^2}$$

$$\rho = 2405\ kg/m^3$$

Rotational Mass Moment of Inertia

(19.11)

$$I = \frac{\pi}{32}\rho\,(\text{Length})\left(Dia_{out}^4 - Dia_{in}^4\right) + wr^2_{drive\ train}$$

$$I = \left[\frac{\pi}{32}\right]\left(2405\frac{kg}{m^3}\right)(0.300m)\left(0.750^4 - 0.182^4\right)m^4 + 0.08$$

$$I = 22.41\ kg-m^2$$

Acceleration Rate

$$(19.12)\quad a = \frac{dv}{dt}$$

$$a_{decel} = 40\ \frac{\frac{m}{min}}{sec}\left[\frac{min}{60\ sec}\right] = 0.667\ m/sec^2$$

$$\alpha_{accel} = \frac{a}{r} = \frac{0.667\ m/sec^2}{0.750m/2} = 1.778\ rad/sec^2$$

$$\alpha_{decel} = 1.333\ rad/sec^2$$

$$\alpha_{estop} = 3.33\ rad/sec^2$$

Torque Inertial

$$(19.13)\quad Torque = I\alpha$$

$$Torq_{accelmax} = \left(22.41\ kg-m^2\right)\left(1.778\ \frac{rad}{sec^2}\right)\left[\frac{N}{kg-m/sec^2}\right]$$

$$Torq_{accelmax} = 39.8\ N-m$$

$$Torq_{decelmax} = 29.9\ N-m$$

$$Torq_{estopmax} = 74.6\ N-m$$

Tension Torque

(19.14)
$$Torq_{tens} = (Tension)(Width)(Diameter/2)$$

$$Torq_{mintens} = \left(238\frac{N}{m}\right)(0.3m)(0.75m/2) = -26.8\ N-m$$

$$Torq_{maxtens} = \left(1690\frac{N}{m}\right)(0.3m)(0.75m/2) = -190\ N-m$$

Torque Run

The maximum torque requirements during the steady state run are quite easy. It is merely the maximum tension torque of -190 N-m calculated above, where the minus sign indicates a braking direction. Obviously, lower torques will be required when the tensions, widths and/or diameters are reduced.

Torque Acceleration

It takes 39.8 N-m to accelerate a full diameter (and full width) roll. Obviously, a sufficiently high web tension torque, such as the maximum at 190 N-m will be sufficient and braking action is all that would be needed for those grades. However, a minimum web tension will only be able to brake at -26.8 N-m. Thus, a motor must provide the difference of up to 13.0 N-m. The only way we could get by with strictly a brake would be if our acceleration requirements were reduced or we raised our minimum tensions.

Torque Deceleration

During deceleration, we must brake at as much as -190 N-m to provide maximum tension and another -29.9 N-m to slow down the roll for a total of -219.9 N-m. However, this does not mean that we must be able to continuously dissipate the power indicated by this torque (x speed) because this is only for a short period. In fact, many devices can take 50-100% more than rated power for periods up to a minute because of thermal inertia.

Torque E-Stop

During an emergency stop, the drives must brake sufficiently, 74.6 N-m, to bring the roll down at least as fast as the given E-stop rate or time without attempting to pull or control tension. From the above calculation, however, a brake sized for decel is adequate for E-stop (not always the case) as long as tension is not being pulled (web break or downstream roller decelerating faster). In an E-stop situation, all drives brake at maximum available torque

RPM's

(19.15)
$$RPM = \frac{0.318 \text{ x } (\text{Speed } MPM)}{(\text{Diameter } M)}$$

$$RPM = \frac{0.318 \text{ x } (180)}{(0.182)} = 314\ RPM$$

This is rather low and might require a gearbox to better match motor torques and speeds.

Transpositions of Inertias

(19.16)
$$I_1^2 = I_2^2 \left(\frac{N_2}{N_1}\right)^2$$

Design Summary

Since a motor is needed, it could be used to provide the entire range of torques (to 220 N–m). Since the RPM's are quite low, a gearbox or timing belt reduction might be used to trade motor RPM's for torque as given in above (which may give 5-10% additional brake capacity because of mechanical losses). The device must be rated -1.52 kW continuously, and slightly more during a short deceleration. Alternatively, we could use a 15 N-m motor and a 200 N-m brake combination. However, a mechanical brake is seldom used with a motor except to provide extra E-stop torque because of the difficulty in coordinating the controls. If this application did not require a motor, we might use a pneumatic or electric brake. We would need to check the maximum torque, horsepower and RPM's as well as whether the device can hold low tensions suitably on a narrow light web during acceleration (torque range).

Bibliography

1. Roisum, David R. *The Mechanics of Winding.* TAPPI PRESS, Atlanta, 1994.

2. Smith, R. Duane and Roisum, David R. et. al. *Roll and Web Defect Terminology.* TAPPI PRESS, Atlanta, 1995.

3. Hadlock, Alan H. *The Principles of Winding.* PIMA, vol 61, no 2, pp 13-16, February 1979.

4. Smith, R. Duane. *Consistently Winding Good Rolls Provides Challenge.* Paper, Film & Foil Converter, vol 65, no 4, April 1991.

5. Smith, R. Duane. *The Art of Winding Quality Rolls.* TAPPI Finishing and Converting Conf. Proc., Houston, pp 163-172, October 6-10, 1991.

6. Frye, Kenneth G. *Winding.* TAPPI Press, Atlanta, 1990.

7. Burns, J.W. *Rho Hardness Meter.* Tappi J., vol 61. no 1, pp 91, January 1978.

8. Campos, S. *Analysis of the Main Causes of Irregular Thickness Profiles and of the Consequent Increase in Reel Defects.* Papel, September 1990.

9. Frye, Kenneth G. *Winding Variables and Their Effect on Roll Hardness and Roll Quality.* Tappi J., vol 50, no 7, pp 81A-86A, July 1967.

10. Roisum, David R. *Winder Vibration Can Reduce Operating Efficiency and Increase Maintenance.* Tappi J., vol 71, no 1, pp 87-96, January 1988.

11. Eriksson, Leif G. *Deformations in Paper Rolls.* Winding Technology Conf. Proc., Swedish Newsprint Research Center (TFL), Stockholm, March 16-17, 1986.

12. Merin, P. *How Mechatronic Engineering Led to Intelligent Paper Roll Clamps.* TAPPI Finishing and Converting Conf. Proc., New Orleans, pp 63-70, October 24-27, 1993.

13. Skinner, J.R. *Newly Designed Automatic Slip Controlled Paper Roll Clamp.* TAPPI Finishing and Converting Conf. Proc., Chicago, pp 37-42, October 2-5, 1994.

14. Good, J. Keith and Fikes, M. *Predicting Internal Stresses in Center-Wound Rolls with an Undriven Nip Roller.* Tappi J., vol 74, no 6, pp 101-109, June 1991.

15. Good, J. Keith and Wu, Zan and Fikes, M.W.R. *The Internal Stresses in Wound Rolls with the Presence of a Nip Roller.* J. of Applied Mechanics, vol 61, no 1, March 1994.

16. Hakiel, Zig. *Nonlinear Model for Wound Roll Stresses.* Tappi J., vol 70, no 5, pp 113-117, May 1987.

17. Hakiel, Zig. *On the Effect of Width Direction Thickness Variations in Wound Rolls.* 1st Int'l Conf. on Web Handling, Oklahoma State Univ., pp 79-98, May 19-22, 1991.

18. Hakiel, Zig. *From Predictive Models to Profitability in the Web Handling Industry.* 3rd Int'l Conf. on Web Handling, Oklahoma State Univ., June 18-21, 1995.

19. Roisum, David R. *How to Measure Roll Quality.* Tappi J., vol 71, no 10, pp 91-103, October 1988.

20. Lucas, Robert G. *Internal Gearing in Paper Rolls.* TAPPI Finishing and Converting Conf. Proc, October 1974.

21. Lucas, Robert G. *Winder Crepe Wrinkles -- Their Causes and Cures.* TAPPI Finishing and Converting Conf. Proc., October 1981.

22. Roisum, David R. *Nip Induced Defects of Wound Rolls.* TAPPI Finishing and Converting Conf. Proc., Chicago, pp 89-98, October 2-5, 1994.

23. Holmberg, Michael W. *Theoretical and Experimental Studies of Air Entrainment in Wound Rolls.* MS thesis, Web Handling Research at Oklahoma State Univ., May 1992.

24. Good, J. K. and Holmberg, M.W. *The effect of Air Entrainment in Centerwound Rolls.* 2nd Int'l Conf on Web Handling, Oklahoma State Univ., June 6-9, 1993.

25. Bouquerel, F. and Bourgin, P. *Irreversible Reduction of Foil Tension Due to Aerodynamic Effects.* 2nd Int'l Conf on Web Handling, Oklahoma State Univ., June 6-9, 1993.

26. Knox, K. L and Sweeney, T. L. *Fluid Effects Associated with Web Handling.* Ind. En. Chem Process, vol 10, pp 201-205, 1971.

27. Tajuddin. *Mathematical Modeling of Air Entrainment in Web Handling Applications.* MS Thesis, Web Handling Research Center at Oklahoma State Univ., December 1987.

28. King, S. L. and Func, B. A and Chambers. Frank W. *Air Films Between a Moving Tensioned Web and a Stationary Support Cylinder.* 2nd Int'l Conf. on Web Handling, Oklahoma State Univ., June 6-9, 1993.

29. Good, James K. *Air Layer Control During Winding/Handling.* AIMCAL 1995 Fall Tech. Conf. Proc., Atlanta, October 2-4, 1995.

30. Lee, Bang-Eop. *Buckling Analysis of Starred Roll Defects in Center Wound Rolls.* Oklahoma State Univ., WHRC, Ph.D. Thesis, May 1991.

Chapter 20

Maintenance

In this chapter, written primarily for the maintenance department, we focus on how to check the condition of rollers, and restore them as needed. However, we also discuss how to evaluate a machine for possible initial design deficiencies and their correction.

Introduction

Rollers and related components do not last forever. Roller surfaces may abrade away to unacceptable cylindricity tolerances. Alignment will be lost due to stress relaxation and overload of frames as well as foundation settling. Brakes, bearings, couplings and mechanical drives will, given sufficient time, eventually wear out. Tension transducers and electrical drive components, especially those with analog parts, will go out of calibration.

The useful life or mean time between maintenance can vary enormously with the situation. At one extreme, a nip roller surface may need reconditioning in as little as a week. At the other extreme, a roller with no more attention than routine lubrication may outlast the useful life of a machine which can be more than a half century.

There are three factors which govern how long an item will be in service. The first is the rate at which tolerance is lost. A good example is the wear of a brake pad in mm/week. The second is the available tolerance. In our example this would be the brake pad thickness. The last is desired reliability of the component or process. Since all brake pads don't wear at precisely the same rates, we will need to change many before they actually score the brake drum, usually during a previously scheduled maintenance down.

Achieving a high process or machine reliability means replacing items before they wear out so they do not fail in service and unexpectedly shut down a line. This, however, comes at the cost of more parts, labor and machine downtime.

If we try and cut it too close however, we will also increase the cost of parts (a failed part can take out a good one) and have more machine downtime (part failure between scheduled downs). Thus, there is an economic optimum replacement time for every component. We can extend the life of a well maintained component either by reducing overall process reliability, or by more frequent checks of condition.

One of the indicators of a superb maintenance department is the tracking of service and life of components which are either expensive or need frequent replacement. This allows them to schedule replacement in a preventative rather than reactive fashion. Tracking life can be done with no more sophistication than log sheets in a file, though some large departments will computer database key records. All process rollers should have their own ID (identification) number stamped on the shell and an associated file folder in which goes the roller's entire manufacturing and rework history.

The other indicator of a superb maintenance department is scheduled checks of all serviceable items on a machine as indicated by the builder or by good practice. Unfortunately, many builders only provide the more obvious schedules, such as for lubrication. Only the rare and well written manual will detail tolerances such as for alignment, cylindricity and so on.

Thus, it is not wise to assume that if the manual did not mention a service check that it needn't be done. It may only mean that the builder did not know any better or does not write thorough manuals. Similarly, it is not wise to assume that the builder got the initial design right. I recall one machine which had gearboxes that needed to be rebuilt every other week. The vast majority of web machine parts should last many years if not decades.

Alignment

Roller alignment is perhaps the single most overlooked maintenance item on web machinery. Indeed, even most of the builders do not know that rollers must be optically aligned during installation and then periodically every few years.[1,2] Roller misalignment can cause a variety of web problems including stretching, breaking,[3] wrinkling,[4] registration difficulties and so on[5] as given in Chapter 2.

The simplest indicator of misalignment on light webs is the walking diagonal wrinkle (crossing a roller) or mild diagonal troughing (in the open web span). Do not be deceived by the deceptively benign appearance of the slight diagonal, it is an indicator of a very severe web stress distribution. Other evidence is a web that is consistently tight or loose on one side of one span.

Rollers which are grossly out of level can be detected with very accurate machinist's (or better yet master) levels. The accuracy is usually written on the level or case as something like 0.001"/ft which means that one mark off center is out of level by 1 mil per foot (truly gross). Before you use the level, however, the bubble needs to be calibrated to center. This is done by setting it on a very level surface and flipping it 180 degrees and making sure the bubble gives the same reading. Levels should be placed at the center of simply supported rollers and at the ends of cantilevered rollers. If the roller surface is uneven, several readings should be taken across the width and an average reported.

Rollers that are even mildly misaligned in parallelism, which is the most important direction, can be diagnosed with a tramming stick as shown in Figure 20.1. This stick is easy to make up with a piece of angle iron, a threaded rod and dial indicator. The angle iron base is held tightly against one roller and rotated so that the dial indicator sweeps against the other. The maximum reading during the sweep is recorded. The parallelism error is the difference in readings between the front and back of the roller. If this difference is more than 0.010" (0.3 mm), the roller is almost certainly out of alignment. (Check with the builder for their suggested tolerances.)

Figure 20.1
Alignment Check with a Tramming Stick

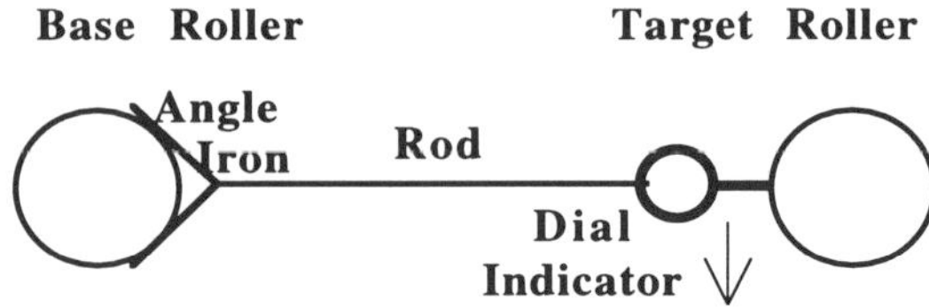

If a roller is out of alignment, do not attempt to correct it manually. First, there are no manual methods of alignment, only of misalignment detection. Second, moving a single roller is likely to just move the wrinkle or other problem into an adjacent span. Only in desperation because product is being destroyed in an area should one attempt to realign a roller manually, and even then one could as easily make the situation worse as better.

The only cure for misalignment is to optically (laser or transit) align all (process and transport) rollers in a section.[6] Optical alignment can be obtained from a number of qualified builders, contractors or engineering firms. There is even one company I know of, Oasis, that specializes almost exclusively in this kind of service. Unfortunately, alignment is an expensive proposition and can be quite difficult to sell to management. However, I consider it to be the first step in solving any wrinkling problem and many registration or web break problems.

Chapter 12 contains a lot of information on how to prepare for an alignment. Briefly, one should first eliminate as many nonessential rollers as possible with the help of a web handling knowledgeable process engineer. It is quite common to have 10-50% of the rollers in a line perform no useful function. Next, all clearances and play (such as on slides, pivots and bearings) must be reduced to an absolute minimum. This is because one can't align rollers to tolerances less than the slop in a machine part. Finally, all roller surfaces must be checked for cylindricity and reworked as necessary. This is because the roller surface is what the optical target sets on.

Balance and Vibration

Excessive vibration of web machinery can significantly shorten the life of mechanical components. Also, vibration can limit the useful speed of a process line or reduce product quality. Thus, it is always desirable to have a smooth running line.[7]

First, we must determine if the vibration of any machine portion is excessive. The most accurate measure is to use an FFT vibration analyzer or simple hand held vibration meter and compare the amplitudes with the suggested maximums given in Chapter 11. If such equipment is not readily available, one can sometimes use a dial indicator mounted on a rigid surface and indicating against a frame or roller member. In general, vibration exceeding about 5 mils on most members or 20 mils near nipped roll(er)s could be cause for concern. Also, one can try the classic nickel method. If a nickel balanced on edge will not stay on the machine, then the vibration might be improved. Finally, if the vibration feels uncomfortable to you, it probably is for the machine and process as well.

When a machine component or area is found to vibrate excessively, the next step is to determine the source(s). The most accurate method again is to use a FFT vibration analyzer. If this equipment is not available, this kind of service can be contracted. Unfortunately, the results are all too often a report filled with charts and readings, but without insight and direction. One should expect that vibration sources, magnitudes and often resonant mode shapes should be clearly identified using this powerful instrumentation.

The two most common sources of vibration both come from rollers. The first is simple roller unbalance. This is identified by a frequency that is 1X or higher multiple of the roller's RPM. Also, the amplitude will usually be highest at the offending roller and get progressively smaller at distances away from the roller. If roller unbalance is the source, the roller must be rebalanced as given in Chapter 11. This can be done in place or on a balancing machine.[8] If there is sufficient need for vibration analysis or balancing, entirely adequate instrumentation can be purchased for under $10,000 US.

Another significant, and often the most severe, source of vibration is from nipped roll(ers). Here, even perfectly balanced rollers will vibrate if the surface has even the tiniest amount of TIR or radial runout. Thus, the first check is to measure the amount of runout and compare it with builder specifications or Chapter 6. If runout is excessive, the roller surface must be reground. If the offending member is a <u>wound</u> roll, there is little that can be done other than change speeds (usually down) when the vibration becomes severe.[9]

In addition to the sources of vibration, there are three other very important considerations. The first are resonances where the machine runs much rougher at certain speeds. Short of an extensive machine rebuild, the only recourse (besides balance and TIR) is to avoid running at those speeds. The second is framework or foundation compliance. Vibration will take advantage of any lack of stiffness in a framework. The most common design error here is to cantilever the mounting of a roller, though more examples are given in Chapter 11. Finally, vibration will take advantage of any clearance in slides, pivots, bearings and mounts. Machine components must be as tight as possible, yet avoid binding.

Vibration may increase with the <u>square</u> of the machine speed. Thus, one might expect that a 10,000 FPM machine will vibrate more than a 100 FPM machine. In practice, however, the amplitudes of vibration of high speed machines are similar to low speed machines. The reason is that high speed machines have been designed robustly and machined with precision so that they can operate quite effectively in their elevated design range. Thus, we should expect that these tools can be also applied to more modest speeds to make them also run smoothly.

As you might have gathered, there is only a limited number of maintenance items associated with vibration. These are: balance rollers, grind rollers to tight TIRs and reduce clearance in mechanical parts such as bearings, slides, pivots and so on. After that, it is the responsibility of a mechanical design engineer to determine what other steps might be taken.

Bearings

There are really only three tasks that a roller bearing must perform. First, it must hold a roller in a fixed x,y position, which means the roller must have no measurable radial play. This can be checked with a dial indicator and hand lifting small rollers or using a wooden prybar on intermediate sized rollers. Second, the bearing must hold a roller in a fixed z position, which means the roller must have no measurable axial play. Again, axial play in small rollers can be checked with a dial indicator and prybar. Third, the roller must rotate with minimal friction. The simplest check of friction is to time how long it takes a roller to coast to a stop. Small and intermediate sized rollers should spin at least 10 seconds with a sharp hand spin.

However, a roller also must have a cost effective life. It is good practice to use a (L10) design life of 40 years, even if the machine's useful life is expected to be less, to ensure a high overall process reliability. Thus, bearing failures should be relatively rare.

The most expensive bearing of all is one that seizes unexpectedly during service and shuts down a process line. Thus, the emphasis here will be to estimate bearing condition so that they can be replaced near the end of their life, but well before catastrophic failure.

The most reliable way of determining bearing condition is to pull it off, clean it up and inspect the rollers, races, cages and clearance. Even the tiniest pits or scratches or other visible flaws mean that the bearing is very near the end of its life. The usual progression of failure is inner race, ball or roller, outer race and finally the cage. Unfortunately, it is not always practical to pull bearings off for inspection.

One way to get a reasonably reliable bearing condition evaluation is to use an FFT vibration analyzer which looks at key frequencies corresponding to the bearing geometry. This can be done with portable equipment,[10] or with permanently mounted continuous monitors[11] if process reliability is at a premium. These are just sophisticated versions of listening to a bearing through a wooden handled screwdriver.

Also, one can have the lubrication analyzed to determine if bearing metal particulate is present. This is also useful to diagnose certain causes of premature bearing failure due to water or dirt contaminants, due to thermal degradation of the lubrication and so on. However, experienced maintenance personnel can sometimes also diagnose these types of problems by sight, feel or smell of the lubricant.

Finally, many bearings tend to run hotter toward the end of their lives. The point here is not whether the bearing temperature is in spec. Rather, it is to determine if the bearing temperature is warmer than in the past or when compared with its neighbors. This is perhaps the easiest check because all one has to do is to instruct the oiler to touch and monitor bearing housing temperatures during his normal duties.

If a bearing fails in less than 10 years of service, its reason for premature failure should be determined. This is not to satisfy an academic curiosity. Rather, it is to determine if anything can be done to prevent it or its cousins from failing again and possibly shutting down a line. Since bearing failure analysis is beyond the scope of this short section, the interested reader should look through the many guides and articles written by bearing manufacturers for more information.

In general, premature bearing failure can be the result of inappropriate design or maintenance. Design problems include: radial or axial overload, overspeed, no provision for thermal expansion of the free bearing, poor fits, and inappropriate seals or lubrication. Maintenance problems include: inappropriate lubrication or lubrication schedule, misalignment, unbalance, poor fits, impact loading of the bearings during handling and so on.

In summary, it is the role of the maintenance department to prevent unexpected bearing failure by appropriate practices and continuous checking. If bearing life is short, the maintenance department should also be the vanguard of problem solving.

Cylindricity

All rollers in a web line must be very true cylinders as given in Chapter 6 and elsewhere in the book. If not, the stresses in a web will be skewed causing a variety of problems such as wrinkles, puckers, baggy lanes, web breaks and so on. Cylindricity must be specified initially and checked periodically. Cylindricity is usually measured as TIR (Total Indicator Reading), maximum diametral variation (taper) and/or station-to-station variation.[12] Additionally, cover hardness needs to be uniform to maintain a predictable geometry under loading.

Specifications for ingoing (at purchase or rework) and outgoing (for rework) cylindricity may be given by the builder. If not, they will need to be developed by a web expert based in part upon the needs of the particular application. The ingoing tolerances for cylindricity may be as tight as 0.1 mil for hard process nips or as much as 10 mils for very large rollers used on very flexible webs. Outgoing tolerances may be 2-4 times ingoing tolerances.

The easiest measure of cylindricity is TIR, as shown in Figure 20.2, because it only requires a readily available dial indicator. TIR needs to be checked as several CD positions or stations. Excessive radial runout can cause problems such as web flutter and nip roller vibration.

Equally important is the maximum diametral variation and particularly the maximum variation between two adjacent stations (CD positions). These can only be measured in situ with a precision caliper as shown in Figure 20.3. PI or flat tapes, especially those lacking a vernier scale, are simply not accurate enough for most applications. Diameter variations should be checked as a function of rotation and CD position.

The life between regrindings can vary enormously. The shortest are measured in days for some applications of covered calendering, coating and printing. The longest life is well over a half century where webs are not abrasive and the shells are ferrous.

Figure 20.2
TIR Measured With a Dial Indicator

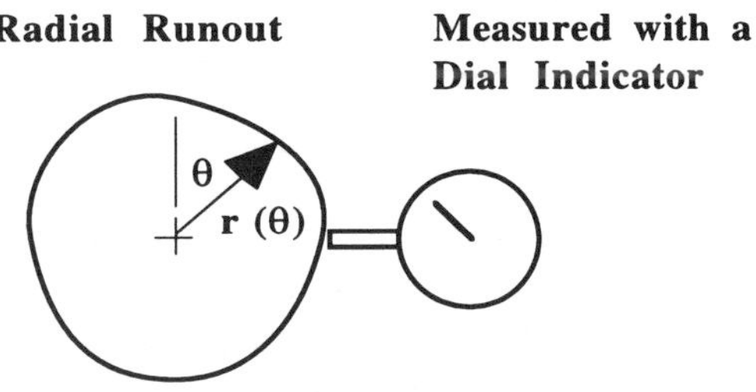

Figure 20.3
Diameter Measured With a Micrometer

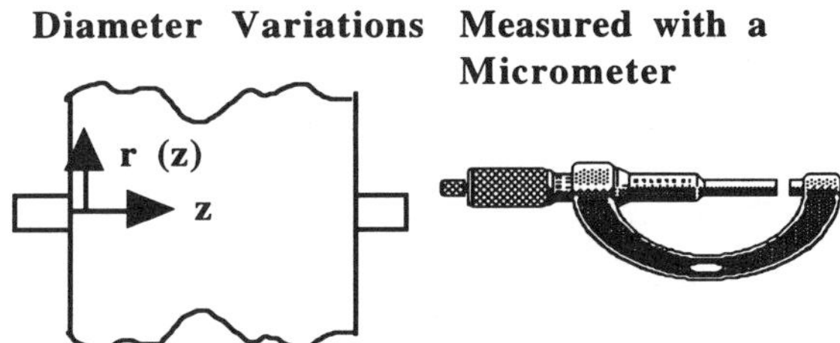

Sometimes life is short due to the process (application, loads, speeds, etc.), an abrasive web (mineral fillers in paper, fiberglass, etc.), doctors and so on. Then, it may be desirable to treat roller surface with hard coatings such as anodize, chrome or tungsten carbide. Cover materials also may need reformulation to give adequate life. If life still remains short, it is important to make sure that roll changeout is smooth and easy. Also remember to maintain alignment during roller changeout. Since roller cylindricity can be a regular maintenance item, it is important to keep track of the rework history for each roller and position.

Finally, no roller can be cylindrical if it deflects excessively. A rough rule of thumb is that the slenderness (length/diameter) should be 10-15 for nipped rollers and 15-20 for unnipped rollers. However, an engineer should verify deflection as calculated in Chapter 3.

Drive Controls

The primary purpose of a drive is to create an appropriate tension setting in each section, and to maintain an appropriate tension tolerance despite internal and external upsets. Thus, our most useful diagnostic tool will be load cells.

As discussed in Chapter 2, the setpoint tension should be between 10 and 25% of the breaking strength of the web in that section. Exceptions to this are rare but include extremely strong materials which may not need as much tension to stay flat and in control, and drawing where tensions must be similar to the yield point of the material in order to permanently elongate it. Tension sensitivity is often an indicator of poor process (material or machine) health.

Also, we discussed tension control tolerances which should not allow variations of more than 5% of the setpoint during steady state and 10% of the setpoint during transients (speed changes etc.) as read by fast acting load cells without significant damping or filtering. There is no good substitute for tension readouts as a barometer of process health. For example, motor speed or load readings do not solely reflect tension changes. Similarly, the absence of dancer movement could mean it is bound up with excessive friction instead of tightly controlling tension, while movement of the dancer could imply an appropriate compensating response to tension fluctuations or that the tension indeed fluctuates undesirably.

Perhaps the greatest drive tension control challenge is during speed changes. Thus, it is imperative to check the acceleration and deceleration rates. Acceleration rates can be calculated as the change in speed (meter) divided by the time in seconds (stopwatch). As a rough guideline, many machines which are intended to run continuously will accelerate at about 10 ft/min/sec (3 m/min/sec). Slower rates may mean an excessive time to get to speed while faster rates may put unnecessary demands on the drives. However, faster machines which run in a discrete start-stop manner, such as winders, may accelerate and decelerate at around 100 ft/min/sec (30 m/min/sec). The fastest acceleration I am aware of is almost 1000 ft/min/sec.

Sometimes it is necessary to check the shape of the acceleration and deceleration curves for the master speed reference using a fast strip chart recorder or similar means. The curve should have a small amount of rounding on both ends of a very linear section. During speed changes, ammeters and other load measurements should move firmly, smoothly and without overshoot. Finally, one may need to check to make sure all drives track together and that the gain (of fixed rollers) and inertia compensation (winding and unwinding rolls) are appropriately tuned. Once tuned, modern digital drives should never drift, but this is a common problem with analog drive controllers.

There are three common drive sizing errors. First, the drive can be undersized so that it can't follow a command for high torques. However, this usually results from the plant who decides to run a heavier grade than the machine was built for. Second, the drive can be oversized so that it lacks finesse when lightly loaded. The indicators to look for here are pneumatic devices operating under 10 psi or electric devices operating at a few percent of their rated torque or power. Third, a tension or torque controlled drive can be loaded up with mechanical losses by having too many belts, gearboxes, slaved rollers and so on attached to a single drive point. Doing this is asking for the proverbial "weighing of a pea in a dump truck" problem. Most of the drive's power should be to generate tension rather than overcome mechanical losses. An easy check for this is to compare the load (operating condition) versus no load (no web) readings.

About the only maintenance items that can be performed on drives is to periodically calibrate load cells, check for excessive dancer friction, and check for excessive mechanical losses coming from components such as bearings, gearboxes and so on. If the drive still doesn't hold tensions closely after these items are taken care of, expert drive engineers must be called upon to diagnose and remedy the problem. More often than not, small plants will not have electrical engineers with sufficient drive expertise to tune or suggest modifications to drives so that trusted vendors must be called upon. This is especially so for drives on start stop operations such as winders.

Drive Mechanicals

Mechanical elements transmit torque from an actuator, such as electric motors, pneumatic brakes and so on, to the web. The mechanical elements may include couplings, shafts, belts, gears, bearings and so on as described in Chapter 13. The mechanicals must transmit this torque without excessive losses, torsional play, and with high reliability.

A universal indicator of mechanical losses is heat. Thus, a simple check of all elements can be made by touching them, while the machine is stopped for safety reasons, and noting the temperature. Obviously, the temperature must be within the rating of the component. However, what we are looking for here is a component whose temperature is higher than it used to be or higher than a similar neighbor. Sometimes losses will also be accompanied by noise so that any changes in amplitude, pitch or character should be investigated.

Bearing problems are a common source of mechanical losses. A quick check of idler roller bearing freeness can be made on many machines by noting whether the roller will coast for at least 10 seconds with a sharp hand spin. Friction on larger rollers or driven systems may have to be checked using a torque wrench. The vendors maintenance manual should be consulted to verify the correct operation of gearboxes and transmissions which often have a very high source of mechanical losses.

A drive train must transmit torque from an actuator to a roller without excessive torsional play or windup. Play and windup will make it difficult for a drive to start, stop or change speed smoothly. Torsional play can be checked by locking up the actuator and torquing the roller from one end of the clearance deadband to the other. Some couplings and winder/unwind clutches are notorious for their looseness.

Torsional windup is usually the result of poor design and can't be addressed by maintenance. Torsional windup can result in the necessity to detune (make sluggish) a drive to keep it stable (no overshoot or oscillation). Long belts or inshafts are the most common source of windup.

Failed or worn out mechanical drive components are a common source of machine downtime and should be minimized. However, most components should be capable of years if not decades of trouble free performance if they are properly sized and maintained.

The first step to achieving high machine uptime is good record keeping of what items are failing on a machine. A very visual method is to place stick pins at component locations on a machine elevation drawing. This will vividly demonstrate where, but not how or why. Paper filing using cards or folders can easily keep track of work, but is difficult to search for trends. The most comprehensive systems use computerized databases.

If several locations of a component type are experiencing a short life, it could indicate either an overall design or maintenance shortcoming. If a component has much more trouble in one area than another, it may be due to lack of needed attention there. In any case, the component builder should provide a list of maintenance checks and schedules.

All too often, worn and failed parts are simply discarded. A wealth of information can be obtained from them by cleaning them up and examining them closely. Indeed, the failed part will record the fingerprint of the failure mode under close examination with a magnifying glass. Only by knowing the mode of failure can we begin to understand what caused it and how it can be prevented in the future.

Similarly, the environment around the part should be examined. For example, a magnet will pick up ferrous metal bits. Normal wear will be as fine as dust while abnormal conditions will result in larger grains, chips or slivers. Belts when overloaded will tend to make black powder near the pulleys.

In closing, check drive elements to make sure they are operated within their rated loads and that service specifications such as alignment and balance are closely followed.

Grinding

In this section, we merely highlight some of the more important considerations for precision grinding. More information can be obtained from roller and/or grinding machine suppliers.

Bearings

The best grinding is often done without bearings, so that journal condition should be evaluated. However, if the roller is ground in its bearings to save time, their condition should be evaluated and corrected. Bearings must be protected from grinding debris and lubricant.

Depth

The amount of material to be removed depends on the as received condition of the roller. The coarse grind must take as many passes as necessary to just clean up 100% of the surface that the web will see. However, some application exceptions will allow less metal to be removed if scratches or narrow dents can be tolerated, and more metal may have to be removed if a nipped roller is barred.

Grinding Machine Condition and Maintenance

Alignment of the grinder must be checked periodically. The foundation must be sturdy and free of vibration induced by nearby web machinery, fork trucks and so on. Isolation mounting components should be inspected for condition. There should be no looseness in bearings, drive belts, gears, stands, slides, screws and so on.

Grinding Parameters

Grinding parameters should be selected from a chart or by experience for each shell/cover material and stage of the grinding process. The most important parameters include wheel grit and composition, wheel infeed rates, wheel traverse speeds, wheel speeds and coolant.

Grinding Wheel

Grinding wheels must be dressed and balanced frequently. Grinding wheels should be cleaned and stored with their axes vertical to avoid water collection on one side of the periphery which will upset balance.

Roller Measurements

Tools used to measure the quality of the roll shapc during thc grinding proccss includc: a Pi tape for nominal diameter, saddle micrometers (0.0001" accuracy) for diametral variation and roller shape, a dial indicator (0.0001" accuracy) for runout, and surface roughness gage.

Shell/Cover Condition

Shells and covers should be inspected for condition. Their surfaces often retain a fingerprint of machine or process problems that can be read and corrected. Surface failure or wear modes are discussed in Chapters 6 and 9.

Storage

Rollers are sometimes stored before and after grinding. The storage section of this chapter will give more information on this subject. However, it is very important that the roller awaiting grinding be turned 1/4 revolution every day or so to avoid taking a set. Also, allow 1-2 days of final storage in the grinding room to allow the roller to come to equilibrium temperature.

Temperature

The roller must be allowed to reach room temperature before grinding is started. This may take as long as 2 days for a large heated roller. The grinding room temperature must be consistently maintained to avoid direct sunlight, drafts from open doors and forced ventilation. Even more important is to avoid excessive heat generated by high infeed rates, high speeds or improper coolant or grinding wheel.

Thickness

Always observe the roller manufacturer's specification for minimum shell/cover thickness.

Unbalance

Rollers may need a coarse balance before grinding in addition to a final balance after.

Heated/Chilled Rollers

The application and mechanics of chilled and heated rollers are covered in more detail in Chapter 15. Here, we merely highlight a few important maintenance considerations.

Temperature, Average Setpoint

The average setpoint temperature can be a critical process parameter. This temperature should be specified by the equipment manufacturer or a plant process engineer for each grade. However, maintenance should check the temperatures periodically and verify the calibration of any temperature readouts. The most important, as far as the web is concerned, is the shell or cover surface temperature. However, interior or fluid temperatures should also be checked. Temperature can be read to about a 5°F accuracy with an instrument quality hand held pyrometer, optical pyrometer, embedded thermocouple or mercury or bimetallic thermometer.

Temperature, Uniformity

In addition to setting an average temperature, the temperature must be uniform to within some application specified tolerance. The problem, however, is that most of the chilled or heated rollers can maintain a temperature uniformity closer than can be measured with the above instruments. Thus, more indirect means must be used to infer temperature uniformity. For fluid systems, temperature uniformity will be improved with higher flow rates and reduced difference between ingoing and outgoing temperatures. For electrical induction or resistance heated systems, consistency will be improved by working in the midpoint power range of the heaters. Low power may mean poor consistency, while high power may also mean poor consistency as well as short life.

Startup/Shutdown

Observe all of the manufacturer's guidelines for changing the roller temperature during startup and shutdown. In some cases, this may mean running at a reduced power for a specified period of time. Be especially careful of the cool down cycle which can generate thermal cracks in the shell material. For example, it is seldom permissible to hose a hot roll down with water during a cleanout. Heated roller shells should be checked periodically for thermal cracks with a dye penetrant or other means. Thermal cracks in steam heated or other pressure vessels can be a safety hazard.

Bearings

There are several potential bearing issues with heated/cooled rollers. First, one side of roller (usually the tending side) must have a floating bearing to take up thermal expansion. Also, the bearing must be appropriately positioned within its available travel (e.g., on the inside for a heated roller at room temperature). Second, the bearing may need special clearances for high temperature applications. Third, the seals lubrication must be appropriate for the temperature range.

Balance

Some chilled or heated rollers present special balance problems. These include scale and contaminant buildup, condensate from steam, or air bubbles in liquids.

Connections

Follow the builders recommendations for rotary fluid joints or electrical contacts. These components are often high maintenance items.

Safety

Heated and sometimes cooled rollers can pose additional safety issues. Hot rolls and fluids can burn operators and maintenance people, so that appropriate precautions must be observed. Hoses and fittings which pass hot fluids must be checked periodically for integrity, and no fitting loosened until the pressure is bled off. Electrically heated devices must be checked periodically to make sure connections, wiring and other insulation is sound. Pressure vessels require special treatment that will not be covered here. Always observe recommendations from the builder, your plant safety procedures, codes, regulations and so on.

Nips

Nipped rollers are covered more thoroughly in Chapter 5. Here, we merely highlight a few important maintenance considerations. The objective is to maintain a consistent nip pressure across the entire roller set and at all times.

Nip Load, Average Setpoint

The average nip load should be set according to the builder's or plant process engineer's specification for each grade. The nip load (force per unit width) is a result of both roll weight and cylinder pressure. The builder or engineer will determine the required nip load and translate that into a cylinder pressure. In general, the pressure on one side of the cylinder will be fixed (except for unloading) and the other will vary by grade. In general, absolute cylinder pressures or cylinder pressure operating ranges which are less than 10 PSI pneumatic or 100 psi hydraulic could indicate a poor design.

Nip Load, Uniformity

The quality of the nip load uniformity is largely determined by the amount of mechanical friction in the pivots, slides and cylinder seals. This deadband friction can be measured as the difference between load and unload pressures as a regulator is slowly turned. Load/unload pressure differences in excess of 10 PSI pneumatic or 100 psi hydraulic could indicate mechanical binding and should be checked regularly.

Cylindricity

Perhaps the most important nip quality parameter is the cylindricity of both rollers. Cylindricity is measured as TIR (Total Indictor Reading), maximum diametral variation (taper via precision caliper), and especially station-to-station diametral variation (difference between two adjacent CD positions of 20 measurements across the width). The cylindricity tolerances of nipped rollers can range from 0.0001" for new metal rollers in hard nip to 0.010" for soft covers in need of regrinding or replacement.[13]

Nip Impressions

Dynamic nip impressions should be taken to verify a uniform mating of nipped rolls. This merely requires access to a machine for 15 minutes, nip impression paper, and a procedure as described in Chapter 5. Nip impressions should be carefully documented and stored in maintenance files.

Crowns and Skewing

Some nipped rollers are crowned (barrel shaped) or skewed (misaligned in square) to even out the nip which tends to be high at the ends and low in the center.[14] The crown magnitude and quality should be specified by engineering and checked via nip impression paper at the operating load. Crowns in excess of 0.00015 x width indicate undersized rollers. Roller skewing must be optically aligned for center and adjusted via a labeled vernier.

Drives

In most cases, one of the nipped rollers must be driven. The drive strategy and tuning should be checked by an electrical engineer. In general, however, it is very poor practice to gear, belt, or otherwise speed control both rollers in a nip. If two drives are required, one should be in torque control. Also, it is not a good practice to speed control any soft covered nip roller.

Common Issues

The most common maintenance issues with nipped rollers include shell/cover wear and vibration.

Safety

Nips present one of the largest safety hazards in any web manufacturing or converting plant. Make sure all guards are in place, interlocks are operational, operators and maintenance are trained and all safety instructions, codes, regulations and practices are observed.

Purchasing

The most common component in any web line is rollers. With the original machine purchase, the vendor will have supplied the rollers and presumably complete maintenance instructions. However, there are many other goods and services that may be needed in order to support a web line.

Spares

Most rollers, unless they are unusually expensive, should be spared. Those that wear out frequently (a few months or less) may need more than one spare. Whether to carry spares for any position is a judgment call. If a roller is lost without a spare, the whole line may be down for at least a week, which is very costly. If excessive spares are purchased, inventory and storage costs are increased.

Roller Sources

Very large rollers can be obtained from dozens of shops, while smaller plain rollers can be obtained from hundreds if not thousands of suppliers. The first place to check is with the OEM (Original Equipment Manufacturer). Next, one can check with the larger or more reputable roller manufacturers. Be careful; there can be a large quality difference between suppliers. The best insurance is to be thoroughly informed on the needs of the application, put it into specifications and tolerances, and caveat emptor (buyer beware).

Delivery

The delivery of rollers can vary from a few days for a plain idler roller to the better part of a year for large, process or specialty rollers.

Bearings

Bearings are more perhaps likely to fail than other parts of a roller and are cheaper to spare out. However, the plant's shop must also carry all tools necessary to remove the roller as well as pull and press the bearing.

Instruments

To measure the quality of a roller surface, the following tools may be needed: precision Pi tape, outside micrometers in sizes to match all roll diameters, precision dial indicators and possibly surface roughness meters and cover hardness meters.

Grinders/Lathes

Plants with many machines and thus rollers may consider purchasing their own lathe/grinder to refinish rollers.

Balancers

Rollers can be balanced in the shop or in place without dedicated balancing machines if vibration instruments and software costing less than \$10K are adapted to the application.

Alignment Checks

Alignment checks can be made with long machinists or better yet masters levels, precision Pi tapes and precision dial indicators. Alignment checks need to be performed every time a roller is changed out and periodically every few years.

Alignment

While alignment can be checked with simple hand tools, realignment must be done by optical transit. A transit setup may cost upward of \$50K and requires skilled labor to operate. However, the cost of outside alignment services is also very expensive.

In-Machine Grinding and Coating

In rare occasions, rollers are resurfaced and coated in place. These will invariably be done by an outside service. Check their reputations and references carefully.

Rigging

Rigging is the attachment of devices to lift a roller out of a machine. Here, we merely highlight a few maintenance considerations.

Load Capacity

Never exceed the load capacity of any lifting device including straps, hoists, cranes, lift trucks. Thus, you must know both the weight of the roller assembly and the rating of each lifting component. Also beware of sideload and overhung load capacities. All rigging components should be inspected periodically.

Safety

Always follow all applicable roll removal and handling recommendations by the roller builder, rigging component manufacturers, plant safety rules, codes and so on. Never, place yourself (or even a portion of yourself) under a lifted roller.

Straps and Ropes

Most rollers will be lifted via a pair of straps or wire ropes attached to each of the journals. Dual hooks should be used for rollers more than a few feet wide, and in no case should the strap angle exceed 30° from vertical or rub on the edge of the roller shell. Metal rollers can be rigged with straps on the shell, but covers could be damaged by even very wide straps. Obviously, chains should never contact any part of any roller.

Bearings

It is very difficult to set a roller down on its bearing housing onto a hard surface without risking impact damage to the bearings in the form of brinelling. While this is unavoidable when setting a roller into a machine, impact can be reduced by setting bearing housings on an appropriate wooden block, and eliminated by setting the shell onto v-shaped blocks (metal) or soft roll cradles (covers).

Storage

Rollers may need to set in storage prior to use in machine or awaiting refinishing. Follow all builder recommendations in this regard.

Floor

Never set a roller shell or cover on the floor. If a small amount of temporary storage is needed, soft sling cradles or journal support stands can be constructed. To avoid a tipping hazard, the base of the stands should be at least 1/2 of the height.

Racks

Racks are a good solution for long term storage of many rollers. Racks should safely hold the required loads, have a base at least 1/2 of the height for stability, and have easy access for moving rolls into and out of storage. Lower storage positions should be used before upper.

Rollers may have to be prepared for long term storage such as changing out the lubrication, coating exposed metal surfaces to avoid oxidation and so on as instructed by the builder. Also, some rollers should be turned 1/4 revolution every couple of days to avoid taking a set which will make it run rough during startup in a machine. An alternative to turning is to support the roller 22% of the face inboard of the shell to reduce bending stresses. However, shell support is applicable only for metal shells and if the rack is padded.

Transport

The requirements for transport are similar to storage with one exception. They must be protected from impacts, bumps and scrapes. In most cases, the shell is covered with a cardboard or other protection. The roller is then mounted in a wooden box with appropriate labels for handling (fragile) and rigging (lift here).

Surface Defects

This section describes a few common metal[15] and covered[16,17] roller surface defects that can be found on rollers. These defects are a fingerprint of their causes which should traced down and eliminated.

Barring
Barring is a series of equally spaced longitudinal patterns caused primarily by nip roller bounce. A barred roll must be reground well below the lowest point.

Chipping
Metal-metal nip can chip away the edges of a roller if they are not dubbed (relieved to a smaller diameter).

Corrosion/Rust
Many of the ferrous roller materials are prone to rust due merely to room humidity and can corrode when in contact with various solids, liquids and gases in the environment. Corrosion may indicate the need to change to a different metal such as chrome plating, electroless nickel, stainless or a nonmetal such as carbon fiber.

Cover Blowout
A loose or lost cover can result from the following failure modes: adhesion to the metal, adhesive failure, adhesion to the cover, and cover failure. Find the initiation site to determine the cause.

Cracks
Cracks on metal shells can result from shock cooling of a heated roller or some types of corrosion. The presence of micro-cracks should be checked with a dye penetrant or similar means. Cracks on covered rollers indicate oxidation.

Crease Marks
Crease marks can be made when a web folds over going into a nip.

Dents
Dents are commonly caused by setting a roller on the floor, dropping tools or parts onto a roller, foreign objects passing through a nip, and transport or handling.

Knife Cuts
Knife cuts are one of the most common as well as most avoidable causes of cuts in roll covers and metal shells. Knife cuts are usually made by operators when they try to remove a wrap. A better way is to unwrap the roller by pulling on the tail.

Roughness/Smoothness
The surface roughness of a roller as supplied will normally be 4-32 microinch rms and has a pattern left by the last finishing operation. Excessive web/roller slippage is evidenced by a surface that becomes smoother and shinier.

Scoring
A roller can be scored (annular scratches) due to contact with a doctor or other stationary object.

Spalling
Spalling is a failure that begins as an interior crack parallel to the surface that propagates to the surface leaving a pit. Spalling indicates an excessive contact or nip load stress. Shock cooling can also cause spalling, but this failure begins perpendicular to the surface.

Stickies
Contaminants sometimes stick to a roller instead of the web. Sticking can be reduced by using Teflon impregnated tungsten carbide or other release coatings.

Swelling
Many covers are attacked by solvents which may show as a swelling.

Wear
Metal or covered surfaces that wear out too quickly may not be hard enough for the possibly abrasive web product.

Bibliography

1. Weiss, Herbert. *Yearly Machinery Alignment Controls Quality Loss, Cost.* Paper, Film & Foil Converter.

2. Anon. *Realignment Program Keeps Quality In Check.* Paper Film & Foil Converter, April 1990.

3. LeBel, Gary A. *Improving Dryer Section Runnability through Optical Alignment.* Tappi J., vol 77, no 5, pp 285-288, May 1994.

4. Roisum, David R. *What Causes Diagonal Shear Wrinkles?* Converting Magazine, Web Works column, pp 22, June 1994.

5. Boggs, R.E. *Rotating Machine Misalignment:* The Silent Disease but not the Symptom Tappi J., vol 73, no 12, December 1990.

6. Allen, George Jr. *Optical Tooling for Paper Machines.* TAPPI Engineering Conf. Proc., pp 595-599, November 1994.

7. Wouk, Victor. *Machinery Vibration: Measurement and Analysis.* McGraw-Hill.

8. Wouk, Victor. *Machinery Vibration: Balancing.* McGraw-Hill, 1995.

9. Roisum, David R. *Winder Vibration Can Reduce Operating Efficiency and Increase Maintenance.* Tappi J., vol 71, no 1, pp 87-96, January 1988.

10. McLain, D. and Hartman, D. *New Instrumentation Techniques Accurately Predict Bearing Life.* Pulp & Paper, February 1981.

11. Liddle, Ian and Reilly, Steve. *Expert Systems Offer Precise Analysis, Diagnosis of Mill Rotating Machinery.* Pulp & Paper, vol 67, no 2, pp 53-55, February 1993.

12. Maniatty , G.S. *Roll Grinding Measurements and their Effect on Paper Making.* TAPPI Finishing and Converting Conf. Proc., New Orleans, pp 311-320, Oct. 24-27, 1993.

13. Kueh, Howard E. *Calendering - The Papermaker's Last Chance.* Pulp & Paper, July 24, 1967.

14. Kueh, Howard E. *Calendering - Controlling Roll Deflection.* Pulp & Paper, July 31, 1967.

15. Anon. *Care of Calender Rolls, Repair, Regrind, Defect.* SHW Internal Report, September 1984.

16. Beucker, Albert W. *On-Machine Maintenance of Rubber Covered Rolls.* Tappi J., vol 52, no 7, pp 1283-1288, July 1969.

17. Giguere, P. *Care and Maintenance of Polyurethane Roll Covers.* CPPA Conf. Proc., Montreal, January 31, 1989.

Appendix A - Units and Conversions

Given below is a set of conversion factors for units commonly used in winding and web handling. For further information on paper lab test value conversions, see the TAPPI Technical Information Sheet 0800-01. To convert from metric (right column) to English (left column), divide by factor instead of multiply.

From	Multiply x	To
Area		
in^2	6.452	cm^2
in^2	6.452e-4	m^2
ft^2	0.09290	m^2
Basis Weight		
lb/3000ft^2	1.628	g/m^2
lb/1000ft^2	4.883	g/m^2
Caliper		
mils (0.001")	2.540e+5	angstrom
mils (0.001")	25.40	micrometer
mils (0.001")	0.02540	millimeter
Energy & Work		
in-lb	0.11297	joule
ft-lb	1.3556	joule
BTU	1054.2	joule
watt-hours	3599.	joule
Density		
lb/in^3	27.68	g/cm^3
lb/in^3	27680.	kg/m^3
lb/ft^3	16.019	kg/m^3
Force		
ouncef	0.2780	newton
poundf	4.448	newton
kilogram	9.806	newton
Force/Length		
Tension, Nip, Torque, Strength		
lb/in (PLI)	0.17513	kN/m

From	Multiply x	To
Force/Area		
Pressure, Stress, Modulus		
lb/in^2	6.895	kPa (kN/m^2)
lb/in^2 x 10^3	6.895	MPa
lb/in^2 x 10^6	6.895	GPa
Frequency & Rotational Speed		
CPM, RPM	0.016667	Hz
Rad/sec	0.15916	Hz
Length		
inch	25.40	millimeter
inch	2.540	centimeter
inch	0.0254	meter
foot	30.48	centimeter
foot	0.3048	meter
mile	1609.3	meter
mile	1.6093	kilometer
Mass		
ounce	28.35	gram
pound	453.6	gram
pound	0.4536	kilogram
ton	907.2	kilogram
Power		
horsepower	0.7457	kilowatts
Speed		
ft/min	0.3048	m/min
Temperature		
Celsius = (5/9) (Farenheit - 32)		
Torque & Bending Moment		
lb-in	0.11299	N-m
lb-ft	1.3558	N-m
Volume		
in^3	16.387	cm^3
ft^3	0.02832	m3

Appendix B - Selected Bibliography

Air Flotation

Fraser, W. A. R. *Air Flotation Systems: Theoretical and Practical Applications.* Paper, Film & Foil Converter, vol 57, nos 5-6, May-June 1983.

King, S. L. and Func, B. A and Chambers. Frank W. *Air Films Between a Moving Tensioned Web and a Stationary Support Cylinder.* 2nd Int'l Conf. on Web Handling, Oklahoma State Univ., June 6-9, 1993.

Knox, K. L and Sweeney, T. L. *Fluid Effects Associated with Web Handling.* Ind. En. Chem Process, vol 10, pp 201-205, 1971.

Tajuddin. *Mathematical Modeling of Air Entrainment in Web Handling Applications.* MS Thesis, Web Handling Research Center at Oklahoma State Univ., December 1987.

Alignment

Allen, George Jr. *Optical Tooling for Paper Machines.* TAPPI Engineering Conf. Proc., pp 595-599, November 1994.

Boggs, R.E. *Rotating Machine Misalignment:* The Silent Disease but not the Symptom Tappi J., vol 73, no 12, December 1990.

Dunn, Jim M. and Pepin, S. and Allen, D. and Leonard, W. and Carter, M. *Precision Alignment of Winders.* TAPPI Finishing and Converting Conf. Proc., San Diego, CA, pp 119-124, October 15-18, 1995.

Gehlbach, Lars S. and Good, J.K and Kedl, Douglas M. *Prediction of Shear Wrinkles in Web Spans.* Tappi J., vol 72, no. 8, pp 129-134, August 1989.

LeBel, Gary A. *Improving Dryer Section Runnability through Optical Alignment.* Tappi J., vol 77, no 5, pp 285-288, May 1994.

LeBel, Gary A. *Optical Diagnosis for Misalignment.* Converting Magazine, pp 46, 48,50, November 1993.

LeBel, Gary A. *Paper Machine Alignment.* Papermaker, pp 12-14, December 1994.

Roisum, David R. *What Causes Diagonal Shear Wrinkles?* Converting Magazine, Web Works column, pp 22, June 1994.

Weiss, Herbert. *Yearly Machinery Alignment Controls Quality Loss, Cost.* Paper, Film & Foil Converter.

Balance

Baur, Paul. *Field Balancing of Rotating Machinery.* Power, October 1983.

Schminke, K.H. *The Importance of Roll Balancing on High Speed Paper Machines.* Papier, July 1991.

Wowk, Victor. *Machinery Vibration, Balancing.* McGraw-Hill, 1995.

Control

Weiss, Herbert L. *Control Systems for Web-Fed Machinery.* Converting Technology Co, Milwaukee WI, 1983.

Converting

Satas, Donatas. *Web Processing and Converting Technology and Equipment.* Van Nostrand Reinhold Co. Inc., 1984.

Weiss, Herbert L. *Coating and Laminating Machines.* Converting Technology Co, Milwaukee WI, 1977.

Weiss, Herbert L. *Technically Speaking.* Converting Technology Co, Milwaukee WI, 1978.

Critical Speed

Roisum, David R. *Winder Vibration Can Reduce Operating Efficiency and Increase Maintenance.* Tappi J., vol 71, no 1, pp 87-96, January 1988.

Shelton, John J. *Deflection and Critical Velocity of Rollers.* 3rd Int'l Conf. on Web Handling, Oklahoma State Univ., June 18-21, 1995.

Wowk, Victor. *Machinery Vibration, Measurement and Analysis.* McGraw-Hill, 1991.

Wowk, Victor. *Machinery Vibration, Balancing.* McGraw-Hill, 1995.

Cylindricity & Surface Roughness

American National Standards Institute. *Surface Texture.* American National Standards Institute, ANSI B46.1, 1978.

Knapp, J.K. and Lampman, R.D. *Characterization of Calender Roll Surface Features Using Interferometric Profiling.* TAPPI Finishing and Converting Conf. Proc., San Diego, CA, pp 15-24, Oct 15-18, 1995.

Maniatty, George S. *Roll Grinding Measurements and their Effect on Paper Making.* TAPPI Finishing and Converting Conf. Proc., New Orleans, pp 311-320, Oct 24-27, 1993.

Maniatty, George S. *Roll Grinding Measurements and Their Effect on PaperMaking.* Paper Age.

Twitchen, D. *Paper Machine Roll Grinding Tolerances.* Canadian Pulp & Paper Association, Technical Section, Data Sheet G-20, May 1988.

Dancers

Bak, David J. *Dancer Arm Feedback Regulates Tension Control.* Design News, vol 43, no 6, pp 132-133, April 1987.

Roisum, David R. *How do Dancers Work?* Converting Magazine, Web Works column, pp 16, January 1995.

Weiss, Herbert L. *Getting Better Quality With Dancer Controlled Unwinders.* Package Printing, vol 28, no 12, pp 11-12 103 and vol 29, no 1, pp 14, 50-52.

Deflection

Luthi, Oscar. *Ring Stresses and Deflections of Thin-Walled Rolls and Application.* Tappi J., vol 56, no 8, pp 121-124, August 1973.

Shelton, John J. *Deflection and Critical Velocity of Rollers.* 3rd Int'l Conf. on Web Handling, Oklahoma State Univ., June 18-21, 1995.

Draw Control

Roisum, David R. *How does Draw Control Work?* Converting Magazine, Web Works column, pp 16, February 1995.

Spielbauer, Thomas M. and Walker, Timothy J. *Theory and Application of Draw Control for Elastic Webs with Nipped Pull Rollers.* 2nd. Int'l Conf. on Web Handling Proc., Web Handling Research Center, Oklahoma State Univ., June 6-9, 1993.

Drive, Electric

Anon. *Controlling Power Transmission Systems.* Library of Congress Card #77-85721.

Anon. *TAPPI Paper Machine Drives Short Course Notes.* TAPPI PRESS, Process Control Electrical and Information Division, March 14-17, 1995.

Bose, Binal K. *Adjustable Speed AC Drive Systems.* Wiley and Sons #0-471-09396-3, Library of Congress Card #80-27789.

Brink, Kenneth D. & Van Lieshout, Robert F. *Application and Selection Guidelines for AC System Drives.* Tappi J., vol 76, no 4, pp 143-155, April 1993.

Eikerberry, M.L. and Brink, K.D. *Adjustable Frequency AC Winder Drive System.* Tappi J., vol 74, no 4, April 1991.

Kheboian, G.I. *Drive Systems for Off-Machine Coaters.* Tappi J., vol 73, no 4, April 1990.

Lowe, J.A.. *Considerations for an Economic Evaluation of AC vs DC Technology for Paper Machine Dries.* TAPPI Short Course, March 1991.

McMahon, Michael J. *A simple method for determining drive roll load shares on machines with line shaft drives.* Tappi J., vol 78, no 4, pp 260-261, April 1995.

Puckhaber, Charles F. *Print Registration Control in a Rotogravure Printing Press Combined with Extrusion Coating or Lamination Processes.* Tappi J., vol 78, no 3, pp 144-154, March 1995.

Guiding

Feiertag, Bruce A. *Selecting a Web Guiding System.* Converting Machinery/Materials. 2nd Conf. Proc., Philadelphia, October 17, 1979.

Hopcus, Kenneth I. *Unwind and Rewind Guiding.* 2nd Int'l Conf. on Web Handling, Oklahoma State Univ, June 6-9, 1993.

Shelton, John J. *Lateral Dynamics of a Moving Web.* Oklahoma State Univ., WHRC, Ph.D. Thesis, 1968.

Nips

Anon. *Roll Crown Specifications and Definitions.* TAPPI Technical Information Sheets, TIS 405-10, September 1975.

Deshpande, Narayan V. *Calculation of Nip Width Between Elastomeric Cylinders.* Tappi J., vol 61, no 10, 115-118, October 1978.

Genisot, Tony. *Nip Impressions - a tool for troubleshooting and maintaining quality.* Tappi J., vol 77, no 9, pp 246-250, September 1994.

Moore, Robert H. and Moschel, Charles C. *A Tool for the Engineered Nip.* Tappi J., vol 77, no 3, pp 117-122, March 1994.

Rodal, Jose J. A. *Paper Deformation in a Calendering Nip.* Tappi J., vol 76, no 12, pp 63-74, Dec. 1993.

Roisum, David R. *The Profile of a Nip.* Converting Magazine, Technical Report, pp 44-46, June 1994.

Twitchen, D. *Paper Machine Roll Grinding Tolerances.* Canadian Pulp & Paper Association, Technical Section, Data Sheet G-20, May 1988.

Normal Entry Law

Pfeiffer, J. David. *Web Guidance Concepts and Applications.* TAPPI Finishing and Converting Conf. Proc., pp 23-31, October 1977.

Printing

Anon. *Gravure Process and Technology.* Gravure Association of America, Rochester NY, 1991.

Weiss, Herbert L. *Rotogravure and Flexographic Printing Presses.* Converting Technology Co, Milwaukee WI, 1985.

Rollers, General Articles and Books

Roisum, David R. *A Guide to Roller Sizing and Selection.* Converting Magazine, pp 48-50, March 1995

Roisum, David R. *The 10 Commandments of Web Machine Design.* Proc. 3rd Int'l Conf. on Web Handling, Oklahoma State Univ., June 1995.

Roisum, David R. *The 10 Commandments of Web Machine Design.* TAPPI Finishing and Converting Conf. Proc., October 1995.

Donatas. *Web Processing and Converting Technology and Equipment.* Van Nostrand Reinhold Co, New York, 1984.

Weiss, Herbert L. *Coating and Laminating Machines.* Converting Technology Co, Milwaukee WI, 1977.

Anon. *Gravure Process and Technology.* Gravure Association of America, Rochester, NY, 1995.

Slitting

Schable, Reinhold. *Spotlight on Slitting.* Series of monthly articles in the Paper Film Foil Converter Magazine beginning in January 1992 and ending in December 1993.

Spreading

Delahoussaye, Ronald D. *Analysis of Deformations, Stresses, and Forces in Webs Encountering Spreading Rollers.* Oklahoma State Univ., WHRC, Ph.D. Thesis, 1989.

Delahoussaye, Ronald D. and Good, J. Keith. *Analysis of Web Spreading Induced by the Concave Roller.* 2nd Int'l Conf. on Web Handling, Oklahoma State Univ., June 6-9, 1993.

Delahoussaye, Ronald D. and Good, J. Keith. *Analysis of Web Spreading Induced by the Curved Axis Roller*. 2nd Int'l Conf. on Web Handling, Oklahoma State Univ., June 6-9, 1993.

Lucas, Robert G. *Better Spreading of Web in the Winder - and How to Achieve It.* Pulp & Paper, vol 30, no 4, pp 154-157, April 1977.

Roisum, David R. *The Mechanics of Web Spreading.* 2nd Int'l Conf. on Web Handling, Oklahoma State Univ., June 6-9, 1993.

Roisum, David R. The Mechanics of Web Spreading - Parts I & II. Tappi J., vol 76, no 10, pp 63-70, October 1993 and vol 76, no 12, pp 75-86, December 1993.

Roisum, David R. *The Well-Proven Principles of Web Spreading.* Converting Magazine, pp 60-64, November 1995.

Static Electricity

Anon. *ESD Handbook.* Electrostatic Discharge Assoc, Rome, NY, 1995.
Horvath, T. and Berta, I. *Static Eliminations.* John Wiley and Sons, Letchworth, UK, 1982.
Roisum, David R. *Static Electricity.* Converting Magazine, September 1995.
Wettermann, Robert P. *Static Control Methods in Hazardous areas in Converting Operations.* Tappi J., vol 78, no. 2, pp 192-200, February 1995.

Stresses in Rollers

Fernside, Richard L. *Granite Roll Analysis.* Canadian Pulp and Paper Assoc. Annual Mtg. Proc., January 1990.
Fernside, Richard L. *Yankee Dryer Joint Corrosion: cause, prevention and repair.* Tappi J., vol 78, no 7, pp 144-151, July 1995.
Luthi, Oscar. *Ring Stresses and Deflections of Thin-Walled Rolls and Application.* Tappi J., vol 56, no 8, pp 121-124, August 1973.
Tonglin, Shu et al. *Assembly Stresses and the Strength of Hollow Iron Rolls.* Tappi J., vol 78, no 3, pp 185-190, March 1995.
Unneberg, Bengt. *Hydrostatic Testing of Yankee Dryers.* Tappi J., vol 78, no 9, pp 97-100, September 1995.

Temperature Measurement and Control

Roberts, Jack. *Computer Program Aids Heat Transfer Adjustments.* Paper, Film & Foil Converter, December 1990.
Stenstrom, Stig G. *et al. Equipment for Measurement of Cylinder Surface Temperatures.* Tappi J., vol 78, no 7, pp 245-246, July 1995.
Weiss, Herbert L. *IR Thermometers Invaluable for Accurate Temperatures.* Paper, Film & Foil Converter, vol 68, no 5, pp 166-167, May 1994.
Wilhelmsson, B.I et al. *Condensate Flow Inside Paper Dryer Cylinders.* J. of Pulp and Paper Science, vol 21, no 1, pp J1-J9, January 1995.
Zaoralek, M. *The Application of Different Designs of Heated Calender Rolls.* TAPPI Finishing and Converting Conf. Proc., Kansas City, MO, pp 153-158, October 1-5, 1989.

Tension Control

Anderson, Glenn. *Fundamentals of Fuzzy Logic: Part III.* Sensors, May 1993.

Tension Measurement

Roisum, David R. *How do I Measure Web Tension?* Converting Magazine, Web Works column, pp 16, December 1994.

Tension Setpoints

DuBois, G.B. *Selecting Proper Winding Tension for Various Web Substrates.* Tappi J., vol 76, no 5, pp 91-93, May 1993.
Jones, J.P. *Determining Tension Values.* Converting Mag., pp 64-72, November 1989.

Traction

Daly, David A. *Factors Controlling Traction Between Webs and Their Carrying Rolls.* Tappi J., vol 48, no 9, pp 88-90, September 1965.

Ducotey, Keith S. *Dynamic Coefficient of Friction Including the Effects of Air Entrainment.* MS thesis, Oklahoma State Univ., Web Handling Research Center, December 1987.

Vibration

Blevins, Robert D. *Formulas for Natural Frequency and Mode Shape.* Van Nostrand Reinhold Co., 1979.

Caccase, Vincent. *Dynamic Behavior of Steel Core in the Unwinding Process.* Tappi J., vol 76, no 10, pp 51-61, October 1993.

Daly, David A. *How Paper Rolls on a Winder Generated Vibration and Bouncing.* Paper Trade J., December 11, 1967.

Gerhardt, Terry D. and Staples, John F. and Lucas, Robert G. *Vibrational Characteristics of Wound Paper Rolls: Experiment and Theory.* Tappi J., vol 76, no 6, pp 121-128, June 1993.

Hussain, Mirza S. *Problems Associated with Rollers in Nip.* Tappi J., vol 76, no 1, pp 246-248, January 1993.

Kaufman, A.B. *Monitor Acceleration, Velocity or Displacement.* Instruments and Control Systems, October 1975.

Roisum, David R. *Winder Vibration Can Reduce Operating Efficiency and Increase Maintenance.* Tappi J., vol 71, no 1, pp 87-96, January 1988.

Wowk, Victor. *Machinery Vibration, Measurement and Analysis.* McGraw-Hill, 1991.

Web, Baggy

Roisum, David R. *What Causes a Baggy Web?* Converting Magazine, Web Works column, pp 22, April 1994.

Roisum, David R. *How Can I Fix a Baggy Web?* Converting Magazine, Web Works column, pp 20, April 1995.

Web Breaks

Roisum, David R. *Runnability of Paper.* Tappi J., vol 73, no 1, pp 97-101, January 1990, and no 2, pp 101-106, February 1990.

Web Curl

Hutten, Irwin M. *Linerboard: the relationship between polar angle profile and twist warp.* Tappi J., vol 78, no 4, pp 189-192, April 1995.

Page, D.H. and Seth, R.S. and De Grace, D.H. *The Elastic Modulus of Paper: The Controlling Mechanisms.* Tappi J., vol 62, no 9, pp 99-102, September 1979.

Roisum, David R. *How Can I Reduce Laminator Curl?* Converting Magazine, Web Works column, pp 22, March 1995.

Roisum, David R. *How Can I Fix a Baggy Web.* Converting Magazine, Web Works column, pp 20, April 1995.

Viitaharju, P.H and Niskanen, K.J. *Chiral Curl in Thin Papers.* J. of Pulp and Paper Science, vol 20, no 5, May 1994.

Web Flutter

Chang, Young Bae and Moretti, Peter M. *Interactions on Fluttering Webs with Surrounding Air.* Tappi J., vol 74, no 3, pp 231-236, March 1991.

Pramila, Antti. *Sheet Flutter and the Interaction Between Sheet and Air.* Tappi J., vol 69, no 7, pp 70-74, July 1986.

Web Wrinkles and Related Defects

Benson, Richard C. and Chiu, Han Chieh and LaFleche, John and Stack, Kenneth D. *Simulation of Wrinkling Patterns in Webs Due to Non-uniform Transport Condition.* 2nd Int'l Conf. on Web Handling, Oklahoma State Univ., June 6-9, 1993.

Friedrich, Craig R. *Geometry and Characterization of Wrinkles in Paper.* Tappi J., vol 73, no 8, August 1990.

Friedrich, Craig R. *Stability Sensitivity of Web Wrinkles on Rollers.* Tappi J., vol 72, no 2, pp 161-165, February 1989.

Gehlbach, Lars S. and Good, J. Keith and Kedl, Douglas M. *Web Wrinkling Effects of Web Modeling.* American Control Conf. Proc., Minneapolis, June 1987.

Gehlbach, Lars S. and Good, J.K and Kedl, Douglas M. *Prediction of Shear Wrinkles in Web Spans.* Tappi J., vol 72, no. 8, pp 129-134, August 1989.

Gopal, H. and Kedl, D. M. *Using Finite Element Model to Define How Wrinkles Form in a Single Web Span Without Moment Transfer.* 1st Int'l Conf. on Web Handling, Oklahoma State Univ., pp 291-303, May 19-22, 1991.

Roisum, David R. *Why is My Web Wrinkled?* Converting Magazine, Web Works column, pp 16, February 1994.

Roisum, David R. *What Causes Machine Direction Trough Wrinkles?* Converting Magazine, Web Works column, pp 22, March 1994.

Roisum, David R. *What Causes Diagonal Shear Wrinkles?* Converting Magazine, Web Works column, pp 22, June 1994.

Smith, R. Duane. *Roll and Web Defect Terminology.* TAPPI PRESS, 0101R234, 1995.

Shelton, John. J. *Buckling of Webs from Lateral Compressive Forces.* 2nd Int'l Conf. on Web Handling, Oklahoma State Univ., June 6-9, 1993.

Zak, M. *Statics of Wrinkling Films* J. of Elasticity, 12, pp 51-63, 1982.

Winding

Frye, Kenneth G. *Winding*. TAPPI PRESS, Atlanta, 1990

Roisum, David R. *The Mechanics of Winding.* TAPPI PRESS, Atlanta GA, 1994.

Smith, R. Duane. *Roll and Web Defect Terminology.* TAPPI PRESS, 0101R234, 1995.

Appendix C - Calculations and Formulae

Air Entrainment

see pp 46-47

(A.1)

$$H = 0.643\,r\left(\frac{6\,\mu V}{T}\right)^{2/3}$$

where

$V = V_R + V_W$

$T = T_0 - B\,V_W^2$

and where

H = air film layer thickness

r = radius of roller or bar

μ = absolute viscosity

V_R = speed of roller (or for bar = 0)

V_W = speed of web

T_0 = tension

B = basis weight (mass per area)

Area Moment of Inertia

see pp 32-34, 36, 92, 95, 101

(A.2) $$I_A = \frac{\pi}{64}\left(D^4 - d^4\right)$$

where
I_A = area moment of inertia (m^4) or (in^4)
D= outside diameter (m) or (in)
d= inside diameter (m) or (in)

Balance Grade, ISO

see pp 145

(A.3) $G = e\omega$

where
G = balance quality grade (mm/sec)
e = eccentricity (mm)
ω = speed (rad/sec)
and where
$\omega = 2\,\pi f$ for f(Hz) or
$\omega = 2\,\pi n/60$ for n(RPM)

Bearing Life

see pp 109

(A.4) $$L'_n = a_1 a_2 a_3 \left(\frac{C}{P}\right)^b$$

where
L = life in revolutions with failure rate of n
a's = life adjustment factors
b = 3.0 for ball bearings or 10/3 for roller bearings
C = dynamic basic load rating of bearing
P is the lessor of $P = XVF_r + YF_a$ or $P = VF_r$
F_r is the radial load
F_a is the axial load
X is a radial load factor from bearing tables
Y is the axial load factor from bearing tables
V is the rotation factor (1.0 for inner ring rotation, 1.2 for outer ring rotation)

Bending Stresses

see pp 95, 101

(A.5) $$\sigma_B = k_b \frac{Mc}{I_A}$$

where
σ_B = max bending stress (N/m^2)) or (lb/in^2)
kb = bending stress concentration
M = bending moment (m-N) or (in-lb)
c = radius to outside (m) or (in)
I_A = area moment of inertia (m^4) or (in^4)

Centrifugal Force

see pp 139, 145

(A.6)

$$F\,(\text{lbf}) = 1.77 \times \left(\frac{\text{rpm}}{1000}\right)^2 \times \text{Unbalance (oz} - \text{in) or}$$

$$F\,(\text{N}) = 1.1 \times \left(\frac{\text{rpm}}{10}\right)^2 \times \text{Unbalance (kg} - \text{m)}$$

Critical Speed - see Natural Frequency

Deflection of Cantilever
see pp 30, 35-36 be careful

(A.7) $$\delta = \frac{w L^4}{8 E I_A}$$

where
δ = deflection of end (m) or (in)
w = distributed load (N/m) or (lb/in)
L = length (m) or (in)
E = modulus (N/m^2) or (lb/in^2)
I_A = area moment of inertia (m^4) or (in^4)

Deflection of SS Dead Shaft Roller
see pp 30, 32

(A.8) $$\delta = \frac{5 w F^4}{384 E I_A}$$

δ = deflection of center (m) or (in)
F = face width of roller (m) or (in)
E = shell modulus (N/m^2) or (lb/in^2)
I_A = area moment of inertia (m^4) or (in^4)

Deflection of SS Live Shaft Roller
see pp 30, 32-34

(A.9) $$\delta = \frac{(12B - 7F) w F^3}{384 E I_A}$$

where
δ = deflection of center (m) or (in)
F = face width of roller (m) or (in)
B = distance between bearings (m) or (in)
E = shell modulus (N/m^2) or (lb/in^2)
I_A = area moment of inertia (m^4) or (in^4)

Draw from Speed
see pp 19

(A.10) $$\%Draw_{y/x} = 100 \times (V_y - V_x) / V_x$$

Expansion
see Thermal Expansion

Force on Roller from Web Tension
see pp 13-14

(A.11) $$F_{tension} = 2 T W \sin(\alpha/2)$$

where
$F_{tension}$ = web force (kN) or (lb)
T = web tension (kN/m) or (lb/in or PLI)
W = web width (m) or (in)
α = wrap angle (degrees)

Friction
see pp 43

(A.12a) $$F \le \mu N$$

where
F = friction force (kN) or (lb)
N = normal force (kN) or (lb)
μ = coefficient of friction

(A.12b) $$\mu = \tan(\theta)$$

where
μ = coefficient of friction
θ = incline slide angle

Gear or Pulley Speed Ratio
see pp 168

(A.13) $$N_2 = N_1 \frac{D_1}{D_2}$$

where
N's = rotational speed (RPM)
D = pitch diameter (m) or (in)

Gear or Pulley Torque Ratio
see pp 168, 244

(A.14) $$Torq_2 = Torq_1 \frac{D_2}{D_1}$$

where
Torq's = torques (N-m) or (lb-in)
D = pitch diameter (m) or (in)

Length of Web
see pp 220

(A.15a) $L = n \pi D$

where
n = number of revolutions
L = length (m) or (in)
D = effective wheel or roller diameter
(physical diameter plus web caliper)

(A15b)

$$\frac{L_B}{L_A} = 1 - \left(\frac{\Delta Tension_{A-B}}{E_{11}\ Caliper} + \alpha_1\ \Delta Temp_{A-B} + \beta_1\ \Delta Moist_{A-B} \right)$$

where
A = subscript for measurement 1
B = subscript for measurement 2
L = length (m) or (in)
ΔTension = change in tension (N/m) or (lb/in)
E_{11} = MD modulus (N/m^2) or (lb/in^2)
Caliper = thickness (m) or (in)
α1 = MD coef of thermal expansion
(m/m/°C) or (in/in/°F)
ΔTemp = change in temperature (°C) or (1 °F)
β1 = MD coef of hygro/solvent expansion
(m/m/%) or (in/in/%)
ΔMoisture = change in water/solvent (%) or (%)

Note: This equation can also be applied to width except that the properties now are in the 2 or CD direction and the tension component is multiplied by a (-μ) to account for Poisson necking.

Mass Moment of Inertia
see pp 59, 240, 243

(A.16a) Metric Units

$$I_M = \rho L \frac{\pi}{32} \left(D_O^4 - D_I^4 \right)$$

(A.16b) English Units

$$WR^2 = [0.000682]\ \rho L \left(D_O^4 - D_I^4 \right)$$

where
I_M = mass moment of inertia ($kg\text{-}m^2$)
WR^2 = the mass moment of inertia ($lb\text{-}ft^2$)
ρ = shell density (kg/m^3) or (lb/in^3)
D = diameters (outside & inside) (m) or (in)

Natural Frequency
see pp 135 and cautions on usage

(A.17)

$$f = \frac{\pi}{2} \sqrt{\frac{g E I}{w l^4}}$$

where
where
f = natural frequency (Hz)
g = acceleration of gravity = 386 in/sec^2
E = Young's Modulus (lb/in^2)
I = area moment of inertia (in4) = $\pi(D^4\text{-}d^4)/64$
D = outside diameter (in)
d = inside diameter (in)
w = weight per length (lb/in)
l = length; and where
Critical Speed (ft/min) = 15.71xfxD

Nip Stress
see pp 74, 77

(A.18a) $N = F / L$

where

N = nip load (N/m) or (lb/in)
F = total force between rollers (N) or (lb)
L = length of web or roller (m) or (in)

(A.18b) $\sigma_c = 1.277 \frac{N}{b}$

where
σ_c = peak nip pressure (N/m^2) or (lb/in^2)
N = nip load (N/m) or (lb/in)
b = nip width (m) or (in)

Power from Speed, Tension and Width
see pp 22, 65, 239, 243

(A.19a) Metric Units

$$P = \frac{T \times W \times S}{60}$$

where
P = power (kw)
T = tension (kN/m)
W = width (m)
S = speed (m/min)

(A.19b) English Units

$$P = \frac{T \times W \times S}{33{,}000}$$

where
P = power (hp)
T = tension (lb/in)
W = width (in)
S = speed (ft/min)

Pressure Under Roller
see pp 49

(A.20) $$P = \frac{T}{r}$$

where
P = pressure (N/m^2) or (lb/in^2)
r = radius of roller/bar (m) or (in)

Note that this pressure is reduced at extremely high speeds due to centrifugal forces as given in page 57.

Misalignment (Inplane) Stress Riser
see pp 157-158 and KL caution

(A.21a) $$\frac{T_{max}}{T_{avg}} = 1 + \frac{\theta_i}{\varepsilon_{lavg}\left(\frac{L}{W}\right)}$$

where
T_{max} = high tension side (N/m) or (lb/in)
T_{avg} = average tension (N/m) or (lb/in)
θ_i = inplane misalignment angle (radians)
L = span length between rollers (m) or (in)
W = web width (m) or (in)
ε_{lavg} = average web strain (m/m) or (in/in)

and

(A.21b) $$\varepsilon_{lavg} = \frac{T_{avg}}{c\,E_1}$$

where
ε_{lavg} = average web strain (m/m) or (in/in)
T_{avg} = average tension (N/m) or (lb/in)
c = web caliper or thickness (m) or (in)
E_1 = MD modulus of web (N/m^2) or (lb/in^2)

(A.21c) $$\theta_{icrit} = \varepsilon_{lavg}\left(\frac{L}{W}\right)$$

where
$\theta_{critical}$ = inplane misalignment angle (radians) that would cause the web to just go slack on one side of the upstream roller.

Misalignment (Out-of-Plane) Stress Riser
see pp 159

(A.22a) $$\frac{T_{max}}{T_{avg}} = 1 + \frac{\varepsilon_{omisalign}}{\varepsilon_{lavg}}$$

T_{max} = high tension side (N/m) or (lb/in)
T_{avg} = average tension (N/m) or (lb/in)
$\varepsilon_{omisalign}$ = out-of-plane strain rise (m/m) or (in/in)

ε_{lavg} = average web strain (m/m) or (in/in)

and

(A.22b) $$\varepsilon_{lavg} = \frac{T_{avg}}{c\,E_1}$$

where
ε_{lavg} = average web strain (m/m) or (in/in)
T_{avg} = average tension (N/m) or (lb/in)
c = web caliper or thickness (m) or (in)
E_1 = MD modulus of web (N/m^2) or (lb/in^2)

and

(A.22c) $$\varepsilon_{omisalign} = \frac{1}{8}\left(\frac{Z_o}{L}\right)^2$$

where
$\varepsilon_{omisalign}$ = out-of-plane strain rise (m/m) or (in/in)
Zo = out of plane misalignment distance (m) or (in)
L span length between rollers (m) or (in)

Speed from Diameter and RPM
see pp 19, 135, 136, 139, 168, 240

(A.23a) Metric Units

$$V = 0.031416 \times D \times RPM$$

where
V = speed (m/min)
D = diameter (cm)
RPM = revolutions per minute

(A.23b) English Units

$$V = 0.26180 \times D \times RPM$$

where
V = speed (ft/min)
D = diameter (in)
RPM = revolutions per minute

Strain from Draw
see pp 19

(A.24) $$\varepsilon_y = \varepsilon_x + \frac{\%Draw_{y/x}}{100}$$

Tension from DC Motor Readings
see pp 65

Metric Units

(A.25a)
$$P_R = P_{nameplate}(Efficiency)\left(\frac{Amps_{meter}}{Amps_{nameplate}}\right)$$

where
P_R = power during run (kW)
$P_{nameplate}$ = motor nameplate power (kW)
Efficiency = elect & mech efficiency ≈ 0.90
$Amps_{meter}$ = meter reading during run (amp)
$Amps_{nameplate}$ = motor nameplate current (amp)

(A.25b)

$$T = \frac{P_R \times 60}{W \times S}$$

where
T = tension (kN/m)
P_R = power during run (kW)
W = web width (m)
S = speed (m/min)

English Units

(A.25c)
$$P_R = P_{nameplate}(Efficiency)\left(\frac{Amps_{meter}}{Amps_{nameplate}}\right)$$

where
P_R = power during run (HP)
$P_{nameplate}$ = motor nameplate power (HP)
Efficiency = elect & mech efficiency ≈ 0.90
$Amps_{meter}$ = meter reading during run (amp)
$Amps_{nameplate}$ = motor nameplate current (amp)

(A.25d)

$$T = \frac{P_R \times 33,000}{W \times S}$$

where
T = tension (lb/in)
P_R = power during run (HP)
W = web width (in)
S = speed (ft/min)

Tension from Web Sag
see pp 21

(A.26) $$T = \frac{L^2\ w}{8\ s}$$

where in consistent units
L = span length between roller tangents (length)
w = basis weight (weight/area)
s = sag (length)

Thermal Expansion
see pp 106
Note also that this equation can be written for changes in length, width, and thickness due to moisture, solvents and temperature rise.

(A.27) $\Delta L = \alpha \Delta T L$

where
ΔL = change in length (m) or (in)
α = thermal expansion coef (m/m/°C) or (in/in/°F)
ΔT = change in temperature (°C) or (°F)
L = length

Torque from Acceleration
see pp 58-61, 243

(A.28a) Metric Units

$$Torq = I_M \alpha$$

where
Torq = torque (N-m) or (lb-in)
I_M = mass moment of inertia (kg-m^2)
α = rotational acceleration rate (rad/sec^2)

$$a = r\alpha = \frac{\Delta V}{\Delta t}$$

where
a = acceleration rate (m/sec^2) or (in/sec^2)
V = speed (m/sec) or (in/sec)

(A.28b) English Units

$$Torq = \frac{WR^2 \Delta V}{[80.5] D_O t}$$

where
Torq = torque in ft-lb
ΔV = change in speed in FPM
t = time in seconds.

Torque to Drive Roller
see pp 50, 66, 241, 243, 244

(A.29)

$$Torq_{Drive} = Torq_{Drag} \pm I_M \alpha - \frac{D_0 w}{2}\left(T_2 - T_1\right)$$

where
$Torq_{Drive}$ = torq required by drive (N-m) or (in-lb)
$Torq_{Drag}$ = drag torq on roller (N-m) or (in-lb) from bearings, nips etc
I_M = mass moment of inertia (kg-m^2)
α = rotational acceleration rate (rad/sec^2)
D0 = roller outside diameter (m) or (in)
w = web width (m) or (in)
T_2 = downstream tension (N/m) or (lb/in)
T_1 = upstream tension (N/m) or (lb/in)

Torque versus Accel and Tension
see pp 60, 61, 63

(A.30)

$$Torq = I_M \alpha = \frac{2 \Delta T_I w}{D_O}$$

where
$Torq_{Drive}$ = torque (N-m) or (in-lb)
I_M = mass moment of inertia (kg-m^2)
α = rotational acceleration rate (rad/sec^2)
ΔT_I = inertial tension difference across roller (N/m) or (lb/in)
w = web width
D_0 = roller outside diameter

Torsional Stress and Angle of Twist
see pp 102, 170

(A.31a) $\tau = k \dfrac{\text{Torq}\, D}{2\, J}$

where
τ = shear stress at outside (N/m^2) or (lb/in^2)
k = stress concentration factors (unitless)
Torq = torque (N-m) or (lb-in)
D = outside diameter (m) or (in)
J = polar moment of inertia (m^4) or (in^4)

(A.31b) $J = \dfrac{\pi}{32}\left(D^4 - d^4\right)$

(A.31c) $\phi = \dfrac{\text{Torq}\, L}{J\, G}$

ϕ = angle of twist (radians)
L = shaft length (m) or (in)
G = shear modulus (N/m^2) or (lb/in^2)

Traction from Friction and Wrap Angle
see pp 50, 52-54, 57, 62, 63

(A.32) $\dfrac{T_2}{T_1} \leq e^{\beta \mu_k}$

where
β = wrap angle (radians)
μ_k = coefficient of web/roller kinetic traction

Weight of Cylinder
see pp 33, 95

(A.33) $W = \dfrac{\pi}{4} \rho L \left(D^2 - d^2\right)$

where
W = weight (kg) or (lb)
ρ = density (lb/in^3) or (kg/m^3)
L = length (m) or (in)
D= outside diameter (m) or (in)
d= inside diameter (m) or (in)

WR2
see Mass Moment of Inertia

Glossary

ABEC - Annular Bearing Engineer's Committee of the AFBMA that establishes standards and grades.
Abrasion - The wearing away of roller or other surfaces due to macro or micro slippage.
AC - Alternating Current.
AC Scalar Drive - A simple but loose electric drive composed of a converter bridge, inverter bridge, controller and AC induction motor.
AC Vector Drive - A highly precise electric drive for an AC induction motor.
Acceleration - To increase the speed of a line at a specified rate or curve.
Accumulator - A series of rollers used to store material. See Festoon.
Acrylic - A polymer used for covers.
Adhesion - The sticking together of two materials.
AFBMA - Anti-Friction Bearing Manufacturers Association.
AIMCAL - Abbreviation for the Association of Industrial Metallizers, Coaters and Laminators.
Air Bar - A stationary bar to reduce or eliminate drag by air flotation, or as elements in an air flotation oven.
Air Doctor - To remove excess coating or fluid by blowing air from a slotted nozzle.
Air Entrainment - The forcing of air between a web and roller or between layers of a winding roll on nonporous webs run at higher speeds. Air entrainment on rollers can cause loss of traction. Air entrainment in rolls can cause a variety of defects.
Air Film Thickness - The distance between the average web and roller surfaces which is filled with air pumped into the ingoing tangent.
Air Flotation - The reduction of contact between a web and roller due to air entrainment.
Alignment - The measure of the level (w.r.t. gravity) and squareness (w.r.t a reference) of rollers.
Aluminum Oxide - A ceramic coating used to improve a roller surface's hardness and abrasion resistance.
Anelastic - A material whose stress-strain relations vary with time or other factors. Anelastic materials are prone to creep.
Anemometer - An instrument to measure fluid velocity.
Anilox - An engraved roller which meters ink onto an applicator roller.
Anisotropic - A material whose properties vary with direction. For example, the MD is stronger than the CD. Even isotropic materials such as metal are anisotropic in a wound roll because the ZD stack modulus is less than the metal's nominal modulus.
Anodize - To add a hard aluminum oxide coating to an aluminum shell for improved wear resistance.
Antinodes - Parts of a structure which have maximum movement during resonant vibration.
Anvil Roller - A smooth and very hard roller which serves as the backup roller for score slitting and rotary die cutting.
Armature Current - The control parameter which largely determines the speed and torque of a DC motor.
Armature - The rotating shaft of a DC motor and its windings.
Axial - In the direction of a roller's axis (usually CD or cross machine direction).
Babbitt - A bearing metal or a journal bearing using this metal in the housing.
Back-up Blade - A blade which gives support to a doctor blade.
Back-up Roller - A roller nipped against a work roller.
Backlash - An undesirable rotational play or clearance on clutches, couplings or gears.
Backstand - Slang for an unwind.
Baggy Web - A web with an uneven natural length or residual tension profile variation.
Balance Grade - The quality of balance determined by ISO standards and charts.
Ball Bearing - A bearing type with numerous rolling balls used for lightly loaded rollers.
Ballard Shell - A process used to make a printing roller by adding a thin copper layer to a shell for engraving, which can be easily stripped off.
Banana Roll - Slang for a bowed spreader roller.
Band Brake - A weighted strap or band partially wrapped around a roll or roller to serve as a brake.
Band-Brake Law - A calculation for the maximum web tension differential across a roller.

Barrel Roller - Slang for a roller whose diameter is greater at the center than at the ends. In traction, this roller will tend to center and contract a web. In slippage, this roller will spread the web.
Barring - A series of CD oriented defects equally spaced around a nip roller or on a product due to the bounce of a nipping roller.
Base Cylinder - A roller or shell which is ready for coating or covering.
Base Speed - The RPM of an electric motor under full load.
Basis Weight - The mass per unit area of a web. Common metric units are (g/m^2). English units for the paper industry are often ($lb/3000ft^2$).
Bearing Housing - A cast or machined steel structure to hold a bearing and its lubrication.
Belt - A mechanical drive component used to transmit torques between pulleys via belt/pulley friction. A belt is a wire or web that is endless.
Benchboard - An operator's control console containing switches, lights, dials, meters and other electronics.
Bent Pipe - A type of single element sliding spreader whose construction is well described by its name.
Bent Shaft - A roller shell or journal which has been permanently bend due to a bending overload accident.
Bevel - The angle of grind of a doctor blade.
Biaxial Stress - A state of stress with two nonzero principle stresses. Webs (MD and CD) and rolls (MD and ZD) are usually in a biaxial stress state.
Bimetallic Thermometer - A thermometer composed of two alloys of different thermal expansion bonded together which deflect upon temperature change.
Blocking - The layers in a blocked roll or stack are stuck together due to interlayer pressure, nip, adhesives, etc. A blocked roll will not unwind well because the web will jump or tear at the unwinding tangent.
Blueing - A thin metal coating used by maintenance to scribe upon for locating parts or to check the fit of parts.
Bolt Bound - A mounting which can't be moved because bolts are bearing on the edges of the bracket holes.
Bow - The distance between the chord and axis at the center of a bowed roller (or severely bent slender roller, or a core). On a spreader, the net bow is the effective working bow and is the vector addition of the pre-bow plus tension deflection plus weight deflection.
Bowed Spreader Roller - A curved axle roller with flexible sleeve or segmented shells used for spreading.
Bowtie Roller - Slang for a concave roller.
Braking - A negative external torque on a roller which will tend to increase the web's tension as it goes over that roller and assist with overcoming deceleration inertia. Braking on an unwind provides a starting tension for the first span. Braking takes energy from the web machine system and turns it into heat (mechanical) or electricity (electrical).
Breaking Length - A measure of the structural efficiency of a web material defined as strength ÷ basis weight.
Brinell - A metal hardness test.
Brinell - To dent an object with a load applied over a small (nearly point contact) area.
Brittle - A material which has little deformation or yielding before it breaks, such as glass.
Brush - The rotary connector (often spring loaded carbon blocks) between the armature commutators and the armature control of a DC motor.
Buckling - A wrinkling or crumpling compressive failure.
Buildup - The ratio of the finished wound roll diameter to its core diameter.
Bulk - Slang for the caliper or thickness of a web. In scientific terms, bulk is the inverse of density.
Bump Test - A test to obtain natural frequencies by striking a structure.
Bumper - Adding tape or similar material to a roller to locally increase its diameter. A crude, but sometimes effective, method of reducing wrinkles.
Buna N - See Nitrile Rubber.
Burnishing - An operation similar to polishing to remove a small amount of material from the surface of a roller.
Burr - A thin sliver of metal extending from the surface of a machined or worn part.

Butt Roll - A narrow roll on the ends of a winding set that will be discarded. Butt rolls result when the total width of an order results in a trim that is too wide to be handled by the trim conveying system. Also called a cookie.
Butyl - A polymer used for covers.
Calender - To run a web between nipped rollers.
Caliper - The thickness of a web expressed in micrometers in the metric system and mils (0.001") in the English system.
Caliper - A disc brake pad and actuator.
Caliper Profile - The variation of caliper as a function of cross machine position.
Camber - A measure of web bagginess as the arc or bow its edge makes when laid flat on the floor.
Can - Slang for a large cooling or dryer roller.
Cantilevered - A roller which is supported on only one end.
Capstan Wrap - The physics of a change of web tension from the ingoing to outgoing tangents as described by the band/brake or belt equation.
Carbon Fiber - A high performance (stiffness to weight) roller shell made of carbon fibers in an epoxy matrix.
Carry Out - The outward motion of a web on the wrapped section of a bowed roller or expander roller spreaders.
Cartridge Heater - See Resistance Heater.
Cascading - At intermediate speeds a fluid (water condensate) will climb part way up one side of the inside of the roller and then fall to the bottom.
Catenary - The mathematical description of the shape of a rope, wire or web sagging due to its own weight.
CC Roll - Abbreviation for a controlled crown roller used for gross adjustment of nip profile.
CD - Abbreviation for Cross (machine) Direction, sometimes also abbreviated as XMD, which is the direction perpendicular to the material flow and material plane in a machine. In many plants, this is referred to as TD or Transverse Direction.
Cell - A tiny etched or engraved well in a gravure cylinder that carries the ink.
CEMA - Abbreviation for the Converting Equipment Manufacturer's Association.
Center-Surface Winder - A winder which drives the roll through torque at the core and through a nipping roller.
Centerline - An imaginary MD line equidistant from both edges of a machine roller.
Centerwind - A winder whose roll is driven solely by torque through the core or mandrel.
Ceramic - A product made but the baking, firing or thermal deposition of a nonmetallic mineral (such as oxides of aluminum, chromium or titanium). Ceramic coated rollers have superior hardness and abrasion resistance.
Ceramic Coating - A nonmetallic, inorganic coating most often used to increase abrasion resistance. The most common roller surface ceramic is aluminum oxide, but oxides of chromium, titanium or zirconium are also used.
Chain Marks - Slang for a web corrugation.
Chevron - Grooving oriented at a slight angle from the rollers axis or CD, also called herringbone.
Chill Roll - A roller cooled with chilled circulating water.
Chilled Iron Rollers - Chilled iron is a gray or white (depending on the rate of cooling) cast iron that is commonly used for calender and other process rollers.
Chromium Oxide - A ceramic coating used to improve a roller surface's hardness and abrasion resistance.
Chuck - A device for holding the end of a rotating tool, shaft or wound roll.
Cigar Roller - Slang for a roller whose diameter is greater at the center than at the ends. In traction, this roller will tend to center and contract a web. In slipping, this roller will spread the web.
Circumference - The circular length around a cylinder defining the length of web passage in one revolution.
Clorobutadiene - See Cloroprene Rubber.
Cloroprene Rubber - A polymer used for roller coverings better known by the DuPont trademark of Neoprene.

Clothing - Slang for felts and wires on a paper machine.
CMM - Abbreviation for Converting Machinery and Materials, which is a large trade show held regularly in the U.S. and other countries.
Coefficient of Friction - The ratio of force to slide over the normal (perpendicular) force.
Coefficient of Thermal Expansion - The unit strain dimensional growth of a material for a unit rise in temperature.
COF - An abbreviation for the coefficient of friction. High web/roller friction improves traction and high web/web friction holds a wound roll together.
Cogged Belt - A belt with teeth on the inside (or sometimes inside and outside) that engage with teeth on a pulley. The cogged belt is commonly used where speed control is used.
Coil - A term for metals in a wound roll form.
Commutator - The segmented copper bars on the back end of a DC motor rotor.
Compliance - Springiness or flexibility. Opposite of stiffness.
Compliant Cover Roller - A roller with a very soft cover with undercut grooving purported by some to spread.
Composite Roller - Broadly speaking, a composite is a structure made of more than one material. When applied to rollers, however, it usually implies a high performance carbon fiber shell.
Compression - A push force as opposed to a pull force (tension).
Compression Set - A measure of the amount of irrecoverable deformation after removing the load from a cover material.
Concave Roller - A roller whose edges have a larger diameter than its center and which is a spreader when in traction.
Concentricity - The distance between the centers of circle and a rotating axis, such as between the shell OD and journal.
Cone Pulleys - A pair of cone shaped pulleys with a flat belt running between them which is used for adjusting speed ratios between sections.
Constant Height Region - The air film thickness between nonporous webs and rollers is nominally constant through much of the wrap, but narrows to about 70% of that near the exit.
Constant Tension Winding - Winding with a web tension maintained constant through the buildup of a wound roll.
Constant Torque Winding - Winding tension drops inversely with current diameter. Useful for structuring a wound roll to avoid some defects, and to reduce the torque requirements of a winder drive.
Contact Resistance - The insulation value of the interface between a web and roller.
Contacting Pyrometer - A handheld roller temperature measuring instrument often employing a thermocouple strip as a sensor.
Continuous Lube - A bearing lubrication system which meters oil in the top and drains it out the bottom.
Contraction - Opposite of spreading.
Converting - The processing of a web material from one form to another. Converting processes include calendering, coating, die cutting, embossing, laminating, printing, punching, sheeting, slitting, treating, winding and unwinding.
Cooling Rolls - Rollers cooled below the temperature of the web, usually by circulating water.
Core - A hollow tube, often of fiber, plastic or metal, upon which a roll is wound.
Core Box - Paper industry slang for core chuck or coreshaft slides.
Core Chuck - A winder component which is inserted into a core to guide the edge position of a winding or unwind roll.
Core Shaft - A mandrel upon which rolls are wound.
Core Slides - Linear ways or bearings which guide core chucks or coreshafts during winding from moving sideways and causing roll offsets.
Corrosion - The oxidation or other chemical attack of metals.
Corrugation - A shear defect of light webs caused by an uneven nip resulting in a narrow tire track or roping band oriented in the machine direction.
Couple Unbalance - An unbalance condition where a statically balanced roller will impose dynamic loads to the front and back bearings which are equal in magnitude but 180° out of phase.

Coupling - A flexible connection between a drive (motor) and a driven element (roller).
Covered Roller - A roller whose surface is a second material such as synthetic rubber.
Crawl - A very slow or web threading speed mode.
Crease - A foldover of a web going over a roller or onto a roll.
Creep - The permanent deformation of a body with time subjected to constant load. Creep moves stresses toward zero while elongating tensioned material and shortening compressed material.
Critical Angle - The inplane misalignment angle that just causes the web to go slack on one side.
Critical Speed - A condition where the roller rpm matches its beam bending natural frequency and thus results in large amplitude resonant vibrations.
Cross Direction - Direction perpendicular to the material flow and material plane in a machine often abbreviated CD or XMD.
Crossing - Another name for skewing nipped rollers.
Crown - The difference in diameter between the center and ends of a roller.
Crowned Roller - A roller made intentionally barrel shaped to improve nip uniformity or to act as a (heavy) web centering device.
Crush Cut - Slang for score slitting.
Curl - An (often laminated) web which tends to roll up out of its plane in the MD, CD or chiral (diagonal) directions.
Current - Motor armature current indirectly controls web tension.
Curved Axis Roller - A formal name for a bowed spreader roller.
Cylindricity - The deviation of a roller shell from a true cylinder as measured by TIR runout, diametral variations and so on.
D-bar Spreader - A profile adjustable bent pipe spreader.
Damping - The resistance to motion which dissipates vibratory energy.
Dancer - A pivoting roller sensor used for feedback control of web tension.
Datum - A reference line for the machine direction, often offset in the aisle, defined by two or three plugs (monuments or landmarks).
DC - Direct Current.
DC Drive - The most common drive used on web applications consisting of AC-DC power conversion, controller and DC motor.
Deadband - The small band spanning a control setpoint within which a controller can neither sense nor react.
Deadshaft Roller - A roller with a stationary central shaft (also see liveshaft roller).
Deceleration - To decrease the speed of a roll(ers) at a specified rate.
Deckle - The width of the web in some part of a machine.
Decurler - A small radiused bar or roller upon which a web is dragged to remove curl. The strength of the effect is primarily determined by the web tension, radius of curvature, and to a lesser extent, wrap angle.
Deflection - The elastic (nonpermanent) bending of a roller due to gravity, tension and nip which primarily determines the required diameter of a roller.
Deformation - The change of length, width or shape of a material subjected to loads or stresses.
Degrease - The cleaning of a roller with soap, solvents or acids to prepare the surface for coating or covering.
Delamination - A separation of materials in a roller covering system.
Density - Weight per unit volume. In English units density is expressed as (lb/in^3) and in Metric Units is expressed as (kg/m^3). Density can refer to either a web or a roll.
Dial Indicator - An instrument for measuring linear displacement, such as the runout of a roller surface.
Dimension Series - The standardized sizing for the width and diameter of bearings.
Displacement Guide - A web edge (or center) guide composed of a pair of rollers mounted on a common pivot table or mechanism.
Doctor - To scrape away contaminants or excess material from a roller with a blade.
Doctor - To salvage rewind a roll on a brand of centerwinder called the Roll Doctor.
DOM - Abbreviation for Drawn Over Mandrel, which is a method of producing pipe or tubing of better cylindricities.

Dowel - A (preferably tapered) pin used to preserve alignment between a roller bearing housing and its framework mount.
Drag Belt - A primitive unwind brake comprised of a weighted belt riding on the roll.
Draw - The percent increase (or decrease) of a roller's speed with respect to an upstream roller.
Draw Control - The control of roller speed so that plastic and deformable webs are strained a given amount.
Draw Control - Slang for speed control which is an indirect method of controlling tension or strain by controlling the speed of driven rollers.
Drawing - A process intended to permanently stretch a material to longer lengths, shorter widths and/or thinner caliper. Drawing is accomplished by a speed (strain) controlled section(s) where downstream rollers run faster than upstream rollers.
Dress a Coreshaft - To slip cores onto a coreshaft and lock them in place prior to winding.
Drilled Roll - A heating or cooling roller that has numerous axial passages drilled in the shell for fluid passage.
Drive - Slang for an electric motor, power conversion and controller system
Drive Roller - A roller which is driven or braked by devices such as electric motors, belts, brakes, clutches
Drive Room - An air conditioned room dedicated for several centralized drive controllers.
and so on.
Drum - A specialized roller, usually large, often having a heavier wall, higher precisions and specialized surfaces.
Dual Bowed Roller Spreader - A spreader using two series bowed rollers pointing perpendicular to the incoming and outgoing web runs.
Dualpass - A circulating fluid roller where the fluid enters and exits from the same end, making two trips through the roller shell periphery.
Dub - To undercut the diameter of the edges of a roller slightly smaller to avoid breaking them out when in nip.
Ductile - A material which will yield and deform considerably before breaking.
Durometer - The hardness of a roller covering as measured by the deflection of an indenting probe. Durometer is a crude indicator of the more important property, namely, modulus.
Duty - The fraction of time that a web machine or component is running.
Dwell - The time a web sits on a roller surface or under a nip.
Dwell - The amount of time the material spends under the nip as calculated by nip width divided by web speed.
Dynamic Balance - To add/remove weights on a roller to achieve both static and couple balance.
Dynamic Basic Load Rating - A factor used to calculate a bearing's life.
Dynamic Braking - The routing of DC motor armature current to a set of heating resistor coils to dissipate braking power.
Dynamic TIR - The dynamic displacement of a roller shell at 1 x rpm.
E-Stop - Short for emergency stop. This is a maximum deceleration of a machine to protect an operator, product or machine from further damage after an accident.
Eccentricity - Distance between the axis of rotation and the center of gravity (balance) or center of the shell, commonly abbreviated as e.
Edge Guide - A sensor and controller for positioning one edge or centering both edges of a web or roller to a setpoint.
Edge Pull Web Stretchers - One of two spreaders which widen a web by pulling on its edges.
Effective Diameter - The diameter a covered nip roller acts like when computing surface speed from RPM.
Effective Groove Thickness - The cross sectional area of an annular or spiral groove divided by the pitch.
Effective Texture Thickness - The cross sectional area of texturized rollers converted to an average thickness.
Elastic - Materials that return to the same size and shape when loads are removed (e.g., doesn't creep).
Elasticity - A branch of engineering mechanics that describes the elastic behavior of materials subject to loads.
Elastomeric - A generic term for rubber coverings.

Electrostatic Assist - A system of applying a charge to a roller which improves attraction or transfer of coating, ink or web.
Emboss(ers) - A patterned nip roller used to pattern a web or bond it together in discrete areas.
Encoder - A sensor used to measure web footage or roller rotational speed.
Endurance Limit - The maximum stress at which a material will not experience fatigue failure at an indefinitely high number of load cycles.
Entering Span - The web span immediately preceding a roller.
EPDM - A polymer cover material.
ESA - Abbreviation for Electrostatic Assist.
Etch - To remove material to pattern a metal roller by using an acid or corrosive.
Expander Roller - A spreader whose shell moves outward underneath the web as it transits from ingoing to outgoing tangents.
Face - The effective width of a roller.
Fading - The loss of friction on a brake or clutch due to excessive sliding or temperature.
Fatigue - A structural failure caused by repeated reversing loads. Journals often fail due to the rotating bending loads.
Fatigue Endurance Limit - The stress at which a material can sustain an indefinitely large number of cycles which is a property of only a few materials such as steels and composites.
Fatigue Strength - The stress at which a material will fail at a given number of load cycles.
Feedback - Output from a sensor which is input to a controller for maintaining a control setpoint.
Feedforward - A control technique to improve responsiveness of controls by letting the actuator know a change is needed even before its sensor detects it.
Felt - An endless woven forming belt.
FEM - Abbreviation for Finite Element Modeling which is a numerical /computer technique that can be used to predict the static (stress), dynamic (vibration) or thermal behavior of a machine.
Festoon - A series of rollers mounted on a translating carriage that can payout and accumulate web during zero speed wind or unwind roll changes while maintaining machine speed. Sometimes (inappropriately) used as a tension sensor.
FFT - Is an instrument which uses the Fast Fourier Transform algorithm to convert vibration amplitude versus time into amplitude versus frequency. The FFT can be used to identify which roller is causing vibration, or the condition of bearings and other components.
Field Weakening - A reduction of a DC motor (outer) field to allow it to run above base speed.
Filled Roll - A nipping roller which is made by slipping either textile or paper disks over a mandrel, compressing them axially until bone hard, then finishing on a lathe or grinder.
Film - A thin polymer (plastic) web such as found in stretch wrap or garbage bags. Also, a thin liquid.
Finish - Short for surface finish (smoothness as opposed to roughness) of a roller.
Finishing - A final process in base web manufacturing such as calendering.
Fire Streaking - A high temperature failure of brakes and clutches.
Fit - Difference between the ID of a hole and the OD of a shaft.
Fixed Gap - To set a distance between two rollers in close proximity.
Fixed Load - To set the load between two nipped rollers.
Flange Mount - A bearing mounting style which bolts to the inside of a frame.
Flat Spot - A lengthwise low spot in a roll(er) often caused by storing it on the floor.
Flattening - The removal of troughs and wrinkles by slipping the web over a bar, shoe or roller. While spreading <u>forces</u> a web to be <u>wider</u> than it would normally be, flattening <u>allows</u> the web to be <u>as wide</u> as it would like to be.
Flexible Rotor - A difficult balance condition where the roller is operated above its 1/2 critical.
Flexible Rotor - A balancing technique for slender rollers run near critical speed.
Floating - One of three states of web/roller interaction where the web wraps but does not contact a roller due to the entrainment of air or other fluid. See also traction and sliding.
Flutter - The ZD vibration of a web which acts much like a plucked string or a flag waving in the wind.
Flying Splice - To join a new unwind roll to an expiring one at speed.
Foil - A thin metal web.

Foil - A special shaped bar or shoe.
Folding Bar - A smooth or air floated bar used for folding the web.
Foldover - A crease in the web caused by a severe trough or baggy area going over a roller or into a nip.
Follower - A roller which is slave to run the same (or ratioed) speed of a master roller drive. An example of a follower is a roller which is chain driven from another which is attached to a motor drive.
Footage Counter - A roller equipped with a revolution counter for measuring length and/or speed. Common methods include a tachometer or encode mounted on a roller or on the end of a drive attached to a roller, or by a proximity switch or photoeye sensing a target on the roller, or by a revolution counter driven by a wheel nipped against the web going over a roller.
Footprint - Slang for nip width or nip pressure profile.
Forced Vibration - Vibration of a structure driven at a frequency.
FPA - Abbreviation for the Flexible Packaging Association.
FPM - Abbreviation for speed in English units of feet per minute.
Frequency - A measure of the speed at which a vibration or other oscillation varies as measured in units of Hz (cycles per second), RPM (revolutions per minute) and so on.
Fret - A small slippage or chaffing between metal parts, such as a bearing slipping in its housing, which accelerates wear.
Fretting Corrosion - Corrosion accelerated by slipping or chaffing between mating metal parts.
Friction - A force resisting motion. Friction between webs and rolls is important because it determines whether the web will track or lose traction. Friction between web and web is important because it determines how stable a stack or wound roll will be. Friction on machines is important because it may compromise the sensitivity of measurement or control.
FTA - Abbreviation for the Flexographic Technical Association.
GAA - Abbreviation for the Gravure Association of America.
Gain - The multiplier for a sensor readout, or the aggressiveness of a controller.
Gauge - Alias for web caliper or thickness.
Glazing - The hardening and/or polishing of the outer surface of a cover due to aging and wear.
Grade (Bearing) - The tolerance quality of a bearing.
Grade (Web) - A specific web material. This term may be used broadly as a subclass of a basic material, such as bond is a grade of paper, or much more specifically as a set of process settings and acceptable test results with its own unique ID number.
Grinding - Machining a roller surface with a rotating abrasive wheel which translates across the roller as it rotates in a lathe.
Grit - The coarseness of the grinding wheel abrasive.
Grooving - Regular spaced annular or helixed grooves cut into a roller on a lathe during manufacturing for purposes of reducing air entrainment. Also, annular damage to a rotating roller due to abrasive wear or contact with a stationary member.
Guide - Short for edge guide.
Guide Roller - A roller with one end that can be moved to tighten the front side of a web versus the back and vice versa.
Half Critical - A speed of slightly increased vibration at exactly one half of the critical speed of the roller, caused by varying wall thickness or geometry of a roller shell.
Hard Balance - A balancer which mounts the roller ends very rigidly (as opposed to soft balance).
Hard Edges - Paper industry slang for a high caliper band on the edges of a web.
Hardening - The chemical aging or oxidation of rubber cover materials.
Hardness - The indentation stiffness of a roller cover or shell, or the impact stiffness of a wound roll.
Head - A disk fit into the end of a roller to connect the shaft or journal with the shell.
Heat Pipe - A cavity which contains a liquid and gas state of a fluid which is used for maintaining tight temperature profiles.
Helical Gear - A gear with teeth set on an angle used where noise is an issue, but generates large thrust loads.
Helper Drive - A drive point intended to counteract modest drag and inertial tension components, thus ensuring a downstream tension close to that of the upstream tension.

Herringbone - Grooving oriented at a slight angle from the rollers axis or CD, also called chevron, which is erroneously thought to provide a spreading function.
Herringbone Gear - A gear with side by side helical teeth, which is very expensive but is quiet running and produces no axial thrust.
Homogeneous - A material whose properties are uniform at any position. Porous or laminated materials are non-homogeneous on a fine scale, but may be treated as homogeneous on a large scale for many analyses.
Hooke's Law - A law that states that stresses equal the product of a material constant called modulus (or Young's modulus) and strains.
HP - Abbreviation for horsepower in English units which is the thermal capacity of a brake or motor.
HVOF - A method of thermal coating.
Hydrogen Embrittlement - Embrittlement of a metal caused by absorption of hydrogen from a corrosive environment.
Hygrothermal - Behavior, such as changes in stress or strain, caused by changes in moisture or solvent content or temperature.
Hypalon - A trade name for a chlorosulfonated polyethylene cover material.
Hysteresis - The material behavior where the load and unload curves are not coincident on the stress-strain curve. Nip induced cover hysterisis heats the roller cover and is the primary load/speed limitation.
ID - Abbreviation for inner diameter.
Idler Roller - A roller which is driven by the web rather than by an electric motor, belt or other external means.
Imbalance - An improper term sometimes used for unbalance.
Impression Roller - A friction driven rubber covered roller that is nips the web against the inked engraved cylinder in gravure printing.
Induction Heat - A means of heating a roller by a coil powered by AC current in the vicinity of metal.
Induction Motor - An AC motor consisting of a mechanically stationary but electrically rotating outer field and a rotor consisting of conducting bars.
Inertia - The property of a roller to resist changes in speed as quantified by mass times radius squared.
Inertia Compensated Dancer Roller - A dancer roller sized to minimized the transmission of tension fluctuations.
Inertia Compensation - An automatic adjustment of the gain of a winder/unwind drive controller to avoid sluggish performance at large roll diameters and instability at small roll diameters.
Infeed Rate - The increase in depth or indexing of the lathe/grinder tooling after pass.
Inshaft - A drive shaft oriented in the CD which may, for example, connect a motor to a live shaft roller journal.
Instant Center - The effective location about which a steering guide pivots, typically adjusted to be 1/2 to 2/3 of the way upstream in the entering span.
Interlayer Slippage - The relative motion of adjacent web layers in a wound roll that is the cause of many roll defects.
Inverter (Drive) - See AC Scalar Drive.
Ionizing Bar - An electrical or nuclear power bar used for removing static charge from a web.
IOPP - Abbreviation for the Institute of Packaging Professionals.
ISO - International Standards Organization.
Isolating Nip - A driven nip roller set to create an additional tension zone. In draw control, tension disturbances pass through a nip.
Jacket - The space between concentric shells for a heated or cooling fluid.
Jog - To briefly run a machine at low (thread) speed to position a web or rollers.
Journal - A stub shaft at the ends of live shaft rollers.
Journal Bearing - A bearing comprised of a journal floating on a cushion of oil in a bearing housing commonly found on large rollers on older machinery.
kW - Abbreviation for kilowatts in Metric units which is the power and/or thermal capacity of a brake or motor.
L10 - The life of a bearing in revolutions where 10% of the bearings have failed.

Labyrinth Seal - A noncontacting seal used for low friction or light duty bearings.
Laminate - To bond two or more webs together by running them through a nip roller set.
Land - The highest unmachined portions of the roller surface.
Landmarks - A pair of plugs in the floor, of in the aisleway, that define an offset reference for the machine direction.
Lathe - A machine tool which is most commonly used on roller surfaces comprised of a tool/grinder translating on a bed while the roller is rotated by chucks at the end.
Lay - The direction of machining marks or wear on a roller surface.
Lay-on Roller - Slang for a rider roller.
Level - A roller whose axis is perpendicular to gravity.
Life - The anticipated life of a component at specified conditions and at a specified reliability. For example, the L10 life of a bearing is the number of revolutions which will cause 10% of the bearings under nominally identical duty to fail.
Lineshaft - A shaft oriented in the MD driven by a single motor from which several rollers are driven off of using right angle belts, gearboxes and the like.
Liveshaft Roller - A roller whose central shaft or journals rotate.
Load Cell - An electronic sensor that measures force. On converting machinery, load cells under the ends of an undriven roller are often used to measure web tension.
Lockout - A safety measure which reduces the potential energy of a machine and/or prevents unexpected movement from occurring during access and repair.
Log - A large roll of paper wound at the end of a paper machine on a reel.
Machine Direction - Direction of the material flow through a machine, abbreviated MD.
Macroslippage - The gross slippage between the web and roller where no position tracks speed exactly.
Magnetic Particle Brake or Clutch - An electromechanical device where current flowing through iron particles causes them to bridge between a rotor and stator.
Mandrel - A cylinder upon which a roll is wound.
Manufacturing - The process of producing a base web material.
Matched Steel - A pair of nested embossing rollers made of steel.
Material - Typical web materials include: paper, plastic film, metal foil, nonwovens and textiles.
MD - Abbreviation for Machine Direction, which is the direction of material flow through a machine.
MEK - A very powerful solvent.
Membrane - A body whose thickness is much smaller than its length and width. Most webs are modeled as membranes so that the negligible bending stresses can be ignored.
Metallizing - The process of adding a thin metal coating to a web, often by electrodeposition in a vacuum.
MG Set - Short for motor generator set which is a nearly obsolete means of DC motor control.
Micrometer - A precision instrument for measuring thickness or roller diameter.
Microslippage - The minute web/roller slippage as the web progresses around any roller with a tension differential across it because the web tension/strain changes as described by the band-brake law. This type of slippage is stable and controlled by mechanics and does not cause either loss of CD or MD web control.
Mil - A measure of distance in the English system corresponding to 0.001 inch, which commonly designates roller static and dynamic TIR runout, as well as thicknesses of webs and coatings.
Misaligned - A roller or machine component which is not level and square with a reference. The reference may be with the upstream roller for web diagnostics or with the world for mechanical alignment.
Misregister - An error in register.
Modal Analysis - The analytical or experimental determination of the modes shapes a structure would prefer to vibrate at.
Mode Shape - The dynamic shape of a structure at one of its resonances.
Modulus - A material property measure of spring rate, expressed in kN/m^2 or MPa in the metric system, and PSI in the English system. It is commonly given the symbol E with subscripts 1,2 and 3 for the MD, CD and ZD respectively. Sometimes test results for modulus are improperly given in units of force/length (web physical property) instead of force/length^2 (material property).
Mohr's Circle - A graphical method of combining axial stresses in the x and y directions and shear stresses in the xy direction.

Moment Transfer - Slippage on a roller due to misalignment which will pass the bending moment beyond the next roller upstream.
Monopass - A circulating fluid roller where the fluid enters from one end and exits from the other.
Monument - See Landmark.
Motoring - A positive external torque on a roller which will tend to increase the web's tension as it goes over that roller as well as counteracting roller drag and acceleration inertia. Motoring on an unwind tends to decrease the tension in the first span. Motoring puts energy into the web machine system.
MPM - Abbreviation for speed in the Metric system in meters per minute.
Mt. Hope Roll - An improper use of a tradename to describe a bowed roller spreader.
Natural Frequency - A frequency a structure will tend to vibrate at when forced or struck.
Natural Rubber - A roller cover material produced from the latex sap of the rubber tree.
Necking - The (usually elastic) reduction of web width due to changes in MD tension.
Neoprene - See Cloroprene Rubber.
Nested - A matched male/female embosser roller set pattern.
Nip - To load two parallel roller(s) together. A nip between two rollers is used in calendering, laminating and printing. A nip between a wound roll and a roller (or drum) is one of the three TNT's of winding which control how hard the rolls are wound.
Nip Width - The contact MD length of a nipped roller set.
Nitrile - A synthetic polymer cover material made from butadiene and acrylonitrile.
Nodes - Parts of a structure which do not move during resonant vibration.
Nonwoven - A web material made of (polymer) fibers laid down in a somewhat random orientation.
Normal Entry Law - A fundamental law stating that a web will seek a perpendicular entry (and exit) to the local axis of a roller in traction which is used to describe alignment, guiding, spreading and other web handling areas.
OD - Abbreviation for outer diameter.
Offset - An abrupt change of a wound roll layer edge position.
Oil Canning - Slang for the ring deflection of thin walled roller shells.
Optical Pyrometer - A nonconctacting temperature measurement instrument using an infrared sensor and a graybody emitter principle.
Orthotropic - A material which has three mutually perpendicular directions of elastic symmetry. Most webs are orthotropic with distinct properties in the MD, CD and ZD.
Oscillate - To cycle a web or unwind axially back and forth a short distance to avoided piling high caliper areas of a web on top of each other to form roll ridges and other roll defects. Also used to extend the life of doctor blades and anvil rollers.
Overload - An unintentional exceeding of a components rating such as radial load, torque or power.
Ozone - The chemical compound O_3 which is produced by electrical discharge and oxidizes rubber covers.
P&J Plastometer - A measure of cover hardness as an indentation of a loaded probe.
Pacer - Slang for a speed reference roller.
PAPRICAN - Abbreviation for the Pulp and Paper Research Institute of Canada.
Payout - To drive an unwind very slowly to allow manual threading.
Pendulum - The tendency of an unbalanced roller to stop with the heavy side down.
Permeability - The air flow per unit area per unit pressure through a permeable web or roller. Porous materials such as nonwovens and tissue to not tend to float over rollers as much because much of the entrain air will escape through the web.
Pi Tape - A flat tape measure whose graduations are multiplied by Pi to measure the diameter of a roller via its circumference.
Picco Abrasion - A test for abrasive wear.
Pillow Block - An integral bearing/housing style that mounts via two bolts through its feet.
Pin - Slang for dowel.
Pin-to-Pin - A male-male embosser roller set pattern.
Pitch - The axial distance between roller grooves.
Pitot Tube - A pipe open on one end and a pressure gauge on the other end for measuring fluid velocity.
PIV - A tradename for a variable speed ratio transmission.

Pk - Abbreviation for peak in vibration terminology.
Pk-Pk - Abbreviation for peak-to-peak in vibration terminology.
Plane - A CD position on a roller in balancing parlance.
Plasma Spray - A method of thermal coating.
Plating - To electrochemically coat a roller surface with a material such as copper or chrome.
PLI - Abbreviation for lineal force per unit width in the English system as lb/in, and is used for tension, nip and torque.
Plug Lines - Slang for landmarks used for establishing a datum for roller alignment.
Poisson's Ratio - The ratio of deformation in the transverse direction divided by the axial direction when a body is subjected to a uniaxial (one direction) load. Poisson's ratio describes necking and width changes.
Polyurethane - Perhaps the most common roller cover material.
Porosity - The void fraction of a structure. Also, a common but inappropriate term for permeability.
Pos-Z - A tradename spreader which uses two airfloat bars.
Pot - Short for potentiometer. A variable resistor often used as a sensor for dancer arm position.
Power - The rate of change of energy with time expressed as horsepower in English units and kilowatts in metric units. The power rating of drives and unwind brakes is determined by thermal limitations.
Pre-Entry Span - The span upstream of the entry span which should be the shorter of the two to reduce steering guide problems due to moment transfer.
Process Roller - A roller used to modify the web.
Profile - The variation of some web or machine property, usually with respect to the CD.
Profilometer - An instrument used to measure surface roughness by dragging a sensored stylus across it.
Puddling - At low speeds a fluid (water condensate) will pond or pool at the bottom of the inside of a roller.
Pull Roller - Any roller which is driven.
PWM - Pulse Width Modulation control of AC motors.
Pyrometer - A sensor for temperature measurement.
Ra - Surface roughness defined as the arithmetical average of the absolute value of the distance from the profile to the average. See also RMS.
Race - The inner and outer tracks of a rolling bearing.
Radial - In a direction perpendicular to a roll(er)'s axis.
Radial Vibration - A mode of nip vibration resonance where both ends are in phase and is a common cause of barring.
Ramp - Slang for the speed versus time plot of a drive during acceleration and deceleration.
Reel - An alias for a large wound roll or the machine upon which it was wound.
Regenerative Drive - An electric motor drive, often found on unwinds, that is able to return roll or roller braking energy back to the power lines.
Registration - To align a machine element or web feature to another in the machine and/or cross machine directions.
Release - The ability of a material to not stick to what it comes in contact with. Release properties of roller shells/covers are important in many adhesive and coating applications.
Resilience - The percentage recovery of energy in a load cycle. High resilience covers are desirable in high load/nip applications to avoid heat generation and failure due to hysteresis heating.
Resistance Heater - A means of heating a roller by passing an electrical current through a resistor.
Resonance - A condition of vibration amplification when running at a speed corresponding to a structural natural frequency.
Rider Roller - A nipping roller on a winding roll. Also called a lay-on roller.
Ridges - A roll defect caused by high caliper bands that cause circumferential bands of a slightly larger diameter than the rest of the roll.
Rig - The fitting of chains, straps or wire ropes around a roller for hoisting it into or out of a machine.
Rigid Rotor - An idealization that the roller mounting does not have play and the shell does not deform dynamically (such as above 1/2 critical or at resonance).
Rimming - At very high speeds a liquid inside a roller will be forced to the walls with an approximately uniform thickness.

Ring Deflection - The deformation of a cross section of a thin walled roller shell from a circle, usually due to nip loads.
RMS - Surface roughness defined as the root mean squared value of the profile.
Rocking - An antisymmetric mode of nipped roll(er) vibration.
Rockwell - A metal hardness tester.
Roll Structure - To program the wind of a roll using TNT's from a hard/tight start at the core which smoothly transitions to looser/softer finish at the end of the roll.
Roll - A web in wound roll form. This term is also used in the industry for rollers. However, in this text the use of rolls will be exclusively for wound forms to avoid ambiguity.
Roller - A rotating cylinder used for web transport. Aliases include drums and rolls.
Rope - A threading system consisting of rope(s) and pulleys.
Rope - Slang for the corrugation web defect.
Rotary Joint - A joint used to pass a fluid between a rotating roller journal and a stationary pipe.
Rotor - A rotating member in balancing parlance.
Rotor - The rotating part of an electric or electromechanical motor/brake. See also Stator.
Roughness - The uneven profile of a surface on a microscopic scale.
Rounding - A small 1-5 second curve section between steady-state speed and an acceleration/deceleration ramp.
Roundness - A measure of how close to a perfect circle the roll face is.
RPM - Abbreviation for the rotational speed of a roll(er) in Revolutions Per Minute.
Runnability - A measure of the mean distance between web breaks on a winder or other converting machine.
Runout - The deviation of circularity of a roll or roller measured as TIR.
S Curve - A linear acceleration/deceleration ramp with a small amount of rounding on each end.
S Wrap - A pair of highly wrapped rollers , usually for the purpose of achieving a higher drive traction.
Saddle Micrometer - A micrometer used for measuring the diameter of large rollers.
SBR - A polymer cover material.
Scale - A mineral coating which contaminates the inside of hot or cold circulating water rollers.
Score Slitter - A pointed slitter blade loaded against an anvil roller.
SCR - Silicon controlled rectifier used in the power conversion and control of some electric motors. See also Thyristor.
Seal - A means of keeping bearing lubrication in and contamination out and are classified as contact, labyrinth or clearance.
Segmented Roller - A series of two or more coaxially located rollers. By segmenting rollers, smaller diameters can be used to avoid the requirement of a drive.
Seized Bearing - A bearing whose elements have locked or welded together causing a catastrophic failure.
Sensitivity - The ability to discriminate, measure or control with great precision.
Set - A permanent deformation occurring when a material such as an elastomeric cover is loaded and does not recover completely when unloaded.
Set - Several narrow rolls wound simultaneously.
Setpoint - A control setting (dial position, etc.) for a single process control variable.
Shafted - To wind a roll on a mandrel as opposed to core chucks.
Shaftless - To wind a roll with chuck end support instead of on a mandrel.
Shear Ledge - An L-shaped bracket used to mount bearing housing on which makes re-alignment easier.
Shear Slitter - A pair of thin rotating slitter blades which are slightly overlapped and cut much like a scissors.
Shear Stress - A stress component which changes angles in a body. Shear stresses in webs are important because they may result in wrinkles. Shear stresses between layers in a wound roll caused by nips is important because slippage may result in roll defects. Shear stresses cause many roller cover failures.
Shell - The cylinder forming the body of a roller.
Shim - Thin brass sheeting used to space a bearing mount for roller re-alignment.
Shoot - Slang for taking an optical alignment measurement.
Shore Scleroscope - A hardness test for covers and shells.

Side Register - The control of register in the CD direction.
Sidedness - A roll(er) that has small bumps as imperfections to cylindricity.
Sidelay - To move an unwind sideways for purposes of aligning a web with downstream components.
Silicone - A polymer used for high temperature roller covers.
Simply Supported - A roller which is supported by on both ends.
Single Plane Balance - Measuring and adding balance weights in a single plane at a time, sometimes synonymous with static balance.
Siphon - A pipe on the inside of a roller used to remove steam condensate.
Skewing - A poor man's crown compensation by misaligning a nipped roller from square to increase the nip at the center versus edges.
Slat Expander - An expander spreader made with overlapping slats for the shell.
Sleeve - A cover or shell which is removable from a mandrel.
Slenderness Ratio - The length of a roller divided by its diameter, or L/D ratio.
Slides - Slang for linear ways or bearings which guide core chucks or coreshafts during winding from moving sideways and causing roll offsets.
Sliding - One of three states of web/roller interaction where the web slips but is in continuous contact with the roller. See also traction and floating.
Sling - To support a roll or roller by a belt, strap or apron.
Sling - To lift, transport or store a roll(er) by straps.
Slip - A mismatch in speed between a web and roller or between a belt and pulley or between a motor and its base speed.
Slip - The difference between synchronous and rotor speeds for an AC induction motor.
Slip Clutch - A clutch used to prevent torque overloads.
Slitter - A device used to cut the web lengthwise, often composed of two rotating cutting wheels.
Slitter - An (unfortunate) alias for a rewinder equipped for slitting.
Soft Edge - Slang for a low caliper area on the edges of a roll.
Soft Balance - A balancer which mounts the roller ends in linear slides (as opposed to hard balance).
Specific Heat - The energy required to raise a unit mass a unit degree of temperature.
Speed Control - The control of motor RPM such that roller surface speed maintains a setpoint or setpoint difference with respect to another roller. Also known as draw control.
Speed Differential - The (controlled) difference in surface speed between two rollers, often expressed as a % difference.
Speed Reference - The roller in a line that is chosen to set the pace for the rest of the machine. There must be one and only one speed reference in any line.
Speed Regulation, Percent - The measure of the accuracy of a drive as how closely it can hold a setpoint speed.
Spherical Roller Bearing - A bearing type used for highly loaded rollers.
Splice - To join together two webs end to end by adhesive tape or other means.
Spline - A drive shaft with a star shaped cross section used on clutches and couplings.
Spoiler Bars - Bars on the inside of a steam condensate dryer shell to enhance heat transfer.
Spool - A mandrel upon which a roll is wound. Sometimes specifically refers to a mandrel with sides.
Spread - To make a web wider at or near a spreader.
Spreader - A device used to remove wrinkles in a web or separate slit webs.
Springback - The partial return of a web or structure to its previous shape after yielding and load release.
Spur Gear - A most common gear, but is noisy and has limited pitch line velocities.
Square - A roller whose axis is perpendicular to a reference such as the centerline of a machine.
Squaring Roller - A roller misaligned in the direction of the wrap bisector to even out the web tensions of the front with respect to back.
Stack - Short for a calender stack composed of three or more nipping rollers arranged in a vertical plane.
Stall Tension - The ability of a drive to hold or create a web tension when a machine is stopped.
Static Balance - A roller which is statically balanced will not tend to pendulum to the heavy side.
Static Basic Load Rating - The maximum stopped or storage load a bearing can take before it permanently dents its ball or race.

Static Electricity - The electrical discharge arc between a web, roller or other object which have become charged due to relative motion.
Static TIR - The TIR runout as measured by slowly rotating a roller against a dial indicator and recording the maximum excursion of the needle.
Static Unbalance - A roller whose center of gravity does not lie on its axis of rotation. **Station-To-Station Diametral Tolerance** - The difference in diameter between two adjacent CD positions on a roll.
Stator - The (usually outer) stationary part of an electric or electromechanical motor/brake. See also Rotor.
Steady-State - A period of time when no pertinent setpoint changes are made such as speed or nip loading.
Steering Guide - An edge guide which pivots one roller in the plane of the incoming web to steer it.
Strain - The change in length, width or angle of a material subjected to stresses. Commonly expressed in units of δlength/length or %.
Strain Gauge - A thin flat electrical transducer for measuring strain that is bonded to a body of interest.
Strap Brake - Alias for band brake.
Stress - Force divided by area expressed in psi in English units and kPa in metric units. Pressure is one type of stress.
Stretch - The ultimate strain of a material expressed in %.
Sunday Drive - A slowing turning drive for a roll or roller in storage to keep it from sagging or taking a set.
Supercalender - A large offline calender stack used in the paper industry to make magazine and other glossy papers.
Surface Tension - The measure of the wettability of a roller surface as measured by the contact angle of a droplet.
Sweat - Condensation on a water cooled drum or roller which is cooler than the dew point of the room.
Swelling - The expansion of cover materials due to the absorption of water, chemicals or solvents which is an indicator of cover/application incompatibility.
Taber Abrasion - A test for abrasive wear.
Tachometer - An RPM sensor for measuring speed. The most common tachometer is a DC generator mounted to the end of a drive motor where the output voltage is proportional to speed.
Tangent Modulus - A measure of material stiffness which is the slope of the stress strain curve at a specified stress or strain.
Tape - Sometimes rollers are taped to increase traction or affect local spreading.
Taper - Slang for the maximum diametral variation of one end of a roller with respect to the other.
TAPPI - Abbreviation for the Technical Association of the Pulp and Paper Industry.
Tare - The component of web tension in the sensing direction of a load cell.
TD - Abbreviation for Transverse Direction. In this book we will use the more common CD or Cross Direction.
Tendency Drive - An idler roller which is driven through bearing friction which is intended merely to counteract the bearing drag tension component.
Tension - Stresses or forces which pull on the material, usually in its MD. Lineal tension for webs is expressed in lb/in in English units and kN/m in metric units. Stress tension for materials is expressed as psi in English units and kPa in metric units.
Tension Zone - A length of machine under nominally the same tension, usually, between driven rollers.
Theodolite - An optical alignment transit.
Thermal Coating - The process of melting materials and propelling them in a molten or softened state onto a roller target. Common thermal coatings include ceramics and tungsten carbide.
Thermal Expansion - The increase in length, width and thickness of a material at increasing temperatures.
Thermal Shock - The potential for tensile cracking of materials due to excessively high cooling rates.
Thermocouple - A thermometer which generates a small voltage at the junction of two dissimilar metals.
Thiokol - A polymer cover material.
Thread - To pass a web through a machine prior to starting.
Thrusting - The axial push of resulting from two roll(er)s that are nipped together.

Thyristor - Silicon controlled rectifier used in the power conversion and control of some electric motors. See also SCR.
Time Constant - The time to effect a change which is 67% of the way from an initial state to a final state. For example, the time constant of a web in a draw zone is span length divided by web velocity.
Timing Belt - A tradename for a cogged belt which can provide torque transmission between cogged pulleys without the slippage associated with flat belts or V-belts.
Tinsel - A grounded metallic tinsel or garland strand placed in proximity to a web to reduce static charge.
TIR - Abbreviation for Total Indicated Runout, is the change of radius as a function of rotational position on a roll or roller, and is a measure of their roundness.
Titanium Oxide - A ceramic coating used to improve a roller surface's hardness and abrasion resistance.
TNT - Short for Tension, Nip and Torque controls of winding tension.
Toluene - A very powerful solvent.
Topology - The description of the surface of a roller which may include roughness, waviness and grooving patterns.
Torque Control - Brakes, clutches and other friction devices whose output is a torque (force times radius).
Torque Differential - The sharing of drive torque between a driven winding roll and a driven nipped roller or between two driven drums upon which is nipped a winding roll. Torque (or speed) differential is one of the three TNTs of winding. Often torque differential is referred to as merely torque.
Torsion - A twist or torque on a motor, brake, coupling shaft or driven roller.
Torsional Resonance - The oscillating twist resonance between a motor/brake and a driven roller, usually associated with very flexible couplings or inshafts.
Tracking - Slang for a condition of traction or control of the edge position or path of a web.
Traction - One of the three possible web/roller interactions where there is no relative movement between a web and roller or between and roll and roller. See also sliding and floating. Traction also refers to the maximum tension differential that can exist across a roller without slippage.
Tram - Slang for parallelism and/or alignment.
Transducer - A sensor.
Transit - A small telescope used for aligning machinery.
Transport Roller - A roller merely used to move a web through a machine rather than modify (process) it.
Transverse Direction - In this book we will use the more common term of CD or Cross Direction.
Traverse Speed - The CD speed of the lathe bit or grinder wheel.
Treater - A process of modifying a web by coating or other means. In film, this refers to bare roll, corona treat or flame treating of the surface to make it easier to print on.
Triboelectric Series - An ordering of materials based on the preference they would have for taking on either positive or negative charges. Static generation is more pronounced when the web and roller are far apart on this series.
Trim - The width of a web.
Trim - A narrow slit on the edges of a web which will be discarded.
Tripass - A circulating fluid roller where the fluid enters and exits from the opposite ends, making three trips through the roller shell periphery for improved temperature uniformity.
Trough - A mild u-shape out-of--plane deformation of a web oriented in the MD. A severe trough may cause a crease or foldover going over a roller.
Tungsten Carbide Coating- A thin flame deposition of a very hard material on a drum or roller. It can be used to enhance wear resistance, or in varying surface roughness, to increase web/roller or roll/roller traction.
Tunneling - Slang for troughing.
Turnbar - A smooth or air floated bar for turning the web, usually at a right angle.
Turnover - Slang for foldover.
Two Plane Balance - Measuring vibration and adding correction weights to dynamically (static & couple) balance a roller.
U-Joint - A yoke and pin coupling capable of very high misalignments.
Ultimate Strain - The maximum elongation (m/m) of a material in uniaxial loading. This is called stretch in web tests and expressed in %. The ultimate strain is usually less in the CD than in the MD.

Unbalance - A roller whose center of gravity does not lie on its axis of rotation is in static unbalance. More complicated unbalances include dynamic and flexible rotor.
Unstable - A dynamic system (vibration or control) whose amplitude naturally increases or diverges.
V-Belt - A commonly used belt/pulley system that transmits torque through the sides of the belt.
Vacuum Bar or Box - A bar which has a porous (drilled) surface to which an internal vacuum is applied for the purposes of holding a web or to draw fluids through a web.
Vacuum (Pull) Roller - A roller which has a porous (drilled) surface to which an internal vacuum is applied for the purposes of traction enhancement, removal of liquids or web drying.
Vickers - A metal hardness test.
Viscosity - The measure of a fluid's resistance to flow.
Waviness - The deviation from cylindricity on a scale between that of roughness and TIR or diameter.
Wear - The abrasive removal of a roller's surface material due to the inevitable slippage that occurs.
Weave - The variable (with time or machine position) path of a web.
Web Handling - The science of web transport at maximum speed and minimum web damage .
Web - A web is structure which is long, thin and flexible. Common web materials include: paper, film, foil, nonwovens and textiles.
Weight - Short for basis weight.
Wheel Speed - The RPM of a grinding wheel.
WHRC - Abbreviation for the Web Handling Research Center at Oklahoma State University.
Wire - An endless web forming belt made of woven polymer or wire strands.
Wire Arc and Wire Combustion - Methods of thermal coating processes.
Worm Rollers - Rollers with raised threads occasionally used on paper machine clothing that is mistakenly though to provide a spreading action.
WR2 - Abbreviation for the rotational inertia of a roll, roller or drive component.
Wrap Angle - The angle between the ingoing and outgoing tangent of a web on a roller, or equivalently, the angle the web deflects as it goes over a roller. High wrap angles help ensure web/roller traction.
Wrinkle - A generic term for a variety of out-of-plane web buckling defects. Wrinkles oriented in the MD are caused by CD strain, and wrinkles at a slight angle are caused by shear stresses.
Yankee Dryer - A very large steam heated roller used for drying of tissue paper.
Young's Modulus - A measure of material stiffness, which is the slope of the stress-strain curve.
ZD - Abbreviation for the Z Direction of a material. In a web, it is normal to the plane of the web. In a roll, this is the radial direction.
Zirconium Oxide - A ceramic coating used to improve a roller surface's hardness and abrasion resistance.
Zirk - The fitting through which grease is pumped into a bearing.

Index